Sustainable Agriculture

Suhaib A. Bandh
Editor

Sustainable Agriculture

Technical Progressions and Transitions

Springer

Editor
Suhaib A. Bandh
Department of Higher Education
Government of Jammu and Kashmir
(Governemnt Degree College DH Pora)
Kulgam, Jammu and Kashmir, India

ISBN 978-3-030-83065-6 ISBN 978-3-030-83066-3 (eBook)
https://doi.org/10.1007/978-3-030-83066-3

This Springer imprint is published by the registered company Springer Nature Switzerland AG
The registered company address is: Gewerbestrasse 11, 6330 Cham, Switzerland

Contents

List of Figures

List of Tables

Chapter 1
Understanding Sustainable Agriculture

Nafeesa Farooq Khan and Sumaiya Rehman

Abstract Agriculture has seen enormous transformations, particularly after the conclusion of World War II. Agricultural production rose dramatically due to new technology, mechanisation, significant chemical usage and government policies that supported output maximisation. While these modifications have had several beneficial consequences and decreased several dangers in agriculture, they have also incurred enormous expenses in topsoil depletion, groundwater pollution, the collapse of family farms and persistent disregard of living and working conditions. Globally, food and agricultural production systems are confronted with unprecedented problems due to the increased demand for food due to population growth, increased hunger and malnutrition, unfavourable climate change consequences, overexploitation of natural resources, loss of biodiversity and food loss and waste. These obstacles may jeopardise the world's ability to satisfy its food demands in the present and future. Agriculture must be sustainable if it fulfils the needs of future generations while also assuring profitability, environmental health and social and economic equality. Today, the notion of sustainable agriculture is gaining greater recognition and support within conventional agriculture. Not only does sustainable agriculture solve several environmental and social challenges, it also creates novel and economically feasible options for producers, labourers, consumers and policymakers across the food supply chain. The present chapter will attempt to cover the fundamentals of the idea using the content format provided.

Keywords Green Revolution · Sustainable agriculture · Organic farming · Micro-irrigation

---

N. F. Khan (✉) · S. Rehman
Biological Invasion Lab, Department of Botany, University of Kashmir, Srinagar, Jammu & Kashmir, India

S. A. Bandh (ed.), *Sustainable Agriculture*,
https://doi.org/10.1007/978-3-030-83066-3_1

1.1 Introduction

The pre-Green Revolution period was a period of scarcity and shortage of food. On account of these scarcities and deficits, the sustainability of life had become difficult. Less production was the prime feature of cultivation. The use of traditional types of crops and methods of cultivation was creating a vicious circle. The manifold increase in the problems of people was putting too much pressure on agriculture.

Consequently, there was a need for hybrid seeds yielding more than traditional ones. Likewise, there was a kind of urgency to use scientific methods of increasing production, such as chemical fertilisers and technologies. All eventually found a more meaningful expression when the Green Revolution occurred, stressing the importance of high-yielding crops, chemical fertilisers, scientific land use and other things involved in boosting agriculture.

1.2 Global Impact of Green Revolution on the Environment

Green Revolution set a tone for greater production. Unlike the past, people began to use new varieties of crops and hybrid seeds. To ensure increasing production, chemical fertilisers and newly discovered technologies were used. An increase in production remained the sole purpose of agricultural activities. To achieve this objective, chemical fertilisers were used, which deteriorated the quality of land in the long run. Without considering the negative impacts of chemical fertilisers, the Green Revolution era, tilling through new technologies and increasing use of inorganic substances, caused innumerable environmental problems. The things that ensured the Green Revolution, in the long run, proved least environmentally friendly. The quality of soil deteriorated. Loss of moisture of cultivable land and water quality degenerated. The land became increasingly less fertile, possessing inadequate and imbalanced proportions of different minerals. Due to pesticides, a more significant number of pro-peasant organisms vanished. In the pace of the Green Revolution, the environment underwent more significant changes, and most of them proved very harmful.

Developing economies like India, by resorting to enhanced agronomic practices, including better agricultural practices, HYVs (high-yielding varieties), plant defence procedures, chemical fertilisers and modernisation, attained higher production of crops. Crops like maize and rice witnessed a boom using enhanced techniques, popularly identified as the "Green Revolution technology". As the total cereal production increased, the Green Revolution technology transformed India into a secure food country. By the 1980s and 1990s, it could rarely export any food (Anderson & Hazell, 1989). Paradoxically, however, the increased production created increased agrarian instability. Several studies show that the improved flux in food grain manufacture in India can be credited to the extensive acceptance of Green Revolution tools (Mehra, 1981; Singh, 1998).

On account of the Green Revolution, selective cropping of wheat and rice became a widespread phenomenon. The other crops received the least attention. Rice and wheat became more lucrative for farmers as yields increased than many other crops. The environmental implication of the Green Revolution includes agricultural increase into peripheral lands, alarming rates of cutting down trees and grazing in dried areas. All this is ultimately having more opposing ecological effects. It is expected that soil and water loss and damage to organic matter may lead to land deterioration and degradation, which eventually will make it more challenging for upcoming generations to meet their requirements for food, fuelwood and other agronomic and forest yields. New technology facilitates the exploitation of land base due to the enhanced yields in prevailing cultivated lands. Hence, increasing the growth and using yield-enhancing machinery lessen the burden on new agricultural lands to fulfil growing demands for nutrition and other agronomic products. The expansion and use of yield-increasing crop varieties, amplified use of chemical fertilisers, improved manufacturing ways and other yield-enhancing elements have triggered land deprivation.

The use of a vast amount of pesticides might disturb the ecosystems badly in a variable number of ways. A decrease in plant genetic diversity is another significant ecological threat linked to it. Once the diversity declines in farms, actual steps should be taken to warrant that the plant genetic substance is preserved somewhere else. Ecological threats related to the use of fertilisers appear to be somewhat unimportant, though excessive usage rates and reduced social practices may cause eutrophication of water bodies.

1.3 Sustainable Agriculture

Sustainable agriculture is indefinitely a comprehensive idea with numerous delineations (Lockeretz, 1988). Summing it up in a single frame is very challenging. Primarily it denotes a variety of approaches for talking about the number of difficulties that agricultural science faces. These difficulties comprise loss of soil production, plant nutrient loss and water pollution by extensive use of pesticides, chemical fertilisers and deposits. Sustainable agriculture is the efficacious organisation of means for agronomy to fulfil varying human requirements through sustaining or improving natural means and circumventing ecological deterioration (TAC-CGIAR, 1988). Sustainable farming is a scheme of farming devoted to preserving and maintaining the cultivation base of atmosphere, soil and water safeguarding the upcoming generation's capability to nourish themselves by a sufficient stock of harmless and healthy diet (Grace, 1990). It is an arrangement that could indeterminately fulfil the desires for food and fibre at reasonable financial and environmental costs (Crosson, 1992).

Sustainable agriculture aspires to create an agricultural environment that is both sustainable and prosperous while conserving the natural resource base. Sustainable agriculture includes protection of the environment and an increase in production, which is economically beneficial. In short, sustainable agriculture means a kind of financially and environmentally sustainable agriculture over a more significant period.

1.3.1 Advantages

Sustainable agriculture seems to be advantageous in multiple ways:

- It causes low or no pollution to the environment.
- Being environmentally friendly, it has the least possible effects on different forms on earth.
- It crucially acts as an active indicator in maintaining the earth's biodiversity.
- It promotes rich biodiversity as well as natural elements and their recycling.
- It emphasises the growth of plants on account of their original potential.
- The conservation of natural resources and their judicious use figures at the centre of sustainable agriculture.
- It helps in maintaining the economic sustainability of farm operations.

1.3.2 Principles of Sustainable Agriculture

There are five fundamental concepts for increasing farm production and sustainability on a global scale. These ideals seek to establish a production structure that is both environmentally friendly and meets human needs. In the first place, sustainable agriculture aims to improve resource utilisation efficiency, which is possible by strictly adhering to sustainability principles. The second principle involves protecting and enhancing natural resources that help maintain favourable conditions for fruitful production and productivity. The third principle envisages healthy human-environment relationships by improving people's resilience, communities and ecosystems. The fourth principle develops a healthy equation between native means of production by expressing faith in a rural lifestyle and social welfare acts. The fifth and final principle brings in the concept of welfare economics wherein the role of governance comes to play a pivotal part.

1.3.3 Goals of Sustainable Agriculture

The following are a few significant goals of sustainable agriculture:

1. It expects a productive and profitable production. It chiefly aims for surplus production and crop varieties to make the human resource equation and equilibrium healthy.

2. The ultimate goal of sustainable agriculture is ever-increasing production and profit with the least harmful impacts.
3. It is driven by the desire to set in a kind of balance between man and the environment.
4. The prime concern of sustainable agriculture is the conservation of resources and protection of health and the environment.
5. Low-input methods, skilled management and recycling of natural elements and less hazardous production methods like novel ideas run central to it.
6. It puts forward the cause of N fixation, pest-predator relationships and nutrient cycling into agriculture procedures.
7. Sustainable agriculture seeks to avoid harmful substances in raising multiple crops.
8. It discourages the use of those off-farm inputs that are non-renewable and external, potentially harming the environment.
9. It prefers only those inputs which are less harmful and renewable and help minimise variable costs.
10. Sustainable agriculture takes farmer and rural life at the centre of its edifice and proclaims a proactive role for peasants. Its goal is to involve farmers, and with their full participation, it assigns a significant place to them in all processes of problem analysis. Besides, it would be preferred to prefer technology to develop, adopt and extend benefits.
11. The whole idea of sustainable agriculture is progress in production, profitability and equitable access to predictive resources and opportunities.
12. To promote agriculture along suitable and sustainable lines, there is a greater need to benefit from the productive use of plant and animal species' biological and genetic potential.
13. An increase in self-sufficiency amongst growers and people of rural areas.
14. Cost-effective and well-organised making with stress on combined practice managing and preserving biological resources, energy, soil and water.

1.4 Farming Systems and Agriculture Sustainability

The farming system could be a complicated inter-linked matrix of animal's implements, soil, plants, capital, labour, power and alternative inputs organised partly by farmhouse people and prejudiced by variable degrees of institutional, political, economic and social forces working at different stages. Specifically, it is a well-defined, exclusive and judiciously constant procedure of field enterprises that a family accomplishes with its physical, biological, socio-economic and cultural situation in agreement with the family's objectives, likings and incomes. Theoretically, it denotes a group of components or parts that remain interconnected amongst themselves. Agricultural initiatives comprise apiary, crops, dairying, fishery, piggery, poultry, sericulture and tree crops. A grouping of one or many enterprises with cropping, when cautiously selected, planned and implemented, provides larger dividends than a solitary enterprise, particularly for minor and peripheral farmers. Farmhouse as a component is thought about and intended to incorporate the initiatives to be shared with crop production activities. The end product or final product of one enterprise is used efficiently as raw material for the other. For instance, the absolute wastes of dairying like urine and dung are used to make compost that functions as an input or manure in the cropping system. Similarly, the grass attained from cereals (maize, sorghum and rice) has been used as cattle fodder. Sustainability is the goal of the agricultural system where the manufacturing process is improved by practical application of efforts without interfering with the excellence of environs with which it cooperates on the one hand and efforts to reach the national objectives on the other hand.

1.4.1 Principles of Farming System

- Ecological equilibrium.
- Opportunity for job creation.
- Increased input performance.
- Integration of two or more enterprises.
- Maximum productivity and profitability.
- Minimisation of risk.
- Optimum use of subsistence systems.
- Recycling of wastes and residues.
- The use of final yields from one individual enterprise as raw material for another enterprise.

1.4.2 Aims of Farming System

1. Food security: In agricultural schemes, varied enterprises yield diverse food sources, viz. fats, minerals, proteins and carbohydrates, from the same land unit, which helps resolve the undernourishment problems predominant amongst the peripheral and sub-peripheral rural families.
2. Employment generation: Numerous farmhouse enterprises like livestock, crop production or some other similar enterprise in the agricultural system may upsurge employment avenues considerably and may help overcome unemployment.
3. Farming system increases input use efficiency by enabling proper resource utilisation with tangible outputs.
4. One of the crucial aspects of the farming system is protecting and preserving the environment by providing different biocontrol measures and pest and disease control mechanisms.
5. It has a more significant advantage because it is eco-friendly, significantly reducing environmental pollution.

6. Farming system possesses a greater potential to prevent a farmer from market fluctuations and benefit from the market by considering the current demands and preferences.
7. It helps in- and outflow of money by selling different products such as eggs, milk, fish, silkworm cocoons, honey, etc. That has increased farmers' purchasing power and opportunity to invest and take advantage of improved technologies.
8. It helps produce both food and fodder simultaneously, enabling farmers to enhance their livestock and income from means other than mere acts of cultivation. Making farmers capable of using proper resource utilisation in the farming systems equips them even for the lean agriculture season.
9. The farming systems help maintain the rich humus and the allied soil potential by minimising synthetic fertilisers and advocating the cause of natural organic materials, manures and wastes. There is a way to prevent soil pollution and the death of pro-farmer organisms in the farming system.
10. The farming systems provide scope for multiple crops in the least expensive and farm-efficient conditions. Less reliance on chemical fertilisers, pesticides and other such things in the farming system has increased opportunities to increase monetary gains per unit area per unit time by increasing crops and connected initiatives.
11. The cost-efficient production, greater profitability and other benefits of farming systems work on the principle of no wastage.
12. Additionally, it saves energy by providing effective recycling of organic wastes.
13. The growth and development of agro-based industries become possible by the hands of IFS.

1.4.3 Organic Farming

Organic farming "is a method of agriculture practice that prevents or substantially eliminates the use of synthetically compounded fertilisers, pesticides, growth regulators and livestock feed additives". Organic agricultural schemes extensively use crop rotation, crop residues, livestock compost, legumes, green manure, off-farm organic wastes and biological pest management (USDA, 1980). The concept of the soil as a biological environment that must be "nursed" in a way that does not restrict the activities of beneficial living organisms necessary for nutrient recycling and humus production is critical to this explanation. "Organic agriculture is a holistic production management system that promotes agro-ecosystem health, including biodiversity, biological cycles and soil biological activity. It emphasises the use of management practices in preference to off-farm inputs, considering that regional conditions require locally adapted systems. This is accomplished by using wherever possible, agronomic, biological, and mechanical methods, as opposed to using synthetic materials, to fulfil any specific function within the system (Di Caracalla, 1997).

1.4.4 Principles of Organic Farming

1. Organic farming is the quest for quality foods with high nutritional values.
2. To develop a healthy man-environment relationship.
3. Creative and constructive use of resources with the most negligible degeneration of life, natural cycles and systems.
4. Maintaining and encouraging organic cycles and preserving the wealth of pro-farming microorganisms, soil fertility and varied life forms with careful mechanical intervention.
5. To prevent soil pollution and water pollution and promote healthy living conditions on earth and water.
6. Conservation of soil and water quality is a priority.
7. To use locally organised agricultural systems with a significant reliance on renewable energy.
8. Making a broad use of organic matter and nutrient elements.
9. Scope for enlarging the livestock conditions.
10. Upholding rich biodiversity on the surface of the earth.
11. Organic production, cultivation of quality life and creation of a safe working environment.
12. Production of food and fodder with the help of renewable resources.
13. Growth and development of crop production and animal husbandry.
14. Promotion of love for nature and output of harmony.
15. A system of just production and distribution.

1.4.5 Relevance of Organic Farming

Attention towards organic farming practices is developing rapidly, particularly in parts where the current modern agricultural system has unchecked several agricultural and ecological issues, both going on and off the farmhouse, threatening food safety. The following are a few examples of the repercussions of modern agriculture that have made organic farming relevant in contemporary times:

(a) Loss of soil fertility.
(b) Contamination of food, soil and water with insecticides and chemical fertilisers.
(c) Health effects on agriculturalists, workers, farm families and rural populations due to the insecticides, pesticides and other deadly chemicals.
(d) Resistance of insects and pests to these chemicals.

1.4.6 Precision Agriculture

Precision agriculture (PA) is one of the emerging concepts towards achieving sustainable agriculture. This farming concept is based on proper measurement, observation and well-responding mechanism to intra- and inter-field inconsistency in crops. This concept's overall goal is to optimise earnings on investments while conserving the resources (McBratney et al., 2005). It aims at increasing production by incorporating technologies and principles to manage environmental quality. The precision agriculture concept works quite well in enhancing crops like coffee, tea, sugar beet and sugarcane. Despite being full of potentials, the idea has not been sufficiently addressed.

The need for precision farming production is an outborn of scarcity-like conditions. Less production and shortage of food created necessary conditions to develop this model. The concept is premised on increasing production through a well-coordinated system of efforts and ingredients like irrigation, fertilisation, pesticides and high-yielding crops. The design invokes a desire to increase production by keeping the growing size of the human population without any degradation of natural resources; by using high-yielding varieties, avoid negative ecological/environmental consequences like waterlogging and chemical deterioration (salinisation and loss of nutrients), respectively.

It involves a systematic way of farming, starting with selecting crop varieties, fertilisers in proper doses, exact types of herbicides and pesticides and adequate means of irrigation to ensure optimum growth and development. Precision agriculture promotes maximum utilisation of advanced technologies such as tools, machinery, satellite-based information and pro-peasant agrarian measures. Precision agriculture by the exact and appropriate quantities in the proper time helps reduce the cost of several agrochemicals used in crop production. This well-coordinated system based on accurate equipment information, land preparation, seeds, fertilisers, pesticides and herbicides, irrigation and post-production activities makes precision farming significant and such anthropogenic activities in precision farming affects the environment the least.

Precision farming comprises many steps that aim to utilise information in agriculture to enhance production by all means available to achieve expected yields. As a system, it makes it imperative to assess the variability for proper management and evaluation.

Assessing variability helps manage the set of factors and the processes playing a pivotal role in crop performance in space and time. The measurement of the inconsistency of factors and procedures determining spatiotemporal variations in crop yield helps precision agriculture farming use the same to achieve maximum production. Assessing spatiotemporal variability statistics makes precision farming highly rewarding. The concept of variability are creatively used in precision soil fertility management by accurately identifying and interpreting the influence of yield, quality and crop-specific environment. However, more important in precision agriculture are the evaluation of data and the analysis of profitability. Proper use of data reduces agrochemicals, enables higher nutrient use efficiencies, helps in managing inputs and maximises production. Due to all these benefits, precision agriculture continues to develop (Pierce & Nowak, 1999).

1.4.7 Climate-Resilient Crop Varieties

Climate-resilient crops have become the need of the hour to endure the drastic climatic change. They can tolerate erratic rains, heat, floods, droughts, chilling and salinity stresses. They are essential ingredients of sustained agriculture, capable of yielding high production and coping with the challenges of climate change. Climate-tolerant crops are evolving as the best potential solutions for addressing climate change problems, meeting the needs and demands of a rising population and enabling competent peasants to be less reliant on what nature provides. Climate-resilient crops are, to a greater extent, capable of coping with abiotic stresses like droughts, heat and cold, the significant factors that adversely affect plant growth and productivity (Maheswari et al., 2012). To produce stress-tolerant varieties, it is critical to recognise the characteristics that support and foster plant growth and development during times of stress (Maheswari, 2017; Maheswari et al., 2016; Shanker et al., 2014).

There are currently many programmes that aim to improve productivity and tolerance to physical stresses. Their availability needs to be encouraged to sustain the production system and meet the increasing demand for food grains. On account of being high yielding and increasingly tolerant to delayed monsoons and droughts, climate-resilient crops are essential for sustainable farming practices.

Climate-resilient crops are likewise capable of enduring heat stress as well as cold stress. Heat stress or exposure of plants to heat hampers agricultural production by causing morphoanatomical, physiological and biochemical changes in plants. The heat-tolerant varieties counter production losses. Many plants, especially those native to warm habitat, exhibit injury symptoms, including reduction of leaf expansion, wilting, chlorosis and necrosis, when subjected to low non-freezing temperatures. Primary injury is the initial rapid reaction to chilling stress resulting in plant dysfunction but is easily reversible when the temperature is elevated to non-chilling conditions (Kratsch & Wise, 2000). To overcome this problem, climate-resilient crops are the only remedy as they are, to a greater extent, capable of enduring even the salinity stresses. So, these crops help meet the twin challenges. They are the key varieties in agricultural production to meet the greater demands of the growing population. Likewise, they protect agriculture from the adverse impacts of climate change. These crops are helpful to overcome the adverse effects of climate change by helping to lower the yield losses under stress condition. For further advancement, the climate-resilient varieties of crops could be utilised as potential genetic resources.

1.4.8 Micro-Irrigation

The idea of sustainable agriculture seems to be incomplete without the concept of micro-irrigation. Micro-irrigation, described as the specific application of water on or below the soil surface at low pressure through small devices that spray, mist, sprinkle, or drip water, is gaining popularity (Hla & Scherer, 2003) and is regarded as critical for increasing irrigation efficiency (Hla & Scherer, 2003) (FAO, 2003). It seems to be instrumental in increasing yields and reducing the rate of salinisation, especially when the crops are not overtly sensitive to salinity (Cetin, 2004).

Micro-irrigation, a concept based on new technologies, is a strategy to overcome the shortage of water. It provides usage of water only as a requirement without least wastage. Thus, it offers a continuous water supply in the crop zone that gives a higher crop yield. According to Postel (2000), it has doubled crop yield per water unit in many applications. It has proven much effective in crops like vegetables, sugarcane, cotton and orchard and vineyard crops. Shah (2011) also claims yield increases and water savings due to micro-irrigation. It additionally saves labour costs and improves crop quality (Madramootoo & Rigby, 1991).

Significantly micro-irrigation makes it possible to reuse wastewater with a high salt content that helps nitrogen in the wastewater be easily absorbed by plants and less likely to pollute groundwater. Micro-irrigation maintains root zone moisture content throughout the season. Micro-irrigation methods have proven too much effective during the dry season in humid areas or arid climates. It offers the best mechanism to make the best use of fertilisers and pesticide residues. It further promotes more efficient use of nutrients, better and longer moisture retention in the root zone, fewer pests and weed invasion and plant diseases and reduced pesticide/herbicide use (Varma et al., 2006).

1.4.9 Tillage Management for the Effectiveness of Fertilisers and Pesticides

As a component of sustainable agriculture, tillage is characterised as a mechanical activity performed on the soil and crop residues to create a suitable seedbed for crop seeds to be sown. This is accomplished by using a variety of farming tools, ranging from the basic digging stick to the paddle-shaped spade that can be pulled by humans or animals (Lal et al., 2007). Tillage impacts weed species by destroying weed seedlings mechanically and altering the vertical distribution of weed seeds in the soil (Peigné et al., 2007). Additionally, it modifies the soil climatic conditions that influence weed dormancy, germination and development. Reduced tillage intensity usually results in a rise in weed concentration in the topsoil (Vasileiadis et al., 2007).

The tillage system needs to be scientific, and conservation tillage needs to replace conventional tilling methods because the scientific conservation tillage helps redistribute organic carbon in the soil (Tebrügge & Düring, 1999). In general, the organic carbon content of surface soil increases as a result of the mulch's presence and decomposition and eventually declines with depth (Six et al., 1998).

1.5 Soil and Its Sustainability

In the scientific literature, the words "soil health" and "soil quality" are used synonymously, and some people assume they are functionally interchangeable. Preferentially, scientists prefer soil quality (Ritz et al., 2009), while farmers prefer soil health for monitoring soils. Soil structure is characteristically essential as it is recognised through values that characterise and classify its health. This substrate contributes to the sustainability of plant and animal production, the preservation or improvement of water and air quality and the overall well-being of plants and animals, thus improving an ecosystem's functioning. However, the soil quality remains an external property of soils that varies according to the intended use of soils. Its quality is compromised by features like soil sodicity, proton concentration, ground compaction, nutrient depletion, low microbiota/biomass, moisture deficit and disrupted nutrient cycling. Methods of unrestricted use of agrochemicals and ill land use practices worldwide, with their disastrous effects on individuals and their surrounding things, have led to substantial changes in people's approach in recent times. Often a belief prevails that plants require nitrogen, potassium and phosphorus, eventually termed limiting nutrients, to improve the soil's water-retaining ability while increasing its fertility. But elevated plant growth fertiliser applications usually include N, P and K, which can cause imbalance and even reduce soil fertility. The imminent fear of chemical persistence in the soil for generations and its consequences raises food safety issue amongst the local masses. Thus, researchers have now shifted their focus on developing an alternative option to chemical fertilisers, which would be more reliable and eco-friendlier. Scientists have made recent advances as a significant contribution to world cropland sustainability by converting theoretical information about soil structure into realistic strategies that help growers assess the sustainability of specific management strategies. This includes total secondary nutrient recovery and the use of soil acids to create organic substance. Soil conservation ensures that nutrients do not become insufficient or harmful to plants and that the right minerals add to the food chain. The sustainability of soils becomes progressively relevant in the years ahead, given the expected rise in world population and the resulting need for intensification of food security. More systematic studies need to be done to avoid further soil contamination due to erosion or pollution and produce enough safe and nutritious food for healthy diets.

1.5.1 Soil and Plant Environment as a Sustaining Environment for Microbial Life

Plants below-ground parts are home to a diverse and abundant microbiota, including algae, bacteria, fungi and protozoa (Glick, 1995; Müller et al., 2016). These microbes can either develop into pathogens or beneficial microbe but are frequently linked to optimised nutrient supply and plant growth roles. The vitality of soil microorganisms is influenced by various factors, including nutrient supply, climate, water and pH. They establish a complex microbial consortium around the rhizospheric soil-root continuum, which has multiple microorganisms living on or inside root cells that improve their growth and survival. Additionally, rhizobia pick up iron (Fe3+), which is present in trace amounts in the rhizosphere, through secreted iron-binding ligands or chelators known as siderophores, which have an affinity for sequestering minimal iron from the microenvironment (Saha et al., 2016). Some other strains can persist and grow at a temperature of 45 °C with a pH 4 by mineralising insoluble phosphates. Rhizodeposition is the primary carbon source for microbial life throughout the soil, supporting various organisms and microbe consuming forms (Nguyen, 2009). In agriculture systems, soil with a dynamic and good load of microbiota produces good crop yield. The decomposition rate is also enhanced through the microbial communities, which produces enzymes that control the dynamics of plant nutrients.

For example, soil microbes, including arbuscular fungi (AMF), aggressive bacteria and helpful nematodes, were closely correlated with crop productivity, fruit development, water-retaining capacity and nutrient availability which play vital roles in plant health and soil fertility. Rhizospheric microbes and connected novel pathways, mechanisms and metabolites have been used to help soils operate more efficiently and sustainably in agricultural production. Biofertilisers are currently utilised as naturally efficient strains of microorganisms or modified organisms that reduce the incidence or severity of diseases caused by plant pathogens. Thus, these microorganisms are also known as biocontrol agents when they exhibit an antagonistic activity towards a particular phytopathogen (Beneduzi et al., 2012). The current understanding of biological nitrogen fixation technology where rhizobium is used as alternative fertilisers is now displaying successful results as biocontrol agents. From an ecological viewpoint, living plant and soil surfaces are nutrient-limited. They could form a particular relation in competition or predation or the possibility of establishing a mutual relationship with indigenous microbes (Avrani et al., 2017).

1.5.2 Mechanisms and Application of Plant Growth-Promoting Microbes in Agricultural Soils

Plant life is often impaired by multiple stressors, such as hydrogen ion toxicity, nutrient deficiencies, toxic effects and imbalances. Plant growth-promoting (PGP) microbes have shown a promise as long-term plant growth modulators and may be capable of dealing with various environmental stressors. Plant growth-promoting factors in the rhizospheric microbiota have a great deal of significance for long-term crop yields. There's a lot of interest in finding out how to use them best under different stress conditions, droughts and high salt contents (Rosier et al., 2018). Microbes contribute to plants' overall strength and health by directly or indirectly controlling the growth of pathogens. The manipulated invasion of plant roots by rhizobia also activates an increased resistance that systematically triggers the plant's complete downstream defence response (Tonelli et al., 2020). Plants respond to several biochemical signals induced by soil- and plant-associated microbes. However, only a few diverse classes of gram-negative, nodule-forming bacteria have been identified that emerge in the rhizosphere and develop a mutually symbiotic relationship with leguminous plants. In their initial interactions with the host, they are considered foreign particles by plants followed by a series of defence mechanism against them by exopolysaccharide secreted by some bacteria (Scheidle et al., 2005). The bacteria overcome the defence by suppressing the plant defence system and acquiring entrance into the root tissues' interior (Cao et al., 2017; Zamioudis & Pieterse, 2012). Once successfully established around the rhizosphere, these bacteria grow fast and multiply (Scheidle et al., 2005). Thus, plant defence mechanism plays a key role, starting from the entry of beneficial microbes to total inactivation of harmful pathogens around its rhizosphere (Yu et al., 2019). Usually, the plant's environment is influenced by soil fertility, and its root exudates which is rich in organic and inorganic compounds such as nutrients, amino acids, hormones, fatty acids and phenols (Gopalakrishnan et al., 2015). The enormous richness of minerals and nutrients around the root attracts other microbial flora capable of establishing an uninterrupted interaction ranging from mutualistic to pathogenic (Patil et al., 2017). For example, rhizobacteria directly or indirectly promote the growth of beneficial organisms and impede the multiplication of various phytopathogens by acting as biocontrol agents (Munees & Mulugeta, 2014). Additionally, rhizobia cause systemic tolerance, which is a beneficial mechanism when contemplating various plant diseases (Fernandez-Göbel et al., 2019; Yu et al., 2019). A plant responds to several biochemical signals induced by soil- and plant-associated microbes. The strength and stability of its cross-talk signal play a key role in determining the quality of resistance against pathogens. The interactions with these microbes can be in the form of different possible relationships (symbiosis, mutualism competition, predation, commensalism, etc.). Following a sequence of events, the hypersensitive response gets active at the initial stage – a mechanism used by plants to prevent the spreading of local infections by microbial pathogens. The plant directs different defence responses around the whole plant system by signalling molecules such as salicylic acid, reactive oxygen species (ROS), nitric oxide, jasmonic acid or ethylene at the onset of systemic hypersensitive response (SAR) (Dong, 1998). This systemic acquired resistance (SAR) is mediated by salicylic acid (SA), a compound that is commonly released in the aftermath of pathogenic infection and reliably results in the expression of pathogenesis-related (PR) proteins. Furthermore, a few downward induced responses mediated by jasmonic acid (JA) and ethylene amplify bacteria's infection risks. However, the SA- and

JA-dependent defence pathways may sometimes be mutually exclusive or antagonistic, and certain strategic pathogens can exploit this process to circumvent systemic acquired resistance. The sensitisation of defence responses by invaders such as rhizobia further activates and stabilises the whole plant system, a process known as priming, which categorically apprises the whole plant system (Mauch-Mani et al., 2017). The activation of systemic expression of pathogenesis-related (PR) genes codes for proteins and enzymes with antimicrobial activity (Pieterse et al., 2014). These PR proteins include various enzymes ranging from chitinase, β-1,3-glucanases, protease and lipase, which extract solute particles from the pathogenic plasma membrane or strengthen the plant's cell wall boundaries to resist infections. Plant microbes are recognised as foreign agents by their perception of unique microbial molecules or effector molecules referred to as microbe-associated molecular patterns (MAMPs) and plant-associated molecular patterns (PAMPs). Few researchers have regularly published on effectors such as flagellin protein, elongation factor Tu (EFTu), bacterial lipopolysaccharides or peptidoglycans and even fungal chitooligosaccharides (Saijo et al., 2018). These PAMPs and MAMPs bind to pattern recognition receptors or PRRs associated with host cells to induce plant defence. Rey et al. (2013) confirmed that *Medicago truncatula* mutants without detecting Nod factors were more susceptible to attacks by *A. euteiches* and *Colletotrichum trifolii* than wild-type plants, implying rhizobia-plant communication. In addition, bradyrhizobial strain, defective in node factor production, exhibits higher prevalence and extremity of the stem rot diseases caused by *Sclerotium rolfsii* in peanut plants (Figueredo et al., 2017). In certain instances, the use of inoculants such as plant growth-promoting rhizobacteria (PGPR) was successful in managing complex diseases such as anthracnose (*Colletotrichum* spp.), angular leaf spot and bacterial wilt (*Erwinia tracheiphila*) (Ongena et al., 2005; Van Loon et al., 1998). Recently, rhizobial inoculation in *Medicago* plants was correlated with priming for salicylic acid (SA) accumulation and salicylic acid (SA)-mediated defence against the powdery mildew (Smigielski et al., 2019). Moreover, a fast systemic oxidative change was observed after inoculation with *Bradyrhizobium japonicum* in soybean roots (Fernandez-Göbel et al., 2019).

1.5.3 Microbial Disease-Suppressive Agents

Prolonged monocultures or minimal crop rotations constitute crop cultivation systems, often resulting in soil deterioration and accumulation of soil-borne infections. Infection thrives in soils where various biotic and abiotic settings favour pathogens. Suppressiveness of diseases due to soils is often attributed to the collective intervention of microbial groups (Mazzola, 2002; Weller et al., 2002). Often soil management techniques are successful in the prevention of soil-borne diseases. A specific disease inhibition can be caused by the bacterium *Pseudomonas* (genus), which grows on root surfaces intensely (Kloepper et al., 1988). Usually, most microbes can synthesise the secondary metabolites in antibiotics for their survival (Gopalakrishnan et al., 2015).

There is a distinct mode of action to counter each type of pathogens. Mostly antibiotics produced by useful microbe's act on harmful microbes of the rhizosphere, e.g. the mode of action of beta-lactam antibiotics like penicillin and vancomycin directly disrupts the mature peptidoglycan molecule on the plasma membrane, which causes cell death due to disruption of osmotic pressure (McDermott et al., 2003). While antibiotics produced from other microbes such as streptomycin specifically targets 16S and 23S rRNA of the 30S subunit of the bacterial translation unit (Wiener, 1996), chloramphenicol has the affinity towards peptidyl transferase of the 50S ribosomal subunit of 70S ribosomes. It inhibits the protein elongation process during protein synthesis. In addition, antibiotics such as fluoroquinolones, bleomycin, cause double-stranded DNA breaks during replication of DNA/RNA (Drlica & Zhao, 1997). There are many reports of antibiotic production by rhizobial strains during their symbiotic relationship with the plant (Bardin et al., 2004a; Chandra et al., 2007; Deshwal et al., 2003). The production of a narrow-spectrum peptide antibiotic trifolitoxin (TFX) by Rhizobium leguminosarum in bean plants changes the overall microbial diversity around its rhizosphere (Triplett and Barta (1987). The resistance of soy bean crop against infection *M. phaseolina* correlated with the direct action of antibiotic rhizobitoxine produced by *B. japonicum* (Chakraborty & Purkayastha, 1987). Bacteriocins are proteinaceous or peptidic toxins that suppress or destroy bacteria belonging to the same or closely related genus (Salto et al., 1979; Tagg et al., 1976). Bacteriocins formed by *Rhizobium* spp. have been classified as phage-like, protease-sensitive or protease-resistant substances with antimicrobial activity (Schwinghamer et al., 1973). Hirsch et al. (1980) discovered that *R. leguminosarum* harbours the symbiotic plasmid pRL1J1, which encodes genes needed for nodulation and nitrogen fixation, as well as bacteriocin development determinants. *R. leguminosarum* bv. viciae, *R. Leguminosarum* bv. trifolii, *R. meliloti*, *R. trifolii*, *Sinorhizobium meliloti* and *Bacterium japonicum* have been reported to secrete antibiotics and cell wall-degrading enzymes that inhibit the phytopathogens (Bardin et al., 2004b; Siddiqui et al., 2000). Recently antibiotic phazolicin class of peptide, previously an unknown compound, was found in the root nodules of wild beans (*Phaseolus vulgaris*), and this compound was produced by a symbiotic soil bacterium that fixes nitrogen for the plants and keeps control over harmful microbes around its rhizosphere (Travin et al., 2019).

1.5.3.1 Siderophore

Siderophores (Greek: "iron carrier") are low-molecular-weight, high-affinity, iron-chelating and water-soluble organic compounds manufactured by many bacteria capable of sequestering ferric ion (Fe^{3+}), fulfilling iron deficiency in plants. Under aerobic conditions, iron undergoes oxidation and forms insoluble complex hydroxides and oxy-hydroxides (Arora et al., 2001). The oxidation state of iron (Fe^{2+}/Fe^{3+}) is regulated by both pH and Eh-activity of electrons (redox potential) of the soil and the accessibility of other minerals (Bodek, 1988). Rhizobia have been reported to produce several siderophores (Gupta et al., 2018; Srinivasan, 2017), but only a few have been structurally characterised until now. These include anthranilate, citrate, rhizobactin and other carboxylates, vicibactin, as well as unidentified catechols and hydroxamates (Carson et al., 2000). Rhizobia develop siderophores in response to iron deficiency, providing a benefit that results in pathogen exclusion due to iron deficiency. The siderophores hinder the development of various phytopathogenic fungi, including *Fusarium oxysporum*, *Phytophthora parasitica*, *Pythium ultimum* and *Scleritinia sclerotiotrum* (Arora et al., 2001). Chelation of bacterial siderophores increases their solubility and availability to plants, which aids in the alleviation of biotic and abiotic stresses (O'Brien et al., 2014).

1.5.3.2 Phytoalexin

Plants need to promptly recognise pathogen attacks to activate their defence mechanism that protects the infection process. Low-molecular-weight secondary metabolites trigger a cascade of antimicrobial and antioxidative events produced in response to stresses, collectively named phytoalexins. Phytoalexins are mostly lipophilic compounds that readily cross the hydrophobic plasma membrane. It is believed that phytoalexins play a major role in plant resistance against pathogenic fungus and bacteria. Phytoalexins were significantly present in higher amounts at the sites of infection compared to healthy plant tissues (Arruda et al., 2016). Pisatin was the first chemically characterised phytoalexin to be purified and chemically identified (Perrin & Bottomley, 1962). Plants can recognise molecular signals originating from both microbes and their cell, which elicit phytoalexins as part of their defence response (Mishra et al., 2012). The pathogens produce virulence factors called effectors/elicitors facilitating pathogen attachment and entry into the host cell and triggering the first line of defence known as PAMP-triggered immunity (PTI), also called as basal or non-host defence and "effector-triggered immunity" (ETI) (Akira, 2009). This is based on the particular interaction between the exogenous or endogenous elicitors and the products of its R gene (Boller & Felix, 2009). Thus, transcription of specific R gene leads to the induction of distinct signalling pathway which can be divided into induced systemic resistance (ISR), activated mainly through abiotic stress and mediated by jasmonate and ethylene, and systemic acquired resistance (SAR), triggered by biotic stress, and is a salicylate-dependent induction (Zhang et al., 2015). The aromatic amino acid phenylalanine acts as a precursor for synthesising phytoalexins via the phenylpropanoid biosynthetic pathway.

1.5.4 Impact of Microbes in Enhancing Soil Fertility, Health and Plant Growth Attributes

Plant growth-promoting microbiota (particularly arbuscular mycorrhizal fungi) boosts farm crop production and returns under natural and stressful environments. Growth regulators, the development of various metabolites and the direct and indirect conversion of atmospheric nitrogen into ammonia are some of the mechanisms involved in growth promotion. It also offers biotic resistance (pathogen) via induced systemic resistance (ISR) and systemic acquired resistance (SAR). Host-microbe interactions help promote plant growth and disease control in a changing climate, allowing for more sustainable farming without jeopardising ecosystem services. The dependence on microbes as biological control agents gives us complete access to the natural system rather than human-made ones. This would also avoid the harmful interference with indigenous microflora and maintain a balanced relationship between human health and its beneficial organisms. Therefore, the industrial application of biocontrol agents will require a deeper understanding of molecular and their biochemical mechanism in particular to combat the pathogen and its interactions with plant and environment. Conservation tillage has the potential to improve grower profitability by lowering production and labour costs. At the same time, organic farming may incur additional maintenance costs due to the increased labour requirements for weeding and pest control, as well as for fertiliser inputs (particularly nitrogen-based), which usually lack the consistency and stability of synthetic fertilisers. Thus, using advantageous plant-microbe interactions, such as those between plant growth-promoting rhizobacterium (PGPR) and arbuscular fungi, assists in achieving long-term agriculture efficiency under normal and challenging environments.

1.6 Conclusions

In the existing paradigm, much emphasis has been placed on implementing sustainable agriculture, in which crops' increased productivity is achieved by using their intrinsic biological capacities, with minimal environmental disruption. Furthermore, the rapid industrialisation and urbanisation releasing effluents to nearby areas have intensified the need for soil quality assets in cities and towns. Recycling, landfilling, incineration and pyrolysis are currently running to eliminate harmful particulate matter concentrations from polluted areas. Still, they all have detrimental environmental implications, creating much more reactive and highly toxic transitional compounds. On the other hand, natural organisms are now being built to degrade chemical contaminants such as plant rhizospheric microbes. Microbes may explicitly enhance plant growth, perhaps by biological nitrogen fixation, hormones, nutrient absorption or indirectly plant protection against biotic and abiotic factors. The majority of previous research has reflected the use of beneficial microbes to improve agricultural production. Besides, these techniques are costly and difficult to adopt for soil. Thus, the hour is to find technical differences and future implementations of microbial inoculation technologies for long-term growth and environmental management against alarming rising population needs and declining productivity issues.

References

Akira, S. (2009). Pathogen recognition by innate immunity and its signaling. *Proceedings of the Japan Academy. Series B, Physical and Biological Sciences, 85*(4), 143–156. https://doi.org/10.2183/pjab.85.143

Anderson, J. R., & Hazell, P. B. (1989). *Variability in grain yields: Implications for agricultural research and policy in developing countries.*

Arora, N. K., Kang, S. C., & Maheshwari, D. K. (2001). Isolation of siderophore-producing strains of Rhizobium meliloti and their biocontrol potential against Macrophomina phaseolina that causes charcoal rot of groundnut. *Current Science*, 673–677.

Arruda, M. P., Lipka, A. E., Brown, P. J., Krill, A. M., Thurber, C., Brown-Guedira, G., … Kolb, F. L. (2016). Comparing genomic selection and marker-assisted selection for Fusarium head blight resistance in wheat (Triticum aestivum L.). *Molecular Breeding, 36*(7), 1–11. https://doi.org/10.1007/s11032-016-0508-5

Avrani, S., Bolotin, E., Katz, S., & Hershberg, R. (2017). Rapid genetic adaptation during the first four months of survival under resource exhaustion. *Molecular Biology Evolution, 34*(7), 1758–1769. https://doi.org/10.1093/molbev/msx118

Bardin, S. D., Huang, H. C., Pinto, J., Amundsen, E. J., & Erickson, R. S. (2004a). Biological control of Pythium damping-off of pea and sugar beet by Rhizobium leguminosarum bv. viceae. *Canadian Journal of Botany, 82*(3), 291–296. https://doi.org/10.1139/b04-003

Bardin, S. D., Huang, H. C., & Moyer, J. R. (2004b). Control of Pythium damping-off of sugar beet by seed treatment with crop straw powders and a biocontrol agent. *Biological Control, 29*(3), 453–460.

Beneduzi, A., Ambrosini, A., & Passaglia, L. M. (2012). Plant growth-promoting rhizobacteria (PGPR): Their potential as antagonists and biocontrol agents. *Genetics and Molecular Biology, 35*(4(Suppl.)), 1044–1051. https://doi.org/10.1590/s1415-47572012000600020

Bodek, I. (1988). *Environmental inorganic chemistry*. Pergamon Press.

Boller, T., & Felix, G. (2009). A renaissance of elicitors: Perception of microbe-associated molecular patterns and danger signals by pattern-recognition receptors. *Annual Review of Plant Biology, 60*, 379–406. https://doi.org/10.1146/annurev.arplant.57.032905.105346

Cao, Y., Halane, M. K., Gassmann, W., & Stacey, G. (2017). The role of plant innate immunity in the legume-rhizobium symbiosis. *Annual Review of Plant Biology, 68*, 535–561. https://doi.org/10.1146/annurev-arplant-042916-041030

Carson, K. C., Meyer, J., & Dilworth, M. J. (2000). Hydroxamate siderophores of root nodule bacteria. *Soil Biology and Biochemistry, 32*(1), 11–21. https://doi.org/10.1016/S0038-0717(99)00107-8

Cetin, Ö. (2004). Role of the micro irrigation on sustainability of soil and water resources. *Proceedings of the International Soil Congress on Natural Resource Management for Sustainable Development*, E 57–65, Erzurum, Turkey.

Chakraborty, B. N., & Purkayastha, R. P. (1987). Alteration in glyceollin synthesis and antigenic patterns after chemical induction of resistance in soybean to Macrophomina phaseolina. *Canadian Journal of Microbiology, 33*(10), 835–840.

Chandra, S., Choure, K., Dubey, R. C., & Maheshwari, D. K. (2007). Rhizosphere competent Mesorhizobiumloti MP6 induces root hair curling, inhibits Sclerotinia sclerotiorum and

enhances growth of Indian mustard (Brassica campestris). *Brazilian Journal of Microbiology, 38*(1), 124–130. https://doi.org/10.1590/S1517-83822007000100026
Crosson, P. 1992. "Sustainable Agriculture", *Resources, 106*, 14–17
Deshwal, V. K., Dubey, R. C., & Maheshwari, D. K. (2003). Isolation of plant growth-promoting strains of Bradyrhizobium (Arachis) sp. with biocontrol potential against Macrophomina phaseolina causing charcoal rot of peanut. *Current Science*, 443–448.
Di Caracalla, V. D. T. (1997). Codex Alimentarius Commission: Procedural Manual.
Dong, X. (1998). SA, JA, ethylene, and disease resistance in plants. *Current Opinion in Plant Biology, 1*(4), 316–323. https://doi.org/10.1016/1369-5266(88)80053-0
Drlica, K., & Zhao, X. (1997). DNA gyrase, topoisomerase IV, and the 4-quinolones. *Microbiology and Molecular Biology Reviews, 61*(3), 377–392. https://doi.org/10.1128/mmbr.61.3.377-392.1997
Fernandez-Göbel, T. F., Deanna, R., Muñoz, N. B., Robert, G., Asurmendi, S., & Lascano, R. (2019). Redox systemic signaling and induced tolerance responses during soybean–Bradyrhizobium japonicum interaction: Involvement of nod factor receptor and autoregulation of nodulation. *Frontiers in Plant Science, 10*, 141. https://doi.org/10.3389/fpls.2019.00141
Figueredo, M. S., Tonelli, M. L., Ibáñez, F., Morla, F., Cerioni, G., del Carmen Tordable, M., & Fabra, A. (2017). Induced systemic resistance and symbiotic performance of peanut plants challenged with fungal pathogens and co-inoculated with the biocontrol agent Bacillus sp. CHEP5 and Bradyrhizobium sp. SEMIA6144. *Microbiological Research, 197*, 65–73. https://doi.org/10.1016/j.micres.2017.01.002
Glick, B. R. (1995). The enhancement of plant growth by free-living bacteria. *Canadian Journal of Microbiology, 41*(2), 109–117. https://doi.org/10.1139/m95-015
Gopalakrishnan, S., Sathya, A., Vijayabharathi, R., Varshney, R. K., Gowda, C. L., & Krishnamurthy, L. (2015). Plant growth promoting rhizobia: Challenges and opportunities. *3. Biotech, 5*(4), 355–377. https://doi.org/10.1007/s13205-014-0241-x
Grace, T. (1990). Misperceptions Cloud Students' Opinions of Agricultural Careers. *Choices, 5*(316-2016-7334).
Gupta, S., Kaushal, R., & Sood, G. (2018). Impact of plant growth–promoting rhizobacteria on vegetable crop production. *International Journal of Vegetable Science, 24*(3), 289–300. https://doi.org/10.1080/19315260.2017.1407984
Hirsch, P. R., Van Montagu, M., Johnston, A. W. B., Brewin, N. J., & Schell, J. (1980). Physical identification of bacteriocinogenic, nodulation and other plasmids in strains of Rhizobium leguminosarm. *Microbiology, 120*(2), 403–412.
Hla, A. K., & Scherer, T. F. (2003). *Introduction to micro-irrigation.*
Kloepper, J. W., Lifshitz, R., & Schroth, M. N. (1988). Pseudomonas inoculants to benefit plant production. *ISI Atlas of Science: Animal and Plant Sciences, 1*(1), 60–64.
Kratsch, H. A., & Wise, R. R. (2000). The ultrastructure of chilling stress. *Plant, Cell and Environment, 23*(4), 337–350. https://doi.org/10.1046/j.1365-3040.2000.00560.x
Lal, R., Reicosky, D. C., & Hanson, J. D. (2007). Evolution of the plow over 10,000 years and the rationale for no-till farming. *Soil and Tillage Research, 93*(1), 1–12. https://doi.org/10.1016/j.still.2006.11.004
Lockeretz, W. (1988). Open questions in sustainable agriculture. *American Journal of Alternative Agriculture, 3*(4), 174–181. https://doi.org/10.1017/S0889189300002460
Madramootoo, C. A., & Rigby, M. (1991). Effects of trickle irrigation on the growth and sunscald of bell peppers (Capsicum annuum L.) in southern Quebec. *Agricultural Water Management, 19*(2), 181–189. https://doi.org/10.1016/0378-3774(91)90007-6
Maheswari. (2017). Enhancing tolerance to climatic stresses in rainfed crops: The road ahead. In V. V. Belavadi, K. N. Nataraja, & N. R. Gangadharappa (Eds.), *Agriculture under climate change: Threats, strategies and policies* (pp. 105–111). Allied Publishers. ISBN: 978–93–85926-37-2.
Maheswari, M., Tekula, V. L., Yellisetty, V., Sarkar, B., Yadav, S. K., Singh, J., … Maddi, V. (2016). Functional mechanisms of drought tolerance in maise through phenotyping and genotyping

under well watered and water stressed conditions. *European Journal of Agronomy, 79*, 43–57. https://doi.org/10.1016/j.eja.2016.05.008

Maheswari, M., Yadav, S. K., Shanker, A. K., Kumar, M. A., & Venkateswarlu, B. (2012). Overview of plant stresses: Mechanisms, adaptations and research pursuit. In *Crop stress and its management: Perspectives and strategies* (pp. 1–18). Springer.rdrecht.

Mauch-Mani, B., Baccelli, I., Luna, E., & Flors, V. (2017). Defense priming: An adaptive part of induced resistance. *Annual Review of Plant Biology, 68*, 485–512. https://doi.org/10.1146/annurev-arplant-042916-041132

Mazzola, M. (2002). Mechanisms of natural soil suppressiveness to soil-borne diseases. *Antonie van Leeuwenhoek, 81*(1), 557–564.

McBratney, A., Whelan, B., Ancev, T., & Bouma, J. (2005). Future directions of precision agriculture. *Precision Agriculture, 6*(1), 7–23.

McDermott, P. F., Walker, R. D., & White, D. G. (2003). Antimicrobials: Modes of action and mechanisms of resistance. *International Journal of Toxicology, 22*(2), 135–143. https://doi.org/10.1080/10915810305089

Mehra, S. (1981). *Instability in Indian agriculture in the context of the new technology, 25*. International Food Policy Research Institute.

Mishra, A. K., Sharma, K., & Misra, R. S. (2012). Elicitor recognition, signal transduction and induced resistance in plants. *Journal of Plant Interactions, 7*(2), 95–120. https://doi.org/10.1080/17429145.2011.597517

Müller, D. B., Vogel, C., Bai, Y., & Vorholt, J. A. (2016). The plant microbiota: Systems-level insights and perspectives. *Annual Review of Genetics, 50*, 211–234. https://doi.org/10.1146/annurev-genet-120215-034952

Munees, A., & Mulugeta,K. (2014). Mechanisms of application of plant growth promoting rhizobacteria: current perspective. *Journal of King Saud University Science, 26*(1), 1–20..

Nguyen, C. (2009). Rhizodeposition of organic C by plant: Mechanisms and controls. *Journal of Sustainable Agriculture*, 97–123.

O'Brien, S., Hodgson, D. J., & Buckling, A. (2014). Social evolution of toxic metal bioremediation in Pseudomonas aeruginosa. *Proceedings of the Royal Society of London. Series B, 281*(1787) PubMed: 20140858.

Ongena, M., Duby, F., Jourdan, E., Beaudry, T., Jadin, V., Dommes, J., & Thonart, P. (2005). Bacillus subtilis M4 decreases plant susceptibility towards fungal pathogens by increasing host resistance associated with differential gene expression. *Applied Microbiology and Biotechnology, 67*(5), 692–698. https://doi.org/10.1007/s00253-004-1741-0

Patil, G., Mian, R., Vuong, T. et al. (2017). Molecular mapping and genomics of soybean seed protein: a review and perspective for the future. *Theoretical and Applied Genetics, 130*, 1975–1991. https://doi.org/10.1007/s00122-017-2955-8

Peigné, J., Ball, B. C., Roger-Estrade, J., & David, C. J. S. U. (2007). Is conservation tillage suitable for organic farming? A review. *Soil Use and Management, 23*(2), 129–144. https://doi.org/10.1111/j.1475-2743.2006.00082.x

Perrin, D. R., & Bottomley, W. (1962). Studies on phytoalexins. V. The structure of pisatin from Pisum sativum L. *Journal of the American Chemical Society, 84*(10), 1919–1922. https://doi.org/10.1021/ja00869a030

Pierce, F. J., & Nowak, P. (1999). *Aspects of precision agriculture advances in agriculture*.

Pieterse, C. M., Zamioudis, C., Berendsen, R. L., Weller, D. M., Van Wees, S. C., & Bakker, P. A. (2014). Induced systemic resistance by beneficial microbes. *Annual Review of Phytopathology, 52*, 347–375. https://doi.org/10.1146/annurev-phyto-082712-102340

Postel, S. L. (2000). Entering an era of water scarcity: The challenges ahead. *Ecological Applications, 10*(4), 941–948. https://doi.org/10.1890/1051-0761(2000)010[0941:EAEOWS]2.0.CO;2

Rey, T., Nars, A., Bonhomme, M., Bottin, A., Huguet, S., Balzergue, S., ... & Jacquet, C. (2013).
NFP, a L ys M protein controlling N od f actor perception, also intervenes in M edicago truncatula resistance to pathogens. *New Phytologist, 198*(3), 875–886.

Ritz, K., Black, H. I. J., Campbell, C. D., Harris, J. A., & Wood, C. (2009). Selecting biological indicators for monitoring soils: A framework for balancing scientific and technical opinion to assist policy development. *Ecological Indicators, 9*(6), 1212–1221. https://doi.org/10.1016/j.ecolind.2009.02.009

Rosier, A., Medeiros, F. H. V., & Bais, H. P. (2018). Defining plant growth promoting rhizobacteria molecular and biochemical networks in beneficial plant-microbe interactions. *Plant and Soil, 428*(1–2), 35–55. https://doi.org/10.1007/s11104-018-3679-5

Saha, M., Sarkar, S., Sarkar, B., Sharma, B. K., Bhattacharjee, S., & Tribedi, P. (2016). Microbial siderophores and their potential applications: A review. *Environmental Science and Pollution Research International, 23*(5), 3984–3999. https://doi.org/10.1007/s11356-015-4294-0

Saijo, Y., Loo, E. P. I., & Yasuda, S. (2018). Pattern recognition receptors and signaling in plant–microbe interactions. *Plant Journal: For Cell and Molecular Biology, 93*(4), 592–613. https://doi.org/10.1111/tpj.13808

Salto, H., Watanabe, T., & Tomloka, H. (1979). Purification, properties and cytotoxic effect of a bacteriocin from Mycobacterium smegmatis. *Antimicrobial Agents and Chemotherapy,15*, 504–509.

Scheidle, H., Gross, A., & Niehaus, K. (2005). The lipid A substructure of the Sinorhizobium meliloti lipopolysaccharides is sufficient to suppress the oxidative burst in host plants. *New Phytologist, 165*(2), 559–565. https://doi.org/10.1111/j.1469-8137.2004.01214.x

Schwinghamer, E. A., Pankhurst, C. E., & Whitfeld, P. R. (1973). A Phage-like bacteriocin of Rhizobium trifolii. *Canadian Journal of Microbiology, 19*(3), 359–368. https://doi.org/10.1139/m73-059

Secretariat, C. G. I. A. R., & CGIAR Technical Advisory Committee. (1988). *Review processes in the CGIAR.*

Shah, T. (2011). *Past, present and the future of canal irrigation in India* (pp. 70–87). India infrastructure report.

Shanker, A. K., Maheswari, M., Yadav, S. K., Desai, S., Bhanu, D., Attal, N. B., & Venkateswarlu, B. (2014). Drought stress responses in crops. *Functional and Integrative Genomics, 14*(1), 11–22. https://doi.org/10.1007/s10142-013-0356-x

Siddiqui, I. A., Ehteshamul-Haque, S., Zaki, M. J., et al. (2000). Effect of urea on the efficacy of Bradyrhizobium sp. and Trichoderma harzianum in the control of root infecting fungi in mungbean and sunflower. *Sarhad Journal of Agriculture, 16*, 403–406.

Singh, I. J. (1998). Farm poverty, household food security and agricultural sustainability in India. *Journal of Rural Development-Hyderabad, 17*, 619–632.

Six, J., Elliott, E. T., Paustian, K., Doran, J. W. (1998). Aggregation and soil organic matter accumulation in cultivated and native grassland soils. *Soil Science Society of America Journal, 62*, 1367–1377.

Smigielski, L., Laubach, E. M., Pesch, L., Glock, J. M. L., Albrecht, F., Slusarenko, A., … Kuhn, H. (2019). Nodulation induces systemic resistance of Medicago truncatula and Pisum sativum against Erysiphe pisi and primes for powdery mildew-triggered salicylic acid accumulation. *Molecular Plant–Microbe Interactions, 32*(9), 1243–1255. https://doi.org/10.1094/MPMI-11-18-0304-R

Srinivasan, T. (2017). Studies on antifungal activity of siderophores produced by rhizobium spp. isolated from groundnut (Arachis hypogaea). *Journal of Agricultural Science and Food Research, 8*(4), 1–2.

Tagg, J. R., Dajani, A. S., & Wannamaker, L. W. (1976). Bacteriocins of gram-positive bacteria. *Bacteriological Reviews, 40*(3), 722–756. https://doi.org/10.1128/br.40.3.722-756.1976

Tebrügge, F., & Düring, R.-A. (1999). Reducing tillage intensity—A review of results from a long-term study in Germany. *Soil and Tillage Research, 53*(1), 15–28. https://doi.org/10.1016/S0167-1987(99)00073-2

Tonelli, M. L., Figueredo, M. S., Rodríguez, J., Fabra, A., & Ibañez, F. (2020). Induced systemic resistance-like responses elicited by rhizobia. *Plant and Soil, 448*(1–2), 1–14. https://doi.org/10.1007/s11104-020-04423-5

Travin, D. Y., Watson, Z. L., Metelev, M., Ward, F. R., Osterman, I. A., Khven, I. M., … Severinov, K. (2019). Structure of ribosome-bound azole-modified peptide phazolicin rationalises its species-specific mode of bacterial translation inhibition. *Nature Communications, 10*(1), 1.
Triplett, E. W., & Barta, T. M. (1987). Trifolitoxin production and nodulation are necessary for the expression of superior nodulation competitiveness by Rhizobium leguminosarum bv. trifolii strain T24 on clover. *Plant Physiology, 85*(2), 335–342. https://doi.org/10.1104/pp.85.2.335
USDA Study Team on Organic Agriculture. (1980). Report and Recommendations on Organic Farming. US Department of Agriculture, Washington, DC. Available at Website http://www.nal.usda.gov/afsic/pubs/USDAOrgFarmRpt.pdf (verified December 20, 2012)
Van Loon, L. C., Bakker, P. A. H. M., & Pieterse, C. M. J. (1998). Systemic resistance induced by rhizosphere bacteria. *Annual Review of Phytopathology, 36*(1), 453–483. https://doi.org/10.1146/annurev.phyto.36.1.453
Varma, S., Verma, S., & Namara, R. E. (2006). Promoting micro irrigation technologies that reduce poverty. *Water Policy Briefing, 23*.
Vasileiadis, V. P., Froud-Williams, R. J., & Eleftherohorinos, I. G. (2007). Vertical distribution, size and composition of the weed seedbank under various tillage and herbicide treatments in a sequence of industrial crops. *Weed Research, 47*(3), 222–230. https://doi.org/10.1111/j.1365-3180.2007.00564.x
Weller, D. M., Raaijmakers, J. M., Gardener, B. B. M., & Thomashow, L. S. (2002). Microbial populations responsible for specific soil suppressiveness to plant pathogens. *Annual Review of Phytopathology, 40*(1), 309–348. https://doi.org/10.1146/annurev.phyto.40.030402.110010
Wiener, P. (1996). Experimental studies on the ecological role of antibiotic production in bacteria. *Evolutionary Ecology, 10*(4), 405–421.
World Health Organization. (2003). *Diet, nutrition, and the prevention of chronic diseases: Report of a joint WHO/FAO expert consultation, 916*. World Health Organization.
Yu, K., Pieterse, C. M. J., Bakker, P. A. H. M., & Berendsen, R. L. (2019). Beneficial microbes going underground of root immunity. *Plant, Cell and Environment, 42*(10), 2860–2870. https://doi.org/10.1111/pce.13632
Zamioudis, C., & Pieterse, C. M. (2012). Modulation of host immunity by beneficial microbes. *Molecular Plant–Microbe Interactions, 25*(2), 139–150. https://doi.org/10.1094/MPMI-06-11-0179
Zhang, D. W., Deng, X. G., Fu, F. Q., Lin, H. H. (2015). Induction of plant virus defense response by brassinosteroids and brassinosteroid signaling in Arabidopsis thaliana. *Planta, 241*, 875–885.

Chapter 2
Biofertilizers: The Role in Sustainable Agriculture

Ishfaq Ul Rehman, Tajamul Islam, Abid Hussain Wani, Irfan Rashid, Ishfaq Ahmad Sheergojri, Maajid Mehraj Bandh, and Summia Rehman

Abstract The global rise in the human population presents a significant challenge to the world's food security. There is a huge gap between production and consumption (7.2 million tonnes nutrient deficit) due to the expanding population and agricultural land's shrinking over time. Therefore, crop production must be significantly boosted in the next few decades to meet the growing population's food demand. Chemical fertilizers have been extensively used to enable the crop outputs to bridge the lacunae between production and consumption, which ultimately seriously damaged both the natural ecosystems and human health. Therefore, biofertilizers' exploitation is, to a certain extent, considered a substitute for chemical fertilizers in the agricultural industry because of their significant potential to improve food production and safety. Biofertilizers are substances that include cells of different varieties of beneficial microorganisms that could become critical components of advanced nutrient management. Organisms typically used as constituents of biofertilizers are nitrogen fixers (N fixer), phosphorus solubilizers (P-solubilizer) and potassium solubilizers (K-solubilizer) or a mixture of fungi and moulds. These possible biofertilizers play a vital role in the production and sustainability of soils and safeguard the environment by being eco-friendly and cost-effective for producers.

Keywords Biofertilizer · Sustainability · Chemical fertilizers · Agriculture · Food security

---

I. U. Rehman (✉) · T. Islam · A. H. Wani · I. Rashid · I. A. Sheergojri · S. Rehman
Department of Botany, University of Kashmir, Srinagar, Jammu and Kashmir, India

M. M. Bandh
Department of Zoology, University of Kashmir, Srinagar, Jammu and Kashmir, India

S. A. Bandh (ed.), *Sustainable Agriculture*,
https://doi.org/10.1007/978-3-030-83066-3_2

2.1 Introduction

Since the evolution of the human race, agriculture has been a primary survival tool for human generations. The agriculture sector generates food, fodder and other non-wood products on which a significant fraction of the world's population relies (Hervé et al., 2016; Kaur et al., 2018, 2018b). Given the population explosion, we need to set up appropriate plans and protocols to promote sustainable agriculture to meet our demands (Bhardwaj et al., 2014; Singh et al., 2014). Traditional methods are sensu stricto just to the farmer's families and the local village communities as it only engages food and feed production at the domestic level (Jehangir et al., 2017; Pandey, 2018). However, with the advent of innovative technologies, the production of agriculture output per hectare increases. Sustainable agriculture is a sensu latu concept to grow crops to their threshold limit while simultaneously protecting the environment (Barragán-Ocaña & Del-Valle-Rivera, 2016). There is a stochastic change in agricultural practice methodology as people have been paying more attention to safeguarding the environment while working to amplify the farm yield. Various hormones, fertilizers and even innovative approaches to fertigation are being used to increase crop production (Campos et al., 2019; Umesha et al., 2018a). Though yield may be amplified with increased application of synthetic chemicals, the overdose of synthetic fertilizers would diminish the utility of living conditions through pollutant and biomagnification or ecological amplification (Uosif et al., 2014). But that would compromise environmental sustainability. So, without compromising the future generation's needs, the current generation could attain sustainability in the agricultural sector by using eco-friendly products and environmentally sound technologies (Calabi-Floody et al., 2018; Umesha et al., 2018b; Wang et al., 2015).

Biofertilizer is one such product that helps us in achieving sustainable agriculture. Biofertilizers are the amalgamation of live or latent cells of competent phosphate-solubilizing strains, N_2-fixing or cellulolytic microorganisms, mainly used to apply to seeds and seedlings, etc. (Agarwal et al., 2018; García-Fraile et al., 2015). They play a significant role in escalating soil fertility by fixing atmospheric nitrogen and converting it into usable products. They also promote root growth by producing necessary hormones and antimetabolites and help in soil mineralization and nutrient decomposition (A, B., Ak, M., M, G., G, G., P, P., … B, J., 2009; Kumar et al., 2017). They are economical and can be used as supplements to synthetic fertilizers. Microflora like bacteria, fungi and blue-green algae are used as the principal ingredients of biofertilizers. To improve their shelf-life, these should be packed in material like peat and lignite powder. This way, biofertilizers have an utmost significance in sustaining agriculture and a safe environment (Agarwal et al., 2018). They can be grouped into different categories (Kumar et al., 2017) based on their service in sustainable agriculture (Table 2.1).

Table 2.1 Categories of biofertilizers

S. no.	Biofertilizers	Examples
1.	**Nitrogen fixing**	*Frankia, Anabaena, Rhizobium* (symbiotic)
		Anabaena, Azotobacter, Clostridium (free-living)
		Azospirillum (associative symbiotic)
2.	**Phosphorus mobilizing**	*Glomus* sp., *Gigaspora* sp., *Scutellospora* sp. and *Sclerocystis* sp. (arbuscular mycorrhiza)
		R. solani (orchid mycorrhiza)
		Laccaria sp., *boletus* sp., *Pisolithus* sp., *amanita* sp. (ectomycorrhiza)
		Pezizellaericae
3.	**Phosphorus solubilizing**	*Bacillus circulans, B. megaterium var. phosphaticum, B. subtilis, P. striata* **(bacteria)**
		Aspergillus awamori, Penicillium sp. **(Fungi)**
4.	**Biofertilizers (micronutrients)**	*Bacillus* sp. (for zinc and silicate Solubilizers)
5.	**Rhizobacteria (plant growth promoting)**	*P. fluorescens* (promoting plant growth)

2.1.1 *Rhizobium*

It is the most extensively studied genus to carrying the function of N_2 fixation (Odame, 1997). This genus's strains *are* symbiotically associated with leguminous crops, the essential food components. Apart from the ingredient of meals, legumes possess the potential to improve soil health via N_2 fixation (Laranjo et al., 2014), with different strains of this genus.

2.1.2 *Azospirillum*

Genus Azospirillum – a free-living genus – fixes the atmospheric nitrogen at a rate of 20–40 kg $ha^{-1}y^{-1}$ (Bashan, 1993). Strains of this genus are used as biofertilizers in various economically important crops like corn, rice, wheat, etc., which are (Döbereiner, 1997; Reinhold & Hurek, 1989; Sundaram et al., 1988). There is a proven fact that there is an increase in agricultural production and soil properties by applying *Azospirillum* strains Motsara et al. (1995).

2.1.3 *Azotobacter*

Azotobacter is an important genus promoting the synthesis of active secondary compounds like heteroxins, vitamins, gibberellins, etc. and thus significantly improves plants' root growth. This genus's species show intolerance to fluctuations of pH, salts and temperature (Jaga & Singh, 2010; Rao, 1986). Growth and crop yield of Triticum aestivum have increased with the augmentation of *Azotobacter* species (Ei-Lattief, 2016). Along with some yeast strains, it shows many improvised results (Ahmed et al., 2011). Some well-known methods of application of *Azotobacter* species are seed sipping and seedling root dipping.

2.1.4 Phosphorus-Solubilizing and Phosphorus-Mobilizing Microbes

Biofertilizers containing P-solubilizing and phosphorus-mobilizing microbes can make the accumulated phosphates readily available for plant growth progress (Goldstein, 1986). PSB also alters the status of soil nutrient structure (Blake, 1993).

2.2 Biofertilizers: Why their Need Is Inevitable?

In contemporary times there is a trend of environmental hazards and threats to sustainable agriculture due to the extensive use of chemical fertilizers. Given this, biofertilizers' continued application proves very economical, eco-friendly, efficient and productive to marginal and small farmers over chemical fertilizers. There are primarily two reasons which push the agriculturists for frequent use of biofertilizers:

- Usage of biofertilizers is a front runner in increased crop productivity.
- Increased usage of chemical fertilizer augments the damage in soil texture with accompanying environmental problems.

2.3 How Biofertilizers Work

Bacteria, fungi and other microorganisms, though distributed heterogeneously, are omnipotent in the world. The microbes that are present in the rhizosphere and can promote growth and development are accordingly termed as plant growth-promoting bacteria (PGPB). These days several PGPBs are available as biofertilizers at a commercial scale (Calvo et al., 2014). PGPB plays a very prominent role in plant growth and soil fertility maintenance in the following ways (Fig. 2.1):

1. Under certain stress conditions, PGPBs may produce and supply important hormones like auxins, cytokinin and gibberellin, which directly regulate plant growth. PGPBs also aid plants by providing essential elements like nitrogen (N) and phosphorous (P) and enhance potassium (K) intake, etc. Since these activities promote plant growth directly, they are called direct ways.

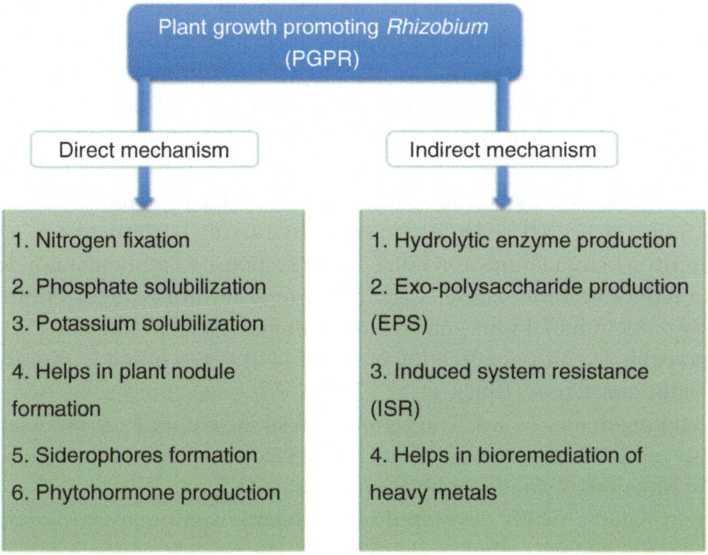

Fig. 2.1 Direct and indirect mode of action

2. PGPBs also promote plants' growth indirectly via different pathways, e.g. protecting against the deleterious effects of plant pathogens (Gamalero et al., 2009).

2.3.1 Direct Way

Nitrogen (N_2) fixation is the best-studied direct way of growth promotion in plants by PGPBs. Nitrogen belongs to the category of essential nutrient element for plants. Though the atmosphere contains 78% nitrogen as dinitrogen (N_2), plants cannot take up and use it in this form. The plant available forms (PAFs) of nitrogen are ammonia and nitrates mainly produced by microorganisms via biological nitrogen fixation (BNF). Plants can assimilate the nitrates and ammonia via assimilation pathways, i.e. ammonium assimilation and nitrate assimilation, respectively (Tairo & Ndakidemi, 2013). The nitrogen-fixing microorganisms (also called diazotrophs) possess a unique enzyme complex known as dinitrogenase which acts on atmospheric nitrogen and converts it into ammonia (Smith et al., 2013). Diazotrophs can be free-living and symbiotic. The symbiotic category includes *Rhizobiaceae* members that share a symbiosis with plants belonging to *Leguminosae* (Ahemad & Khan, 2012a, 2012b, 2012c, 2012d).

Symbiotic nitrogen-fixing rhizobacteria collectively called *Rhizobia*, e.g. *Rhizobium*, *Bradyrhizobium*, *Sinorhizobium*, *Azorhizobium* and *Mesorhizobium*, belong to the *Rhizobiaceae* (*Alphaproteobacteria*) family. They create a symbiotic relationship with their hosts (legumes) by infecting their roots. A complex interplay of chemical signalling between the host and symbiont is required to establish this relationship, which results in the root nodule formation *Rhizobia* resides intracellularly as symbiont (Allito et al., 2015). Simultaneously, the non-symbiotic nitrogen-fixing rhizobacteria form a non-obligatory relationship with the non-leguminous plants (Verma et al., 2010). All diazotrophs carry nitrogen fixation by a complex molybdenum-iron dinitrogenase system, consisting of dinitrogenase reductase with a cofactor of Fe (iron) and dinitrogenase with Fe and Mo (molybdenum) as its cofactors (Smith et al., 2013). However, many free-living bacteria like *A. vinelandii* contain iron-iron or vanadium-iron cofactors in response to molybdenum depletion. Ferredoxin reduces dinitrogenase reductase, which in turn reduces dinitrogenase, followed by reduction of dinitrogen (N_2) to ammonia (NH_3) (Santi et al., 2013). The genes responsible for nitrogen fixation are called Nif genes found in free-living and symbiotic nitrogen fixers (Black et al., 2012). *Nif* genes consist of two types of structural and regulatory genes. The former is responsible for Fe protein activation, biosynthesis of Fe-Mo cofactor and donation of electrons, and the latter accountable for enzyme functionality and synthesis (Ahemad & Kibret, 2014).

Maximum legume plants develop de novo lateral root organs to accommodate the symbiotic *Rhizobium*, called "root nodules". The process of development of root nodules in legumes is an essential feature associated with nitrogen fixation. It involves de-differentiation of differentiated root cortical cells following bacterial infection to the roots. The de-differentiated cells then differentiate into root nodules harbouring the nitrogen-fixing microbes (Suzaki et al., 2015). The schematic representation of plant-bacteria interaction has been presented in Fig. 2.2.

For the proper formation of nodules, two regulatory events, viz. infection of bacteria in epidermis and organogenesis of the cortex's nodule (Suzaki & Kawaguchi, 2014), must be well synchronized. Nod factors produced by rhizobia are chemically lipochoto-oligosaccharides which initiate the symbiotic relation between the

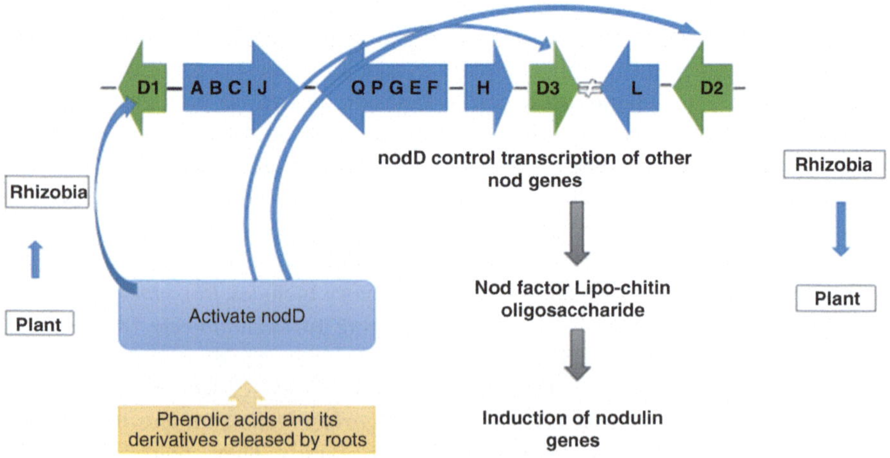

Fig. 2.2 Diagrammatic representation of plant-bacteria interaction

rhizobium and host plants (Maillet et al., 2011). The amount of plant ethylene has been reported to increase after the infection with *Rhizobium* spp., which prevents further rhizobial disease and nodule development (Abeles et al., 2012). Several rhizobial strains restrict the increase in ethylene production by synthesizing "rhizobitoxine", a small molecule that inhibits the function of enzyme 1-aminocyclopropane-1-carboxylate (ACC) synthase – the main regulatory enzyme in the ethylene biosynthesis pathway (Nascimento et al., 2012). The limit on ethylene production because of rhizobitoxine causes an increase in the number of nodules formed on host plant roots and enhances symbiosis (Vijayan et al., 2013). Some rhizobial strains decrease the ethylene concentration by producing the enzyme ACC deaminase, which breaks down some ACC before its conversion to ethylene. This ethylene reduction results in increased formation of nodules and plant biomass by 25–40% (Zahir et al., 2011). About 10% of rhizobial strains in the field naturally have ACC deaminase, so the number of nodules formed by *Rhizobia* strains without ACC deaminase can be enhanced through transformation via genetic engineering (Glick, 2014). The same approach was used in the Sinorhizobium meliloti strain lacking this enzyme which was transformed with the ACC deaminase gene from *Rhizobium leguminosarum* bv. *viciae*. This transformation reportedly increased the number of nodules significantly by approximately 35% and the biomass of alfalfa by 40% compared to the control of wild strain (Glick, 2012, 2015).

Phytohormones produced and supplied by plant-associated microflora can stimulate growth and development in the host plant by modulating its endogenous hormone levels (Gray, 2004; Van Loon, 2007). Auxins (indole-3-acetic acid), gibberellins (GAs) and cytokinins are among the most significant plant growth regulators produced by associated microbes. It is reported that 80% of associated rhizospheric microbes of different crops synthesize auxins as their secondary metabolites (Ahemad & Khan, 2011). They do so through the tryptophan-dependent and tryptophan-independent pathways. The following three tryptophan-dependent auxin synthesis pathways are known:

(i) Indole-3-pyruvic acid (IPyA) pathway present in *Rhizobium*, *Bradyrhizobium* and *Azospirillum*
(ii) Indole-3-acetamide (IAM) pathway found in certain pathogenic bacteria such as *Pseudomonas syringae* and *Agrobacterium tumefaciens*
(iii) Tryptamine pathway found in species like *Bacillus megaterium* and *Bacillus licheniformis*

IAA from a rhizobacterial source has been recognized as the primary causal molecule of pathogenesis or phytostimulation (Ahemad & Khan, 2012b; Mahanty et al., 2017). In addition to IAA, cytokinin modulation is reported to be involved in phytostimulation. For example, enhancement of seedling growth in *Arabidopsis thaliana* and *Proteus vulgaris* has been reported via cytokinin synthesis by *Bacillus megaterium* (Ortíz-Castro et al., 2008). Diverse genera of bacteria like *Azospirillum*, *Bacillus*, *Klebsiella*, *Proteus*, *Pseudomonas*, *Xanthomonas*, etc. include well-characterized cytokinin-producing species. Besides cytokinin and ethylene, gibberellin (GA) production has also been observed in plant-associated bacteria and fungi. GA-producing bacteria are used to boost seed germination rate even though bacterial GA's precise function is not understood (Goswami et al., 2016).

2.3.2 Indirect Way

As mentioned above, PGPB may also indirectly promote the growth of plants with which they are associated, e.g. by reducing the adverse effects of pathogenic fungi or bacteria on plant growth and development. Scientists are also promoting the use of PGPB as an eco-friendly alternative to chemical fungicides.

Antibiotics
PGPBs protect plants from many pathogen attacks by producing antibiotics acting against the pathogens (mainly fungi). They make different antibiotics under different conditions, with many of them studied in detail. Some of these antibiotic-producing strains have also been commercialized. Scientists have also modified these antibiotic-producing strains of PGPBs to produce antibiotics under laboratory conditions (Devine et al., 2017).

Cell Wall-Degrading Enzymes
Some plants defend themselves against the pathogenic fungus by producing enzymes involved in cell wall degradation, e.g. such as chitinase, which hydrolyses chitin in the fungal cell wall. Similarly, some bacteria (used as biocontrol agents) produce enzymes like cellulases, chitinases, glucanases, lipases and proteases, which can degrade many pathogenic fungi's cell walls (Kim et al., 2015). Hence PGPBs control fungal diseases and prevent yield loss which is a promising and eco-friendly way.

Hydrogen Cyanide
Many PGPBs like *Rhizobium*, *Pseudomonas* and *Bacillus* produce hydrogen cyanide to control many diseases. The lower levels of hydrogen cyanide do not allow fungal pathogens to attack plants and enhance plants' resistance against the diseases. The hydrogen cyanide works by inhibiting the cytochrome-c oxidase and its other metabolites. Some bacteria also produce antibiotics and HCN that act synergistically against fungal pathogens and prevent the development of resistant pathogenic fungi (Olanrewaju et al., 2017; Ramette et al., 2006).

2.4 Methods of Application of Biofertilizers to Crops

2.4.1 Seed Treatment

In this method, selected biofertilizers are mixed with water and gently combined with the seed mass with the help of an adhesive like gum acacia, jiggery solution, etc.; however, before sowing, the seeds are shade dried on a clean sheet or piece of cloth.

2.4.2 Seedling Root Dip

In this method, selected biofertilizers are mixed with water, and the seedling roots are dipped in the mixture for about 8–10 hours before transplantation. This method is usually used for transplanting crops such as rice.

2.4.3 Soil Treatment

Recommended biofertilizers and compost are mixed in the ratio of 1:50 by weight and kept overnight. This mixture is mixed with soil at the time of sowing seeds or transplanting the seedlings.

2.5 The Role of Biofertilizers in the Alleviation of Environmental Stresses

Plants, being sessile, are vulnerable to a plethora of stress (abiotic and biotic) at any given instant, interrupting normal metabolism seen as abnormal physiology. During stress, the peculiar features in plants are the excessive formation of reactive oxygen species (ROS), which include both free radicals and non-radical molecules. These extreme ROS molecules damage cellular lipids, hence damaging cells, and cause metabolic disorders and variations in senescence. Several plant growth-promoting rhizobacteria (PGPR) are known to help plants alleviate stress. For example, rhizobacteria maintain the cytoplasmic osmolarity under drought stress by producing various osmolytes like glycine betaine, proline, ectoine, and trehalose. The production of glycine betaine helps plants tolerate droughts, frost and salinity stresses simultaneously.

As mentioned above, several rhizobacterial species may produce growth regulators like ethylene and cytokinin, thus promoting crop plants' growth under stress. *Pseudomonas alcaligenes*, *P. aureofaciens*, *P. aurantiaca* and *P. chlororaphis* have plant hormones under unfavourable conditions in saline arid soils (Verma et al., 2017; Yadav et al., 2018). Further, alleviation of heat stress in plants by cytokinin-producing PGPB was isolated from the rhizosphere's soil (Arkhipova et al., 2007). Liu et al. (2013) also reported that a cytokinin producer, *Bacillus subtilis*, improved the tolerance to drought stress. Forchetti et al. (2010) showed that *Bacillus pumilus* and *Achromobacter xylosoxidans* which are drought-tolerant endophytic bacterial strains produce salicylic acid. Similarly, inoculation with rhizobacterial strains enhances the growth parameters of sunflower under conditions of water stress.

Inoculation of cucumber plants with bacteria like *Pseudomonas fluorescens*, *P. extremorientalis*, *Stenotrophomonas rhizophila* and *Serratia plymuthica* increases dry weight appreciably up to 62% compared to the control. The improvement in growth and salt tolerance has been reported to be the IAA production under a salt environment. Further, the fruit yield of the cucumber was also enhanced under controlled conditions (Egamberdieva, 2011). Timmusk et al. (2014) showed that wheat treatment with PGPB under drought stress increased biomass to 78% higher than untreated plants. Enhanced biomass and root architecture modifications under drought stress were reported when inoculated with PGPB strain (Bresson et al., 2014) *Phyllobacterium brassicacearum* (STM196).

2.6 Some Factors Limiting the Use of Biofertilizers

- Lack of regulatory acts and facilities regarding testing of samples: One of the potential limitations to the use of fertilizers is a scarcity of facilities provided by institutions for testing biofertilizer samples. Further, there is a lack of government involvement in this area which is a potential eco-friendly alternative to chemical pesticides. In biofertilization, future research should be focused on options available to confront the issues and propose valid frameworks for the development of eco-friendly practices that allow advancement on the efficiency and subsequent supply of product for the industry in the global economies. Furthermore, their application's technical tests must authenticate their safe use at the worldwide level.
- Biofertilizers' insufficient popularization and low level of farmer acceptance: Among farmers, biofertilizers have not gained the required popularity. However, it comes with various potential benefits for crops, especially under stresses. This non-acceptance seems to be the lack of awareness among the farmers compared to their synthetic counterparts. Other problems such as lack of timely financing, experts' involvement and biofertilizers' non-availability also hinder their popularity and acceptance.

2.7 Conclusions

Biofertilizers form a significant component of organic farming in modern agricultural practices in terms of being a sustainable alternative to chemical fertilizers, linked with various environmental hazards. However, to popularize the biofertilizers' status, increased demand and awareness about its uses are yet to be created. Biofertilizer technology, a significant part of sustainable agriculture, has to be proper for farmers' and planters' social and infrastructural situations. It should be economically feasible for all farmers, renewable, adaptable to prevailing local conditions and satisfactory from the society's cultural patterns, practically implementable and productive.

Acknowledgement The authors are thankful to the University Grants Commission for providing fellowship to corresponding author and grateful to the Department of Botany, the University of Kashmir, for sharing their facilities and those who made their literature available for the compilation of this manuscript.

References

A, B., Ak, M., M, G., G, G., P, P., ... B, J. (2009). Biofertilizers: A novel tool for agriculture. *International Journal of Microbiology Research, 1*(2), 23–31. https://doi.org/10.9735/0975-5276.1.2.23-31

Agarwal, P., Gupta, R., & Gill, I. K. (2018). Importance of biofertilizers in agriculture biotechnology. *Annals of Biological Research, 9*(3), 1–3.

Ahemad, M., & Khan, M. S. (2011). Effects of insecticides on plant-growth-promoting activities of phosphate solubilizing rhizobacterium Klebsiella sp. strain PS19. *Pesticide Biochemistry and Physiology, 100*(1), 51–56. https://doi.org/10.1016/j.pestbp.2011.02.004

Ahemad, M., & Khan, M. S. (2012a). Effects of pesticides on plant growth promoting traits of Mesorhizobium strain MRC4. *Journal of the Saudi Society of Agricultural Sciences, 11*(1), 63–71. https://doi.org/10.1016/j.jssas.2011.10.001

Ahemad, M., & Khan, M. S. (2012b). Effect of fungicides on plant growth promoting activities of phosphate solubilizing Pseudomonas putida isolated from mustard (Brassica compestris) rhizosphere. *Chemosphere, 86*(9), 945–950. https://doi.org/10.1016/j.chemosphere.2011.11.013

Ahemad, M., & Khan, M. S. (2012c). Ecological assessment of biotoxicity of pesticides towards plant growthpromoting activities of pea (Pisum sativum)-specific rhizobium SP. STRAINMRP1. *Emirates Journal of Food and Agriculture*, 334–343.

Ahemad, M., & Khan, M. S. (2012d). Evaluation of plant-growth-promoting activities of rhizobacterium Pseudomonas putida under herbicide stress. *Annals of Microbiology, 62*(4), 1531–1540. https://doi.org/10.1007/s13213-011-0407-2

Ahemad, M., & Kibret, M. (2014). Mechanisms and applications of plant growth promoting rhizobacteria: Current perspective. *Journal of King Saud University – Science, 26*(1), 1–20. https://doi.org/10.1016/j.jksus.2013.05.001

Ahmed, M. A., Ahmed, A. G., Mohamed, M. H., & Tawfik, M. M. (2011). Integrated effect of organic and biofertilizers on wheat productivity in new reclaimed sandy soil. *Research Journal of Agriculture and Biological Sciences, 7*(1), 105–114.

Allito, B. B., Nana, E. M., & Alemneh, A. A. (2015). Rhizobia strain and legume genome interaction effects on nitrogen fixation and yield of grain legume: A review. *Molecular Soil Biology, 6*. https://doi.org/10.5376/msb.2015.06.0004

Arkhipova, T. N., Prinsen, E., Veselov, S. U., Martinenko, E. V., Melentiev, A. I., & Kudoyarova, G. R. (2007). Cytokinin producing bacteria enhance plant growth in drying soil. *Plant and Soil, 292*(1–2), 305–315. https://doi.org/10.1007/s11104-007-9233-5

Barragán-Ocaña, A., & Del-Valle-Rivera, M. D. C. (2016). Rural development and environmental protection through the use of biofertilizers in agriculture: An alternative for underdeveloped countries? *Technology in Society, 46*, 90–99. https://doi.org/10.1016/j.techsoc.2016.06.001

Bashan, Y. (1993). Potencial use of Azospirillum as biofertilizer. *Turrialba, 43*, 286–286.

Bhardwaj, D., Ansari, M. W., Sahoo, R. K., & Tuteja, N. (2014). Biofertilizers function as key player in sustainable agriculture by improving soil fertility, plant tolerance and crop productivity. *Microbial Cell Factories, 13*(1), 66. https://doi.org/10.1186/1475-2859-13-66

Black, M., Moolhuijzen, P., Chapman, B., Barrero, R., Howieson, J., Hungria, M., & Bellgard, M. (2012). The genetics of symbiotic nitrogen fixation: Comparative genomics of 14 rhizobia strains by resolution of protein clusters. *Genes, 3*(1), 138–166. https://doi.org/10.3390/genes3010138

Blake, F. (1993) In: Cartwright (Ed.). *Organic food production. In-world Agriculture* (pp. 22–24). Hong Kong: Strerling Publication Ltd.

Bresson, J., Vasseur, F., Dauzat, M., Labadie, M., Varoquaux, F., Touraine, B., & Vile, D. (2014). Interact to survive: Phyllobacterium brassicacearum improves Arabidopsis tolerance to severe water deficit and growth recovery. *PLOS ONE, 9*(9), e107607. https://doi.org/10.1371/journal.pone.0107607

Calabi-Floody, M., Medina, J., Rumpel, C., Condron, L. M., Hernandez, M., Dumont, M., & Mora, MDLL. (2018). Smart fertilizers as a strategy for sustainable agriculture. In *Advances in Agronomy* (p. 147). Academic Press. https://doi.org/10.1016/bs.agron.2017.10.003

Calvo, P., Nelson, L., & Kloepper, J. W. (2014). Agricultural uses of plant biostimulants. *Plant and Soil, 383*(1–2), 3–41. https://doi.org/10.1007/s11104-014-2131-8

Campos, E. V. R., Proença, P. L. F., Oliveira, J. L., Bakshi, M., Abhilash, P. C., & Fraceto, L. F. (2019). Use of botanical insecticides for sustainable agriculture: Future perspectives. *Ecological Indicators, 105*, 483–495. https://doi.org/10.1016/j.ecolind.2018.04.038

Devine, A., Harvey, R., Min, A. M., Gilder, M. E. T., Paw, M. K., Kang, J., … McGready, R. (2017). Strategies for the prevention of perinatal hepatitis B transmission in a marginalized population on the Thailand-Myanmar border: A cost-effectiveness analysis. *BMC Infectious Diseases, 17*(1), 552. https://doi.org/10.1186/s12879-017-2660-x

Döbereiner, J. (1997). A importância da fixação biológica de nitrogênio para a agricultura sustentável. *Biotecnologia Ciência*, 2–3.

Egamberdieva, D. (2011). Survival of Pseudomonas extremorientalis TSAU20 and P. chlororaphis TSAU13 in the rhizosphere of common bean (Phaseolus vulgaris) under saline conditions. *Plant, Soil and Environment, 57*(3), 122–127. https://doi.org/10.17221/316/2010-PSE

Ei-Lattief, E. A. (2016). Use of Azospirillum and Azobacter bacteria as biofertilizers in cereal crops: A review. *International Journal of Research in Engineering and Applied Sciences (IJREAS), 6*(7), 36–44.

Forchetti, G., Masciarelli, O., Izaguirre, M. J., Alemano, S., Alvarez, D., & Abdala, G. (2010). Endophytic bacteria improve seedling growth of sunflower under water stress, produce salicylic acid, and inhibit growth of pathogenic fungi. *Current Microbiology, 61*(6), 485–493. https://doi.org/10.1007/s00284-010-9642-1

Gamalero, E., Lingua, G., Berta, G., & Glick, B. R. (2009). Beneficial role of plant growth promoting bacteria and arbuscular mycorrhizal fungi on plant responses to heavy metal stress. *Canadian Journal of Microbiology, 55*(5), 501–514. https://doi.org/10.1139/w09-010

García-Fraile, P., Menéndez, E., & Rivas, R. (2015). Role of bacterial biofertilizers in agriculture and forestry. *AIMS Bioengineering, 2*(3), 183–205. https://doi.org/10.3934/bioeng.2015.3.183

Goldstein, A. H. (1986). Bacterial solubilization of mineral phosphates: Historical perspective and future prospects. *American Journal of Alternative Agriculture, 1*(2), 51–57. https://doi.org/10.1017/S0889189300000886

Goswami, D., Thakker, J. N., & Dhandhukia, P. C. (2016). Portraying mechanics of plant growth promoting rhizobacteria (PGPR): A review. *Cogent Food and Agriculture, 2*(1). https://doi.org/10.1080/23311932.2015.1127500. PubMed: 1127500

Gray, W. M. (2004). Hormonal regulation of plant growth and development. *PLOS Biology, 2*(9), E311. https://doi.org/10.1371/journal.pbio.0020311

Hervé, M., Albert, C. H., & Bondeau, A. (2016). On the importance of taking into account agricultural practices when defining conservation priorities for regional planning. *Journal for Nature Conservation, 33*, 76–84. https://doi.org/10.1016/j.jnc.2016.08.001

Jaga, P. K., & Singh, V. (2010). Effect of biofertilizer, nitrogen and sulphur on sorghum-mustard cropping system. In *Proceedings of the National Seminar on Soil Security for Sustainable Agriculture Held at College of Ariculture, Nagypur (MS on February 27–28, 2010) "XXII SAVETOVANJE O BIOTEHNOLOGIJI"* (p. 1).

Jehangir, I. A., Mir, M. A., Bhat, M. A., & Ahangar, M. A. (2017). Biofertilizers an approach to sustainability in agriculture: A review. *International Journal of Pure and Applied Bioscience, 5*(5), 327–334. https://doi.org/10.18782/2320-7051.5011

Kaur, R., Kaur, M., & Purewal, S. S. (2018). Effect of incorporation of flaxseed to wheat rusks: Antioxidant, nutritional, sensory characteristics, and in vitro DNA damage protection activity. *Journal of Food Processing and Preservation, 42*(4), e13585. https://doi.org/10.1111/jfpp.13585

Kim, J. S., Lee, J., Lee, C. H., Woo, S. Y., Kang, H., Seo, S. G., & Kim, S. H. (2015). Activation of pathogenesis-related genes by the rhizobacterium, Bacillus sp. JS, which induces systemic resistance in tobacco plants. *Plant Pathology Journal, 31*(2), 195–201. https://doi.org/10.5423/PPJ.NT.11.2014.0122

Kumar, R., Kumawat, N., & Sahu, Y. K. (2017). Role of biofertilizers in agriculture. *Pop Kheti, 5*(4), 63–66.

Laranjo, M., Alexandre, A., & Oliveira, S. (2014). Legume growth-promoting rhizobia: An overview on the Mesorhizobium genus. *Microbiological Research, 169*(1), 2–17. https://doi.org/10.1016/j.micres.2013.09.012

Liu, J., Mehdi, S., Topping, J., Friml, J., & Lindsey, K. (2013). Interaction of PLS and PIN and hormonal crosstalk in Arabidopsis root development. *Frontiers in Plant Science, 4*, 75. https://doi.org/10.3389/fpls.2013.00075

Mahanty, T., Bhattacharjee, S., Goswami, M., Bhattacharyya, P., Das, B., Ghosh, A., & Tribedi, P. (2017). Biofertilizers: A potential approach for sustainable agriculture development. *Environmental Science and Pollution Research International, 24*(4), 3315–3335. https://doi.org/10.1007/s11356-016-8104-0

Motsara, M. R., Bhattacharyya, P., & Srivastava, B. (1995). Biofertiliser technology, marketing and usage: A sourcebook-cum-glossary. *Fertiliser Development and Consultation Org.*

Odame, H. (1997). Biofertilizer in Kenya: Research, production and extension dilemmas. *Biotechnology and Development Monitor, 30*, 20–23.

Olanrewaju, O. S., Glick, B. R., & Babalola, O. O. (2017). Mechanisms of action of plant growth promoting bacteria. *World Journal of Microbiology and Biotechnology, 33*(11), 197. https://doi.org/10.1007/s11274-017-2364-9

Ortíz-Castro, R., Valencia-Cantero, E., & López-Bucio, J. (2008). Plant growth promotion by Bacillus megaterium involves cytokinin signaling. *Plant Signaling and Behavior, 3*(4), 263–265. https://doi.org/10.4161/psb.3.4.5204

Pandey, G. (2018). Challenges and future prospects of agri-nanotechnology for sustainable agriculture in India. *Environmental Technology and Innovation, 11*, 299–307. https://doi.org/10.1016/j.eti.2018.06.012

Ramette, A., Moënne-Loccoz, Y., & Défago, G. (2006). Genetic diversity and biocontrol potential of fluorescent pseudomonads producing phloroglucinols and hydrogen cyanide from Swiss soils naturally suppressive or conducive to Thielaviopsis basicola-mediated black root rot of tobacco. *FEMS Microbiology Ecology, 55*(3), 369–381. https://doi.org/10.1111/j.1574-6941.2005.00052.x

Rao, D. L. N. (1986). Nitrogen fixation in free living and associative symbiotic bacteria. In *Soil microorganisms and plant growth* (pp. 116–140). Oxford and IBH Publishing.

Reinhold, B., & Hurek, T. (1989). Location of diazotrophs in the root interior with special attention to the kallar grass association. In *Nitrogen Fixation with Non-Legumes* (pp. 209–218). Springer.

Santi, C., Bogusz, D., & Franche, C. (2013). Biological nitrogen fixation in non-legume plants. *Annals of Botany, 111*(5), 743–767. https://doi.org/10.1093/aob/mct048

Singh, S., Singh, B. K., Yadav, S. M., & Gupta, A. K. (2014). Potential of biofertilizers in crop production in Indian agriculture. *American Journal of Plant Nutrition and Fertilization Technology, 4*(2), 33–40. https://doi.org/10.3923/ajpnft.2014.33.40

Smith, B. E., Richards, R. L., & Newton, W. E. (Eds.). (2013). *Catalysts for nitrogen fixation: Nitrogenases, relevant chemical models and commercial processes, 1.* Springer Science+Business Media.

Sundaram, S., Arunakumari, A., & Klucas, R. V. (1988). Characterization of azospirilla isolated from seeds and roots of turf grass. *Canadian Journal of Microbiology, 34*(3), 212–217. https://doi.org/10.1139/m88-040

Tairo, E. V., & Ndakidemi, P. A. (2013). Possible benefits of rhizobial inoculation and phosphorus supplementation on nutrition, growth and economic sustainability in grain legumes. *American Journal of Research Communication, 1*(12), 532–556.

Timmusk, S., Abd El-Daim, I. A., Copolovici, L., Tanilas, T., Kännaste, A., Behers, L., ... Niinemets, Ü. (2014). Drought-tolerance of wheat improved by rhizosphere bacteria from harsh environments: Enhanced biomass production and reduced emissions of stress volatiles. *PLOS ONE, 9*(5), e96086. https://doi.org/10.1371/journal.pone.0096086

Umesha, S., Manukumar, H. M., & Chandrasekhar, B. (2018a). Sustainable agriculture and food security. In *Biotechnology for sustainable agriculture* (pp. 67–92). Woodhead Publishing.

Umesha, S., Singh, P. K., & Singh, R. P. (2018b). Microbial biotechnology and sustainable agriculture. In *Biotechnology for sustainable agriculture* (pp. 185–205). Woodhead Publishing.

Uosif, M. A. M., Mostafa, A. M. A., Elsaman, R., & Moustafa, E. S. (2014). Natural radioactivity levels and radiological hazards indices of chemical fertilizers commonly used in Upper Egypt. *Journal of Radiation Research and Applied Sciences, 7*(4), 430–437. https://doi.org/10.1016/j.jrras.2014.07.006

Van Loon, L. C. (2007). Plant responses to plant growth-promoting rhizobacteria. *European Journal of Plant Pathology. Dordrecht: Springer, 119*(3), 243–254. https://doi.org/10.1007/s10658-007-9165-1

Verma, P., Yadav, A. N., Kumar, V., Singh, D. P., & Saxena, A. K. (2017). Beneficial plant-microbes interactions: Biodiversity of microbes from diverse extreme environments and its impact for crop improvement. In *Plant-microbe interactions in agro-ecological perspectives* (pp. 543–580). Springer.

Verma, J. P., Yadav, J., Tiwari, K. N., & Singh, V. (2010). Impact of plant growth promoting rhizobacteria on crop production. *International Journal of Agricultural Research, 5*(11), 954–983. https://doi.org/10.3923/ijar.2010.954.983

Wang, H. Y., Liu, S., Zhai, L. M., Zhang, J. Z., Ren, T. Z., Fan, B. Q., & Liu, H. B. (2015). Preparation and utilization of phosphate biofertilizers using agricultural waste. *Journal of Integrative Agriculture, 14*(1), 158–167. https://doi.org/10.1016/S2095-3119(14)60760-7

Chapter 3
Organic Farming for Sustainable Soil Use, Management, Food Production and Climate Change Mitigation

Abdul Kareem, Zia Ur Rahman Farooqi, Amina Kalsom, Waqas Mohy-Ud-Din, Muhammad Mahroz Hussain, Mohsin Raza, and Muhammad Moaz Khursheed

Abstract Organic farming or organic agriculture has been adopted as a new and sustainable agriculture technology due to its environmentally friendly nature and healthy produce. It is an efficient technology to conserve soil fertility and health, improve crop yield and quality and reduce carbon concentrations in the atmosphere. It is more efficient in its use of non-renewable energy and has less detrimental effects on water quality and biodiversity. Different strategies are employed in organic farming for crop production. It uses biosolids in organic manures and amendments and crop rotations, mulching, non-synthetic fertilisers, zero tillage, integrated nutrient, pest management, etc., for crop management and productivity.

Keywords Soil health · Carbon sequestration · Soil fertility · Soil nutrients

3.1 Introduction

Organic farming refers to an agriculture system that relies on green manures, biofertilisers, plant growth-promoting bacteria, integrated pest management (IMP)/biological pest control (integrated pest management (IPM)), integrated nutrient management (INM) and zero or minimum tillage, mulching and crop rotation. It is a practice that entails the routine plant cultivation and animal rearing (Trewavas, 2001). Biosolids like organic manures and amendments along with biopesticides are among the biological materials and fertilisers used in this operation (allelopathic

A. Kareem · Z. U. R. Farooqi (✉) · W. Mohy-Ud-Din · M. M. Hussain · M. Raza · M. M. Khursheed
Institute of Soil and Environmental Sciences, University of Agriculture, Faisalabad, Pakistan

A. Kalsom
Soil Chemistry Section, Ayub Agriculture Research Institute, Faisalabad, Pakistan

S. A. Bandh (ed.), *Sustainable Agriculture*,
https://doi.org/10.1007/978-3-030-83066-3_3

and plant growth-promoting bacteria). It avoids using synthetic chemicals to preserve land biodiversity and ecosystem equilibrium, reducing emissions and waste production (Hussain & Farooqi, 2021). To put it another way, organic farming is a technique that includes caring for and growing of crops without the use of chemical fertilisers and pesticides (Hole et al., 2005) and no production and involvement of genetically modified organisms (Heckman, 2006).

As organic farming relies on livestock manures, off-farm organic wastes, mineral grade rock additives, crop residues, crop rotation, biological system of nutrient mobilisation and plant preservation to the greatest degree possible, rather than conventional inputs (such as hormones, chemical fertilisers, feed additives and so on), it is the most preferred for agriculture nowadays. It has numerous advantages, i.e. it encourages and improves biodiversity, biological cycles and soil biological activity. Crop substitution, renewable fertiliser, organic wastes, natural weed management, mineral and rock treatments are also an example of ecologically balanced farming values. Pesticides and fertilisers are used in organic cultivation if they are deemed safe, while petrochemical fertilisers and pesticides are avoided (Barton, 2018; Palaniappan & Annadurai, 2018).

3.2 Need for Organic Farming

There is an increased need for organic farming due to the following reasons (Lammerts van Bueren et al., 2011; Macilwain, 2004; Stolze & Lampkin, 2009):

1. The organic food market is rapidly expanding and is highly profitable.
2. Food quality and environmental protection.
3. The enhancement of human wellbeing.
4. Organic goods have a rich flavour.
5. Analytical authentication ensures that the commodity is of a high standard.
6. The preservation of agricultural diversity.
7. Non-use of drugs, antibiotics and hormones in agricultural goods.

3.3 Key Aspects of Organic Farming

The following are the key features of organic farming (Altieri et al., 1983; Mäder et al., 2002; Pugliese, 2001) (Fig. 3.1):

1. Use organic content to protect soil health while still promoting biological development.
2. Use of soil microorganisms to provide crop nutrients indirectly.
3. Legumes are used to fix nitrogen in soils.

Fig. 3.1 Benefits and prospects of organic farming

4. Weed and pest management using crop rotation, environmental variation, natural pests, organic manures and appropriate chemical, thermal and biological action, among other approaches.
5. Animal husbandry, including shelter, diet, hygiene, rearing and breeding.
6. Environmental stewardship, including the preservation of natural ecosystems and biodiversity.
7. Development of natural crops.
8. Development of natural animals and poultry.
9. Grass and disease management that is organic.
10. Conservation of soil.
11. Maintenance of biological equilibrium.

3.4 Organic Fertilisers

Organic fertilisers, such as manure and compost, may help improve soil productivity and crop yields by supplementing conventional fertilisers. However, due to their lower C:N ratios, organic fertilisers applied to soils can increase N_2O (nitrous oxide) discharges, possibly contributing to global warming (Shen et al., 2018). When opposed to conventional fertiliser applications, organic fertilisers provide several advantages for farm soils and the climate, as explained below (Shen et al., 2018; Wang, 2014):

1. Increasing the organic matter content in the soils.
2. Improving soil's water-holding capability.
3. Reduction in soil degradation as organic matter increases soil porosity.

Organic fertilisers with high carbon content may stimulate microbial development, resulting in both nitrification and denitrification. Compost and other organic fertilisers have been studied extensively for their effects on carbon-nitrogen dynamics, N_2O productions and soil composition (Shen et al., 2018).

3.5 Principles of Organic Farming

Four primary principles help in organic farming (Luttikholt, 2007):

3.5.1 Principle of Health

Organic farming would benefit the health and wellbeing of the land, plants, livestock, people and the environment. It is the maintenance of personal, economic, environmental and social health. For example, it protects humans from contamination and offers them chemical-free, healthy food.

3.5.2 Principle of Fairness

The maintenance of equality and justice of the everyday world among humans and other living beings demonstrates its principle of fairness. Organic cultivation improves people's lives and aims to alleviate hunger. Natural resources must be utilised wisely and kept safe for forthcoming generations.

3.5.3 Principle of Ecological Balance

Organic agriculture can be based on biological processes. Organic cultivation practices must be compatible with natural ecosystem balances and periods.

3.5.4 *Principle of Care*

Organic farming can be done with caution and responsibility to support the needs of present and future generations and the climate. Organic agriculture, unlike industrial and traditional farming practices, does not depend on synthetic pesticides. It uses sustainable organic methods to improve soil productivity, such as improving plant nutrition by microbial activity. Second, organic farming's multiple cropping increases biodiversity, which improves sustainability and stability while still contributing to a sustainable farming environment.

3.6 Unsustainability of Conventional Farming

There are specific reasons due to which modern farming is considered unsustainable (Kingwell, 2011; Sverdrup et al., 2017):

1. Soil productivity is lost because of improper toxic pesticide usage.
2. Water supplies are contaminated by nitrate drainage after rainstorms.
3. Soil degradation because of deep ploughing and heavy rains.
4. Increased fuel needs for agriculture.
5. Animal cruelty in terms of living, feeding, breeding and slaughtering.
6. Monoculture has resulted in a loss of biodiversity.
7. Invasive creatures and hybrids take up more room than native animals and plants.

3.7 Essentials of Organic Farming

When we begin using/performing organic farming methods, we must ensure the presence of specific essential characteristics and components explained as follows:

3.7.1 *Farmyard and Other Organic Manures*

Farmyard, poultry and other manures are decomposed combinations of farm animals' dung and faeces and debris and leftovers from roughages or fodder. Cow dung, bird droppings, excess grass and other dairy wastes are used to make it. It is highly beneficial, and some of its characteristics are as follows:

1. These are nutrient-dense.
2. As manures can be combined, the plants receive a well-balanced diet.
3. Potassium and phosphorus availability is comparable to that of inorganic sources.
4. Application of manures to the soil increases its productivity.

Manure that strengthens the soil composition is used as a natural fertiliser in agriculture. It expands the soil's ability to hold more water and nutrients. It also boosts the soil's microbial production, which improves mineral availability and plant nutrition. Manures like farmyard manure (FYM) deliver all major nutrients (N, P, K, S, Ca, Mg) and micronutrients for plant development (Cu, Fe, Mn and Zn). As a result, it acts as a mixed fertiliser. Farmyard manure helps improve the physico-chemical and biological properties of the soil. The application of FYM improves the soil composition, which creates a more robust atmosphere for root growth. FYM also upsurges the capacity of the soil to hold water (Tadesse et al., 2013). In the Indian subcontinent, FYM is the most used sustainable manure. It's a composted combination of cow dung, stable bedding and the leftovers of straw and plant stalk fed to cattle. In India, rainfed (non-irrigated) dryland cultivation accounts for roughly 70% of the cultivated land (144 million ha). Dryland agroecosystems have poor crop yields due to low soil moisture and nutrient supply. Due to low soil moisture levels, chemical fertiliser usage is restricted in these agroecosystems. Low-input organic cultivation has been proposed to preserve soil moisture and increase soil productivity to realise the maximum economic capacity of these thinly spread drylands (Ghoshal & Singh, 1995).

3.7.2 Vermicompost

Vermicompost is a nutrient-dense organic fertiliser and soil conditioner that contains water-soluble nutrients. It is a mixture of decomposing vegetable or food wastes, bedding materials and vermicast created by the decomposition process of various worm species, most commonly white worms, rid wiggler worms and other earthworms. The process of vermicompost production is referred to as vermicomposting, and the practice of raising worms is referred to as vermiculture. Vermicast is the final outcome of earthworms decomposing organic matter (also known as worm humus, worm faeces, worm manure or worm castings). The castings are known to have less contaminants and a higher nutrient saturation level than organic materials prior to vermicomposting.

Key Benefits of Vermicomposting The following are the key benefits of vermicomposting:

1. Vermicomposting provides nutrient-rich castings and diverts wastes from landfills.
2. As compared to conventional composting, lesser space is needed in vermicomposting.
3. There is a reduction in the amount of strength needed.
4. Compost processing can be done more quickly.
5. Red wiggler worms consume half of their body weight in food waste every day.

Vermicompost is superior to compost because of its higher N, P and K contents and its potential to enhance soil structure and retain more water. Vermicompost is an excellent organic manure for improving plant growth and yield. Vermicomposting is a basic biotechnological composting technique in which some earthworms improve the waste conversion method and provide a healthier product. In many aspects, vermicomposting varies from composting. It is a mesophilic system that makes use of earthworms and microbes that survive between 10 and 32 degrees Celsius (not ambient temperature but temperature within the pile of moist organic material). The process is faster than composting since the waste is absorbed by the earthworm's intestine. It undergoes a significant but little-known transformation, yielding vermicasts rich in plant growth regulators, microbial activity and pest repellent properties. In a nutshell, earthworms transform waste into gold through biological alchemy (Adhikary, 2012). Vermicompost has the same recorded advantages as traditional composts, such as providing organic matter, improving moisture retention and increasing nutrient absorption and plant hormone-like action. Enhancements in the physico-chemical composition of the growth media have been attributed to the increased plant growth. However, it suggests that using vermicompost impacts plant growth that is not explicitly related to physical or chemical properties. Any growth promotion appears to be attributed to plant hormone-like behaviour associated with vermicomposting microflora and metabolites formed due to secondary metabolism (Bachman & Metzger, 2008).

3.7.3 Green Manuring

Green manuring is the process of incorporating undecomposed green plant tissue into the soil. A green manure crop aims to introduce the organic matter to the soil. Because of the incorporation, the soil's nitrogen intake is improved, and some nutrients become more readily accessible, improving the soil's efficiency. It aids in the preservation of arable soil's organic matter content (Farooqi et al., 2021). It acts as a source of energy and food for microbes that multiply quickly, not only decomposing the green manure but also releasing plant nutrients in usable forms for crop usage. The function of a green manure crop differs depending on the circumstance, but the following are some of the advantages they provide:

1. Increasing soil humus and organic matter.
2. Increasing nitrogen fixation.
3. Defending the soil's crust.
4. Preventing the process of erosion.
5. Keeping or changing the composition of the soil.
6. Access to unavailable nutrients from the lower soil profile.
7. Reducing soil vulnerability to leaching.
8. Ensuring access to readily accessible nutrients for the next generation of crops.

Green manuring is a method of soil enrichment that involves plowing under or adding certain green manure crops into the soil when it is either green or shortly after they begin to bloom. The importance of green manuring resides in the incorporation of organic matter into the soil is widely acknowledged as one of the essential components for proper soil fertility. Green manure crops are cultivated to produce green manure that has only recently become common among our farmers. According to estimates, a green manure crop that is 40–50 days old provides up to 80–100 kg N ha^{-1}. A green manure crop replaces 50–60 kg fertiliser N ha^{-1}, even if half of this nitrogen is crop utilisable. Dhanicha, sun hemp, cowpea, mung, potato, guar and berseem are possible green manuring legumes. These crops grown during the Kharif season have been recorded to contribute 8–21 tonnes of green matter and 42–95 kg of N per hectare. Similarly, during the rabi season, cowpea and berseem contribute 12–29 tonnes of green matter and 67–68 kg of N per hectare (Dubey et al., 2015; Mishra & Nayak, 2004; Sharma et al., 2013).

3.7.4 Organic Matter Application and Restoration

The three major components of soil organic matter (SOM) are small (fresh) plant residues, decomposing (active) organic matter and solid organic matter (humus). Soil organic matter serves as a nitrogen and phosphorus reservoir for crops, enhancing fertiliser exchange, preserving moisture, decreasing compaction, reducing surface crusting and rising water infiltration. SOM does not include plant nutrients on the soil surface, such as leaves, fertiliser or seed debris, which are usually sieved out of soil samples prior to examination using a wire mesh (Farooqi et al., 2018).

Plant residues with a low carbon:nitrogen ratio (high nitrogen content) decompose more rapidly than those with a high carbon:nitrogen ratio (low nitrogen content) and contribute less organic matter to the soil. Excessive tillage destroys soil aggregates and accelerates the decomposition of organic matter in the soil. Healthy soil aggregates increase the amount of usable organic matter in the soil, thus shielding it from microbial deterioration. Accelerating SOM decomposition requires steps to increase soil moisture, temperature and aeration. SOM may be degraded or increased depending on the management measures used on the areas being analysed. The following are some main steps that may help improve SOM (Umar et al., 2020):

(a) Use of cropping systems.
(b) Reducing or eliminating tillage.
(c) Reduce erosion using appropriate measures.
(d) Soil-test and fertilise properly.
(e) Use of perennial forages.
(f) Nutrient supply.
(g) Water-holding capacity.
(h) Soil aggregation.

(i) Erosion prevention.

Organic matter in the soil is diverse and heterogeneous, made up of a variety of organic materials that are preserved to differing degrees through physical isolation from microbial biomass, molecular recalcitrance and direct contact with clay surfaces and inorganic ions (Shepherd et al., 2002).

3.7.5 *Crop Rotation*

Crop rotation is the practice of planting multiple crops on the same plot of land in order to improve soil quality, maximise fertiliser availability and combat pest and weed pressure. Consider a farmer who has cultivated a cornfield. He could plant beans after the corn harvest is over since corn absorbs many N and beans return N to the soil. Simple rotations may consist of two or three crops, whereas complicated rotations may have a dozen or more (Behnke et al., 2018).

3.7.5.1 Principles for Crop Rotation

Crop rotation has several general rules, described as follows:

1. Plant a high N-demanding crop after a legume crop.
2. In a particular region, grow annual crops for just 1 year.
3. Do not plant one crop after another that is directly associated.
4. Use seed sequences that encourage the growth of healthy crops.
5. Implement weed-controlling seed series.
6. Use seasonal crops for more extended periods on sloping land.
7. Use a deep-rooted crop in rotation.
8. Use crops that can leave a large amount of residue in the rotation.
9. When cultivating a diverse range of crops, divide them into blocks based on plant families, crop timing, crop form (leaf vs fruit vs root), fertiliser requirements or crops with common cultural traditions.

3.7.5.2 Steps for Crop Rotation and Planning

In organic farms the following steps are adapted from crop rotation (Mohler & Johnson, 2009):

(a) Determine and prioritise the crop rotation priorities (e.g. environmental enforcement, weed management, soil quality and disease control).
(b) Create a crop mix list.
(c) Look for an overabundance of land in one family.

(d) Prepare a crop rotation plan, noting which beds or fields (or parts of fields) are at risk of affecting such crops.
(e) Find crop couplets and short sequences that are appropriate for your farm (including cover crops).

3.7.6 Mulching

Mulch is essentially a substance that is applied on top of the soil to have a protective layer. Mulches may be inorganic (plastics, bricks and stones) or organic (wood chips, grass clippings, straw and other similar materials) with a number of benefits including the following:

- Protects the soil from erosion and alleviates compaction caused by heavy rains.
- Conserves moisture and eliminates the need for regular irrigation.
- Maintains a more constant temperature in the soil.
- Prevents the proliferation of weeds.
- Maintains the cleanliness of fruits and vegetables.

Chemical mulching has been used to successfully kill weeds and minimise soil degradation in organic growing schemes by directly providing organic C nutrients to the soil. It also helps conserve soil moisture and buffer abrupt shifts in soil temperature, particularly in sandy soils where significant differences in soil temperature and soil moisture are common (Tu et al., 2006). Chemical mulches have a mixed effect on crop production. Mulching enhances plant development, production and consistency of yield. The release of nutrients from decomposing mulches (both quickly and slowly decomposing) may benefit the soil (Sinkevičienė et al., 2009). It also impacts humans in terms of cost, aesthetics and ease of service and weeding. It is the key to any garden or orchard's success, particularly during a drought (Farooqi et al., 2020a, b). In semi-arid and arid areas, it's even great for the greenhouse or orchard. It acts as an insulator, allowing the soil temperature to be cooled and moderated through hot days and cold nights. It boosts the development of beneficial microbes and aids in the prevention of diseases. Additionally, it protects soil moisture from excessive solar rays and air movement through the soil surface, resulting in decreased soil moisture loss (Ranjan et al., 2017).

3.7.7 Integrated Nutrient Management

Integrated nutrient management (INM) is a term that refers to the most environmentally friendly method of disposing of crop residues, producing high-quality compost for soil fertility maintenance and supplying plants with the optimal amount of nutrients available during the life cycle to support the yield (Selim, 2020).

The following elements are the essential components of the INM scheme (Chen et al., 2011; Wu & Ma, 2015):

1. All possible nutrient sources that can be used to develop nutrient input programmes with the goal of increasing nutrient use production and high yield performance must be carefully considered.
2. The forms and quantities of soil nutrients in the root zone, also referred to as soil balance, and their availability to meet cover crop requirements.
3. Identifying and mitigating nutrient losses, especially in intensive agriculture.
4. Taking into account all factors affecting the plant/nutrient relationship in order to achieve high yield production, which is a critical aim and major advantage of integrated nutrient management (INM), water use efficiency, grain superiority, high economic returns and sustainability.

4R's of Nutrient Management
There is a popular nutrient management of 4R technology given below:

1. Right source.
2. Right rate.
3. Right time.
4. Right place.

3.7.8 Zero Tillage

Zero tillage describes arable land that receives no tillage between harvest and sowing. It's a low-tillage method in which the grain is sown directly onto the land that hasn't been tilled since the previous crop's yield. Herbicides and adequate mulching are used to manage weeds, and the stubble is held for erosion control.

There are some advantages of zero tillage which includes:

1. Reduced soil erosion caused by wind and water (because the mulch cover of previous crops covers the soil).
2. Compaction of the soil is reduced.
3. Soils that are more fertile and robust.
4. Moisture evaporation is reduced.
5. Cost savings on diesel and labour.

Land degradation on a physical and chemical level, a deficiency in organic matter, reduced ecological activity in the soil and crop yield loss are all effects of intensive or traditional agriculture. Sustainable agriculture, on the other hand, envisions a prosperous and long-term farming system based on three basic principles: soil-free agriculture, crop rotation and a constant soil surface littered with plan debris and vegetation (Shrestha et al., 2018).

3.8 Benefits of Organic Farming

There are several benefits for organic farming, which are briefly discussed as follows:

3.8.1 Crop Productivity

Crop production is increased as organic farming is practised. To increase productivity and sustainability, existing cropping systems must be transformed to address growing environmental effects and reduced inputs. Since organic farming does not encourage pesticides and is generally thought to focus more on crop variety than its traditional equivalent, it is regarded as one type for improving the resilience of current agriculture and cereal-rich cropping systems (Bedoussac et al., 2015). Increasing crop yields by adding organic matter to the soil is a well-known procedure. According to several studies, the use of organic materials improves rice grain and straw production. Some researchers reported that the spent mushroom and rice straw compost improved rice grain yields by 20% over NPK fertiliser. The benefits of organic farming are established for developed and developing countries in terms of biodiversity conservation, environmental improvement, decreased energy use and carbon dioxide emissions, increased crop production without excessive dependence on expensive inputs and sustainable resource use (Yadav et al., 2013).

3.8.2 Soil Fertility and Biological Parameters

Biological parameters are a significant part of soil quality evaluation. These biological properties are critical when evaluating soil quality because flora and fauna in the soil significantly impact it. Soil microbial biomass and microbial interactions are essential for soil fertility to be sustained. A balanced ratio of microbial biomass and operation in the soil is needed to ensure the continuous release of nutrients to plants. Organic farming has been shown to increase microbial biomass and operation by 20–30% and 30–100%, respectively, as compared to conventional agriculture. A soil with a solid organic matter content has more microbial activity and more soil N, providing strength than a soil with a low organic matter content (which is managed inorganically). Furthermore, soil organic matter can absorb CO_2 from the atmosphere, raising the carbon content of the soil and thereby enhancing microbial biomass and respiration.

3.8.3 Sustainable Soil Management

There are several components of sustainable land management. The approach and its implementations have a fragile yet complicated framework, explaining the variety of ingredients and medications (Kwiatkowska-Malina, 2018).

3.8.4 Water Management

Sufficient moisture in the soil's root area during the growing season of the plant is the most important factor in ensuring steady plant growth in sustainable agriculture. The average rainfall is the first indicator of precipitation. In situations where rainfall is insufficient, irrigation water can be used to provide the required water. A decrease in yield is typically caused by inadequate or excessive soil moisture in the plant root region.

Water resource sustainability is a relational, physical, fiscal and ecological term. Future generations' water uses, drainage, agricultural and recreational water storage and habitat conservation programmes are also covered by sustainable water management. The following factors should be weighed to maintain its long-term viability (Chiappetta,, & K. J. L. R. o. B., 2017):

- The irrigation system should be continuously tracked.
- The pumps should be operating at full capacity, the water volume should be measured and the water distribution should be even.
- The irrigation time and volume should be determined in relation to the plant's water requirements, with the aim of ensuring the most effective water use possible.
- Irrigation should be discontinued during windy conditions and in the middle of the day; instead, it should be done at night, with drip irrigation used as required.
- The facility's load should be kept as high as possible.
- Pipes should be tested regularly, and leakage should be stopped.
- Poisoning of water supply and drainage channels should be avoided in any case.
- To avoid waterborne degradation, it is essential to ensure that water is infiltrated into the soil using agricultural principles and drainage methods.
- Water quantity and distance from water supplies should be considered when preparing output.
- Raw agricultural wastes and wastewater should not be discharged into natural surface waters.
- Measures should be taken to mitigate irrigation's harmful impact on the environment.

Drone and sensor technology may also be used to gather data for the implementation of a successful irrigation methodology (Farooqi et al., 2021):

- Soil moisture meters are used to determine the demand for water in the soil.
- Thermal photographs of soil and crop moisture material collected from drones.
- A multispectral camera can detect nitrogen deficiency.
- A variable-rate irrigation scheme should be designed based on environmental data and forecasts.
- Variable-rate implementations should be performed at the best possible time in areas of different water requirements.

3.8.5 *Pest and Disease Management*

Integrated pest management (IPM) is a foundational technique used in modern agriculture that utilises all available plant defence strategies. IPM involves integrating successful approaches to eliminate risks to human health and the environment by preventing the proliferation of insect species and ensuring that plant management pesticides and other methods of action are used at economically and ecologically justified levels. A well-designed integrated pest management system (IPM) consists of three critical stages for maximum effectiveness and minimum environmental effects in weed, disease and pest management (Tuğrul, 2019):

(i) Identifying rodents, viruses and weeds is the first step for farmers, followed by settling on physical, chemical, biological and regulatory enforcement choices.
(ii) After identifying invasive plants, keep an eye on reproductive rates.
(iii) As the number of invasive organisms exceeds a certain level, several safety options are enabled. The most efficient protection strategy against invasive plants is using chemicals that do the minor environmental damage with all defence strategies. Crop loss may also be reduced by crushing early or through other physical defence techniques. The presence of helpful species should be considered when choosing a defence system, and dangerous species may be battled with pest-fighting species without chemicals.

3.8.6 *Cover Crops and Crop Rotation*

Cover plants may be grown between the central plant rows during off-season cycles when the soil is bare. They reduce the need for herbicides while preventing soil degradation, renewing soil resources, controlling weeds and protecting soil health. It is the method of manufacturing various goods in the sector one after the other year after year. As a result, various portions of the soil are used for different crops, and pests and pathogens unique to each crop are kept at bay. Covered plants contribute significantly to agricultural productivity by protecting the land; maintaining the desired temperature, humidity or illumination; and controlling pests and weeds. The weed problem has emerged as a result of sustainable agriculture's reduced soil cultivation. Clover, vetch, trefoil, oats, wheat and sorghum are only a few examples of plants with a broad range of uses and cultivation purposes. Cereals, for example, are superior at weed management, while legumes excel at nitrogen addition to the crop plant. The most critical thing to remember when growing cover plants is to strike a compromise between the system's expense and benefits.

3.9 The Organic Food System

Organic agriculture is a mode of development that promotes vegetation, ecosystems and human health. It is based on natural processes and methods that have been modified to local environments, rather than the input of any harmful chemical. This system consists of various steps, which are described below:

3.9.1 Classification

There are three separate food system classifications, conventional, transitional and the other brands, including seasonal, renewable, pesticide-free and environmentally safe. The word "clean" refers to plants, goods, processors and other value-added intermediaries in the production-to-consumption chain that has been approved by certifying bodies (CB). The certifying bodies are fee-for-service organisations and are typically supervised by the National Food Inspection Authorities. Organic certification is a lengthy procedure that takes at least 3 years to complete if done on a farm that was already farmed using traditional techniques. This means that all chemicals have been leached from the soil and that time has been allowed for organic amendments to restore soil fertility.

Transitional organic is a regional label for farms that have made the commitment to pursue organic certification. For example, the word "transitional" refers to farms that have converted to certifiable organic practices and are in the 36-month period between the last pesticide application and the time when the soil will be declared chemical-free and the farm certified organic. Small farmers catering to local/regional clients also use labels like natural, local, environmentally sustainable and pesticide-free. Except for selling board-regulated goods such as dairy or chicken, the processing and storage of items marketed under these brands are primarily unregulated, except for governmental entities and district health units. Consequently, the knowledge on farms that are not certified organic is dispersed and incomplete.

3.9.2 Production

Organic farming is a comprehensive system aimed at the growth and fitness of agro-ecosystems such as soil, plants, cattle and humans. The primary aim of organic farming is to create profitable and environmentally friendly (Hamzaoui-Essoussi & Zahaf, 2012).

We may divide organic farmers into three divisions based on how they leverage the supply chain to bring their goods to consumers: big-, mini-, and medium-sized operations. Large farmers can be identified by organic cash crops that are exported or imported immediately after harvesting, as well as livestock and field crops that are more likely to be shipped to dealers and processors for further processing. In this case, most dairy farms will be called significant producers. Medium-sized businesses typically cater to a broader local sector (Hall & Mogyorody, 2001). Due to infrastructure constraints, some of these farmers band together to grow their goods, collaborating with complementary companies to increase their on-farm market's offerings and draw more buyers. Others also formed alliances with small regional processors and manufacturers to hit restaurants and speciality food retailers. Most medium-sized growers have on-farm markets set up as permanent storefronts, with goods offered on consignment or resold to other farmers in the region. Small organic farmers are less likely to employ delivery mediators. Farmers' markets and on-farm markets are where they work on building direct partnerships with customers. They may supply a few restaurants, speciality stores or small grocers, but these are carefully nurtured partnerships that depend on niche marketing and personal connections. Owing to the paperwork and costs involved, small farmers are more likely to abandon organic certification.

3.9.3 Distribution

In recent years, organic food has been a significant segment of the food retailing industry. Natural produce has slowly progressed from specialised markets, such as independent retail shops, to mass markets, such as massive grocery chains (Jones et al., 2001; Tutunjian, 2008). Speciality retailers (95%) accounted for most of the revenue 10 years ago, with mainstream stores accounting for the remaining 5%. The pattern has now been reversed (Monitor, 2010). Farmers' markets and other alternate delivery networks are utilised and feature a clear connection between the manufacturer and the customer (Smithers et al., 2008). Distributors in several countries are marketing their product lines under particular brand names (Rostoks, 2002; Tutunjian, 2004).

3.10 Effect of Organic Farming on Climate Change

3.10.1 Reduction of Greenhouse Gas Emission

Crop and livestock agriculture emissions have increased by more than 14% since 2001, from 4.7 billion tonnes of CO_2, equal to more than 5.3 billion tonnes today. Organic farming aids in the fight against climate change by lowering greenhouse gas emissions. The volume of N fertiliser added to farmland has a direct relationship with N_2O pollution. In the EU, N_2O emissions from controlled soils account for about 40% of the total farm emissions. This is especially significant since 1 kg of nitrous oxide has a 300-fold greater warming effect on the environment than 1 kg of CO_2.

Organic farming does not enable the use of synthetic N fertilisers while concentrating on maintaining closed nutrient cycles and minimising losses by drainage, volatilisation and pollution, resulting in lower N levels per hectare on organic farms than on traditional farms, which lead to a healthy, climate-friendly development method that provides adequate food supplies.

3.10.2 Reducing Energy Use

Synthetic fertilisers and pesticides are widely used in conventional cultivation. The production of these chemicals necessitates a considerable amount of energy. Organic cultivation, however, reduces energy usage per unit of land by 30–70% by replacing the energy used to produce conventional fertilisers and utilising internal field inputs, which reduces transportation fuel consumption as well.

3.10.3 Helping Farmers to Adapt to Climate Change

Organic farming also aids in the fight against global climate change by trapping carbon in the soil. Many organic agricultural management methods, such as minimal tillage, restoring crop residues to the surface, using cover crops and rotations and incorporating more nitrogen-fixing legumes, improve carbon return to the soil. This boosts efficiency while still promoting carbon conservation.

3.10.4 Storing Carbon in the Soil

Organic cultivation aids farmers in adapting to climate change by preventing nutrient and water depletion by high soil organic matter quality and soil cover. Soils become more resistant to hurricanes, droughts and ground erosion because of this. Farmers may develop new cropping systems which respond to climatic changes. Organic farming reduces the risk for farmers by providing healthy agroecosystems and returns and lower production costs (Farooqi et al., 2018; Farooqi et al., 2020a, b).

3.10.5 Advocating for Policy Change

Organic farming can eliminate greenhouse pollution, increase land productivity and strengthen environmental resilience. As a result, we suggest that:

1. Governments recognise organic agriculture as a viable method for reducing greenhouse gas emissions and carbon sequestration.
2. Developing world policymakers should provide policies focused on the ideals of organic agriculture.
3. Appropriate mitigation actions to assist farmers in adapting to climate change through study and extension services.

3.11 Conclusions

The increasing yield loss phenomena, soil structure and quality degradation and greenhouse gas emissions have emphasised converting the current farming methods to sustainable soil use and crop productivity. So, organic farming, which uses many techniques and tools, holds a promise in making agriculture sustainable. It helps protect people's health by providing them the safer foods and protecting the health of the environment by stopping the use of synthetic fertilisers and pesticides, which otherwise damage the environment.

References

Adhikary, S. (2012). Vermicompost, the story of organic gold: A review. *Agricultural Sciences, 03*(7), 905–917. https://doi.org/10.4236/as.2012.37110

Altieri, M. A., Davis, J., & Burroughs, K. (1983). Some agroecological and socio-economic features of organic farming in California. A preliminary study. *Biological Agriculture and Horticulture, 1*(2), 97–107. https://doi.org/10.1080/01448765.1983.9754384

Bachman, G. R., & Metzger, J. D. (2008). Growth of bedding plants in commercial potting substrate amended with vermicompost. *Bioresource Technology, 99*(8), 3155–3161. https://doi.org/10.1016/j.biortech.2007.05.069

Barton, G. A. (2018). *The global history of organic farming*. Oxford University Press.

Bedoussac, L., Journet, E.-P., Hauggaard-Nielsen, H., Naudin, C., Corre-Hellou, G., Jensen, E. S., … Justes, E. (2015). Ecological principles underlying the increase of productivity achieved by cereal-grain legume intercrops in organic farming. *A review. Agronomy for Sustainable Development, 35*(3), 911–935. https://doi.org/10.1007/s13593-014-0277-7

Behnke, G. D., Zuber, S. M., Pittelkow, C. M., Nafziger, E. D., & Villamil, M. B. J. A. (2018). Long-term crop rotation and tillage effects on soil greenhouse gas emissions and crop production in Illinois, USA. *Ecosystems, & environment, 261*, 62–70.

Chen, X. P., Cui, Z. L., Vitousek, P. M., Cassman, K. G., Matson, P. A., Bai, J. S., … Zhang, F. S. (2011). Integrated soil–crop system management for food security. *Proceedings of the National Academy of Sciences of the United States of America, 108*(16), 6399–6404. https://doi.org/10.1073/pnas.1101419108

Chiappetta, & K. J. L. R. o. B. (2017). *Book review: Water: Abundance, scarcity and security in the age of humanity by Jeremy J.. Schmidt.*

Dubey, L., Dubey, M., & Jain, P. (2015). Role of green manuring in organic farming. *Plant Archives, 15*(1), 23–26.

Farooqi, Z. U. R., Ayub, M. A., Nadeem, M., Shabaan, M., Ahmad, Z., Umar, W., & Iftikhar, I. (2021). Precision agriculture to ensure sustainable land use for the future: Precision agriculture and arable land use. In *Advances in Public Policy and Administration* (pp. 210–230). IGI Global. https://doi.org/10.4018/978-1-7998-4372-6.ch011

Farooqi, Z. U. R., Ayub, M. A., Ur Rehman, M. Z., Sohail, M. I., Usman, M., Khalid, H., & Naz, K. (2020a). Regulation of drought stress in plants. In *Plant life under changing environment* (pp. 77–104). Elsevier.

Farooqi, Z. U. R., Sabir, M., Zia-Ur-Rehman, M., & Hussain, M. M. J. C. C. (2020b). Adaptation, A.S., & strategies, M. *Mitigation of Climate Change Through Carbon Sequestration in Agricultural Soils, 87.*

Farooqi, Z. U. R., Sabir, M., Zeeshan, N., Naveed, K., & Hussain, M. M. J. C. C. (2018). Utilization, & sequestration. *Enhancing Carbon Sequestration Using Organic Amendments and Agricultural Practices*, 17–35.

Ghoshal, N., & Singh, K. P. (1995). Effects of farmyard manure and inorganic fertiliser on the dynamics of soil microbial biomass in a tropical dryland agroecosystem. *Biology and Fertility of Soils, 19*(2–3), 231–238. https://doi.org/10.1007/BF00336165

Hall, A., & Mogyorody, V. (2001). Organic farmers in Ontario: An examination of the conventionalisation argument. *Sociologia Ruralis, 41*(4), 399–322. https://doi.org/10.1111/1467-9523.00191

Hamzaoui-Essoussi, L., & Zahaf, M. (2012). The organic food market: Opportunities and challenges. *Organic Food and Agriculture-New Trends and Developments in the Social Sciences.*

Heckman, J. (2006). A history of organic farming: Transitions from Sir Albert Howard's. *War in the soil to USDA National Organic Program. Renewable Agriculture and Food Systems*, 143–150.

Hole, D. G., Perkins, A. J., Wilson, J. D., Alexander, I. H., Grice, P. V., & Evans, A. D. (2005). Does organic farming benefit biodiversity? *Biological Conservation, 122*(1), 113–130. https://doi.org/10.1016/j.biocon.2004.07.018

Hussain, M. M., & Farooqi, Z. U. R. (2021). Allelopathic bacteria as an alternate weedicide: Progress and future standpoints. In K. R. Hakeem, G. H. Dar, M. A. Mehmood, & R. A. Bhat (Eds.), *Microbiota and biofertilizers: A sustainable continuum for plant and soil health* (pp. 211–230). Springer International Publishing.

Jones, P., Clarke-Hill, C., Shears, P., & Hillier, D. (2001). Retailing organic foods. *British Food Journal, 103*(5), 358–365. https://doi.org/10.1108/00070700110396358

Kingwell, R. (2011). Managing complexity in modern farming. *Australian Journal of Agricultural and Resource Economics, 55*(1), 12–34. https://doi.org/10.1111/j.1467-8489.2010.00528.x

Kwiatkowska-Malina, J. (2018). Qualitative and quantitative soil organic matter estimation for sustainable soil management. *Journal of Soils and Sediments, 18*(8), 2801–2812. https://doi.org/10.1007/s11368-017-1891-1

Lammerts van Bueren, E. T., Jones, S. S., Tamm, L., Murphy, K. M., Myers, J. R., Leifert, C., & Messmer, M. M. (2011). The need to breed crop varieties suitable for organic farming, using wheat, tomato and broccoli as examples: A review. *NJAS – Wageningen Journal of Life Sciences, 58*(3–4), 193–205. https://doi.org/10.1016/j.njas.2010.04.001

Luttikholt, L. W. M. (2007). Principles of organic agriculture as formulated by the International Federation of Organic Agriculture Movements. *NJAS – Wageningen Journal of Life Sciences, 54*(4), 347–360. https://doi.org/10.1016/S1573-5214(07)80008-X

Macilwain, C. (2004). *Organic: Is it the future of farming?* Nature Publishing Group.

Mäder, P., Fliessbach, A., Dubois, D., Gunst, L., Fried, P., & Niggli, U. (2002). Soil fertility and biodiversity in organic farming. *Science, 296*(5573), 1694–1697. https://doi.org/10.1126/science.1071148

Mishra, B., & Nayak, K. (2004). Organic farming for sustainable agriculture, Orissa Review. *Monthly journal of the government of Orissa. Published in the month of*, 42–46.

Monitor, O. (2010). *The global market for organic food and drink: Business opportunities and future outlook*. Organic Monitor.

Palaniappan, S., & Annadurai, K. (2018). *Organic farming theory and practice*. Scientific Publishing.

Pugliese, P. (2001). Organic farming and sustainable rural development: A multifaceted and promising convergence. *Sociologia Ruralis, 41*(1), 112–130. https://doi.org/10.1111/1467-9523.00172

Ranjan, P., Patle, G. T., Prem, M., & Solanke, K. R. (2017). Organic mulching-A water saving technique to increase the production of fruits and vegetables. *Current Agriculture Research Journal, 5*(3), 371–380. https://doi.org/10.12944/CARJ.5.3.17

Rostoks, L. (2002). Romancing the organic crowd: This new category may yield plenty of profits for you, if you master the new merchandising rules to attract the organic consumer. *Canadian Grocer, 116*, 22–24.

Selim, M. M. (2020). Introduction to the integrated nutrient management strategies and their contribution to yield and soil properties. *International Journal of Agronomy, 2020*, 1–14. https://doi.org/10.1155/2020/2821678. PubMed: 2821678

Sharma, G., Thakur, R., Raj, S., Kauraw, D., & Kulhare, P. (2013). *Impact of integrated nutrient management on yield, nutrient uptake, protein content of wheat (Triticum astivam) and soil fertility in a typic haplustert* (pp. 1159–1164). Washington, DC: BioScan, *8*(4).

Shen, J., Treu, R., Wang, J., Thorman, R., Nicholson, F., & Bhogal, A. (2018). Modeling nitrous oxide emissions from three United Kingdom farms following application of farmyard manure and green compost. *Science of the Total Environment, 637–638*, 1566–1577. https://doi.org/10.1016/j.scitotenv.2018.05.101

Shepherd, M. A., Harrison, R., & Webb, J. (2002). Managing soil organic matter—Implications for soil structure on organic farms. *Soil Use and Management, 18*(s1), 284–292. https://doi.org/10.1111/j.1475-2743.2002.tb00270.x

Shrestha, K. P., Giri, R., Kafle, S., Chaudhari, R., Shrestha, J. J. F., & Management. (2018). Zero tillage impacts on economics of wheat production in far western. *Nepal, 3*(2), 93–99.

Sinkevičienė, A., Jodaugienė, D., Pupalienė, R., & Urbonienė, M. (2009). The influence of organic mulches on soil properties and crop yield. *Agronomy Research, 7*(1), 485–491.

Smithers, J., Lamarche, J., & Joseph, A. E. (2008). Unpacking the terms of engagement with local food at the farmers' market: Insights from Ontario. *Journal of Rural Studies, 24*(3), 337–350. https://doi.org/10.1016/j.jrurstud.2007.12.009

Stolze, M., & Lampkin, N. (2009). Policy for organic farming: Rationale and concepts. *Food Policy, 34*(3), 237–244. https://doi.org/10.1016/j.foodpol.2009.03.005

Sverdrup, H., Koca, D., & Ragnarsdottir, K. V. (2017). Defining a free market: Drivers of unsustainability as illustrated with an example of shrimp farming in the mangrove forest in South East Asia. *Journal of Cleaner Production, 140*, 299–311. https://doi.org/10.1016/j.jclepro.2015.06.087

Tadesse, T., Dechassa, N., Bayu, W., & Gebeyehu, S. (2013). *Effects of farmyard manure and inorganic fertiliser application on soil physico-chemical properties and nutrient balance in rain-fed lowland rice ecosystem.*

Trewavas, A. (2001). Urban myths of organic farming. *Nature, 410*(6827), 409–410. https://doi.org/10.1038/35068639

Tu, C., Ristaino, J. B., & Hu, S. (2006). Soil microbial biomass and activity in organic tomato farming systems: Effects of organic inputs and straw mulching. *Soil Biology and Biochemistry, 38*(2), 247–255. https://doi.org/10.1016/j.soilbio.2005.05.002

Tuğrul, K. M. (2019). Soil management in sustainable agriculture. In *Sustainable crop production*. IntechOpen.

Tutunjian, J. (2004). Are organic products going mainstream? *Canadian Grocer, 118*(7), 31.

Tutunjian, J. (2008). Market survey 2007. *Canadian Grocer, 122*(1), 26–34.

Umar, W., Ayub, M. A., & Rehman, M. Z. u., Ahmad, H. R., Farooqi, Z. U. R., Shahzad, A., & Nadeem, M. (2020). Nitrogen and phosphorus use efficiency in agro-ecosystems. In S. Kumar, R. S. Meena, & M. K. Jhariya (Eds.), *Resources use efficiency in agriculture* (pp. 213–257). Springer Singapore.

Wang, J. (2014). Decentralised biogas technology of anaerobic digestion and farm ecosystem: Opportunities and challenges. *Frontiers in Energy Research, 2*(10). https://doi.org/10.3389/fenrg.2014.00010

Wu, W., & Ma, B. (2015). Integrated nutrient management (INM) for sustaining crop productivity and reducing environmental impact: A review. *Science of the Total Environment, 512–513*, 415–427. https://doi.org/10.1016/j.scitotenv.2014.12.101

Yadav, S. K., Babu, S., Yadav, M. K., Singh, K., Yadav, G. S., & Pal, S. (2013). A review of organic farming for sustainable agriculture in Northern India. *International Journal of Agronomy, 2013*, 1–8. https://doi.org/10.1155/2013/718145

Chapter 4
The Role of Plant Extracts in Sustainable Agriculture

Aadil Gulzar, Tajamul Islam, and Maroof Hamid

Abstract Environmental and toxicological problems have significantly increased from the last couple of decades due to non-judicial agrochemicals. There is a drastic decrease in the growth and yield of most crops due to diseases and infections, and globally the insects and pests reduce the overall crop production by nearly 20%. In the wake of all this, the novelty and application of biopesticides are increasing tremendously and are effectively used as green pesticides globally. Many plant extracts from various plants like neem (*Azadirachta indica*) showed promising effectiveness as antimicrobials. Found relatively safe, they received greater acceptance from the users. In industrialized countries, botanical pesticides have been actively used in integrated pest management (IPM) for organic farming.

Further, due to the harmful nature of chemical pesticides, demand to use biocontrol agents in the agricultural industry is increasing vigorously. Researchers and scientists are now exploring more and more available alternatives that are relatively more eco-friendly, non-toxic and readily biodegradable. They are focusing on various types of plant extracts which can be applied in green agriculture. Some of the potentially effective plant extracts obtained from the rhizome of ginger (*Z. officinale*), garlic (*A. sativum*), pawpaw (*C. papaya*), neem (*A. indica*), independent weed (*C. odorata*), bitter kola (*G. kola*), miracle tree (*M. oleifera*), etc. are economically efficient, eco-friendly and helpful for sustainable agricultural and horticultural.

Keywords Agriculture · Plant extracts · Chemicals · Fungicides · Disease · Bacterial

A. Gulzar (✉)
Department of Environmental Science, University of Kashmir, Srinagar, J&K, India

T. Islam · M. Hamid
Department of Botany, University of Kashmir, Srinagar, J&K, India

S. A. Bandh (ed.), *Sustainable Agriculture*,
https://doi.org/10.1007/978-3-030-83066-3_4

4.1 Introduction

Naturally obtained plant products have a significant potential and scope in modern agriculture. Some plant products have been reported to contain biochemically active compounds effective against various insects and pests (Aikawa, 2002). The active chemicals extracted from the plants and effectively used in integrated pest management are technically known as botanical pesticides or botanicals. Their use is extensively getting accepted in modern agriculture because of their prominent role in reducing crop pathogenicity (Bruneton, 1999). Besides this, the botanicals are getting degraded more rapidly compared to synthetic chemicals. They are therefore considered as eco-friendly and simultaneously do not kill the beneficial insects in the croplands.

4.2 Commonly Used Botanicals

Plant extracts: Various plants such as neem, garlic, eucalyptus, turmeric, ginger and tobacco are the rich sources of extracts or botanicals.

Essential oils: There are diverse types of essential oils such as nettle oil obtained from *Urtica* spp., thyme oil from *T. vulgaris*, eucalyptus oil from *E. globules*, rue oil from *R. graveolens*, lemongrass oil from *C. flexuosus* and tea tree oil from *Melaleuca alternifolia* (Al-Samarrai et al., 2012), which have some proven potential as botanicals.

Gel and latex: *Aloe vera* is a rich repository of gel and latex.

4.3 Significance of Botanicals

- Sustainable and reliable
- Reduce yield losses
- Environmentally friendly
- Readily degradable
- Extremely useful in green farm practices and are cost-effective
- Essential agents of IPM (integrated pest management)

Plants are the natural laboratories where the biosynthesis of some elite secondary metabolites takes place. They further harbour inbuilt capacities to defend the harmful impacts of their natural enemies. These plants' properties give rise to researchers' scientific curiosity to recognize the potential of bioactive secondary metabolites or compounds against harmful pathogens. For long, many plants and their metabolites have been effectively used in the agriculture and health sector. Using plant extracts to control insects, too, has a long history, and, even in the seventeenth century, the nicotine from tobacco leaves was used to kill plum beetles.

In recent times, researchers have been focusing on chemotherapeutics from the plants as biocontrol agents, as they can be effectively used as agriculture insecticides. Even some essential extracts from different plant species can sometimes be used against pests in sophisticated storage godowns (Ilcim et al., 1997). Some aromatic and volatile oils can also be successfully used to safeguard food items from spoilage and microbial contamination. In most developed countries, biopesticides are widely used in the organic agriculture sector; however, these botanicals can be a promising tool for post-harvest protection of crops in developing countries. Currently, several biopesticide plant extracts are available in the market, which can be categorized into the following classes:

4.4 Plant Extracts Used as Biopesticides (Based on Different Categories)

- **Growth regulators:** Plant extracts that do not permit the insects to achieve their reproductive stage.
- **Feeding deterrents:** These are the biochemical compounds that do not permit the insects and pests to consume anything after entering their bodies. They lead to their starvation that ultimately ends with their death.
- **Repellents:** These extracts can produce such types of odours that repel the pests, e.g. garlic- or pepper-based insecticides.
- **Confusants:** These biochemical compounds puzzle the pests and confine their contact to find the crops (Matsumura, 2001). Most common examples of confusants have been found in pyrethroid and neonicotinoid group of insecticides derived from pyrethrum (*T. cineraria folium*) and tobacco, respectively (Arshad & Parvez, 2010).

4.5 Positives of Biopesticides

- Biopesticides are least harmful than synthetic pesticides.
- Biopesticides are very target-specific against synthetic pesticides, which kill non-target organisms like birds, insects and mammals.
- Biopesticides are readily biodegradable and thus pose less risk of environmental pollution as against synthetic pesticides, which are recalcitrant and non-biodegradable.
- Biopesticides are a very crucial ingredient of IPM programmes and thus diminish the pressure of synthetic pesticides and keep the yield optimal (Ncube et al., 2008).

4.6 Plant Extracts Used as Bioherbicides (Categorized Based on Different Modes of Action)

Plant antibiosis: It is the mechanism where one plant inhibits the growth of another plant in its vicinity, and this function is shown by black walnut trees (*Juglans regia*). They produce an active biochemical called juglone, which can potentially be used in agriculture as an herbicide (Satish et al., 2007).

Plant growth regulation: Some plant oils play a significant role in some specific functions like disrupting cell membranes and essential enzymes in plant tissue and inhibiting the amino acid synthesis. This feature of these oils makes them potentially workable to be used as bioherbicides.

Mechanical control: Some plant extracts like D-limonene can destroy the waxy cuticle of weeds, causing necrosis, dehydration and weed death, thus acting as a potent herbicide.

4.7 Plant Extracts Used as Fungicides and Antimicrobial (Based on Modes of Action)

- **Fungicidal control:** Some botanicals tend to crumble the cell membrane of fungal cells at different stages of their development and hence alter their enzymatic activities and metabolic processes to act as a suitable fungicide.
- **Induced resistance:** Plant extracts can escalate some specific proteins that are capable of inhibiting different fungal diseases. These chemicals enhance the immune system of plants to defend themselves against fungal attacks (Sahan, 2011; Kalkisim, 2012).

Various secondary metabolites, such as alkaloids, phenolics and subclasses, viz. phenolic acids, quinones, flavonoids, tannins and coumarins, possess diverse antimicrobial properties. They perform the function of defence mechanism against various pathogens (B.D. & A.G., 2004), e.g. phenolics are toxic to microorganisms because of their site-specific action and several hydroxyl groups. These extracts or essential oils are often found effective against a broad spectrum of diseases and pathogens (Table 4.1) and are extensively used to eradicate plant pathogens or diseases (Table 4.2).

4.8 Secondary Metabolites and their Mechanism of Action

- **Phenolics**
  - Membrane disruption and deprivation in substrate concentration.
  - Bind to CAMs and form complexes with the cell wall to change its integration.
  - Change enzyme configuration.

Table 4.1 Botanicals and their antimicrobial activity

Name of the plant	Compound/ derivatives	Class/ group	Activity
M. pumila	Phloretin	Flavonoids	Broad range
W. somnifera	Withaferin A	Lactone	Antibacterial and antifungal
A. marmelos	Essential oil	Terpenoids	Antifungal
E. globulus	Tannin	Polyphenol	Antifungal, antibacterial and antiviral
A. cepa and *A. sativum*	Allicin	Sulfoxide	Antifungal and antibacterial
T. vulgaris	Caffeic acid	Terpenoid	Antifungal, antibacterial and antiviral
Curcuma longa	Curcumin	Terpenoid	Antifungal, antibacterial and antiprotozoal diseases
D. stramonium	Hyoscyamine scopolamine	Alkaloids	Antifungal
P. nigrum	Piperine	Alkaloids	Antifungal
R. communis	Ricinine, Ricinoleic	Alkaloids	Antifungal
A. indica	Azadirachtin	Terpenoids	Antifungal and antibacterial

- **Terpenes and Oils**
  - Causes cell wall and membrane destruction.
- **Alkaloids and Flavonoids**
  - Span with cellular membranes.
- **Tannins and Coumarins**
  - Bind to different biological components.
  - Halt enzyme activities.
  - Interact with DNA, lectins and polypeptides.
  - Form disulphide bridges in proteins.

4.9 Plant Extracts with Anti-Parasitic Properties

Traditionally, some plants like garlic (*Allium sativum*), marigold (*Tagetes erecta*) and the goosefoot or epazote (*Chenopodium ambrosioides*) possess the eminent capability to control the parasitic insect population.

4.10 Conclusions

Agrochemicals have created a massive problem for emerging agriculture and horticulture through chemical residue contamination, phytotoxicity, environmental pollution and soil biochemistry alteration. Such problems led to the demand for

Table 4.2 Different botanicals, preparation and effectiveness against diseases/pathogens

Plant	Part/parts used	Preparation	Diseases/pathogens
D. stramonium and *Calotropis* spp.	Whole plant	Crude extract	*C. lunata*
Z. officinale	Rhizome	Crude extract	*P. infestans*, *F. solani*, *P. oryzae*
L. inermis	Leaf	Crude extract	*D. oryzae*
A. indica, *A. squamosa* and *O. sanctum*	Leaf, stem Bark, root	Crude extract	Anthracnose
C. procera	Leaf	Crude extract	Tikka leaf spot disease of groundnut (*C. tikka*)
L. camara	Leaf	Crude extract	Castor grey rot (*Botrytis ricini*)
P. pinnata	Leaf	Crude extract	Leaf blight of onion
A. barbadensis and *N. tabacum*	Leaf	Crude extract	Dry rot of yams *F. oxysporum* and *A. niger*
S. aromaticum and *O. Sanctum*	Leaf, seed and fruit	Crude extract	*Aspergillus flavus*
L. esculentum	Leaf, stem	Crude extract	Bacterial disease on onions
C. japonica and *P. coreana*	Roots, stem	Crude extract	Rice blast and wheat leaf rust
O. gratissimum, *A. melegueta*	Leaf	Crude extract	Post-harvest yam rot
Ocimum spp.	Leaf	Essential oils	*A. flavus*
M. spicata, *S. fruticosa* and *Thymbra* spp.	Leaf	Essential oils	*R. solani* and *S. sclerotiorum*
O. sanctum and *P. persica*	Leaf	Essential oils	Grey mould of grapes
O. Hercleoticum	Leaf	Essential oils	*F. Oxysporum* and *P. tracheiphila*
N. sativa and *F. asafoetida*	Seeds	Essential oils	*F. Oxysporum*, *A. niger* and *A. flavus*
P. nigrum, *S. aromaticum*, *Pelargonium graveolens*, *Myristica fragrans* (*O. vulgare* and *T. vulgaris*)	Seeds	Volatile compounds	Antibacterial
Rubus and *Fragaria* spp.	Fruit	Volatile compounds	Post-harvest decay (fungi)
Cymbopogon spp. *and T. vulgaris*	Leaf, root	Volatile compounds	Black mould disease (*A. niger*)
U. dioica, *A. millefolium*	Leaf	Volatile compounds	*A. Alternate*
Allium sativum	Bulb, leaf	Ethanol extracts	*C. lunata*
C. variegatum	Leaf	Phenolics	*A. Alternate* and *F. oxysporum*

sustainable biocontrol options to deal with the alarming issues associated with synthetic agrochemicals. To deal with plant diseases, these extracts are very effective, cheap, biodegradable, assessable, readily available, target-specific, environment-friendly and harmless to human beings than synthetic agrochemicals. Studies have shown promising results of the increase in crop production and decrease in disease incidence in most of the tested crops. Hence, plant extracts are strongly recommended for the farmer community to achieve sustainable horticultural and agricultural crop production.

References

Aikawa, Y. (2002). Topical preparations containing mango seed kernel oils. *Jpn Kokai Tokkyo Koho, 5*, JP20002322074.

Al-Samarrai, G., Singh, H., & Syarhabil, M. (2012). Evaluating eco-friendly botanicals (natural plant extracts) as alternatives to synthetic fungicides. *Annals of Agricultural and Environmental Medicine, 19*(4), 673–676.

Arshad, A., & Parvez, Q. (2010). Bio-efficacy of some plant leaf extracts against mustard aphid, *lipaphiserysimi*kalt. On Indian mustard, *Brassica juncea*. Rizvi, khan, F.R. *Journal of Plant Protection Research, 50*(2), 130–132.

B. D., & A. G. (2004). Antimicrobial activity of some Turkish medicinal plants. *Pakistan Journal of Biological Sciences, 7*(9), 1559–1562. https://doi.org/10.3923/pjbs.2004.1559.1562

Bruneton, J. (1999). *Pharmacognosy, phytochemistry, medicinal plants*. Lavoisier Publishing.

Ilcim, A., Dıgrak, M., & Bagci, E. (1997). *The antimicrobial effect of Juniperus drupaceous Lab. Morus nigra and Jasminum fruticansL* (pp. 116–121). Kizilirmak Science Congress.

Kalkisim, O. (2012). In vitro antifungal evaluation of various plant extracts against walnut anthracnose (*Gnomonialeptostyla* (Fr.) Ces et de Not.). *Journal of Food, Agriculture and Environment, 10*(4), 309–313.

Matsumura, Y. (2001). Transparent solid soap using hardened castor oil. *Jpn Kokai Tokkyo Koho, 4*, JP2001152197.

Ncube, N. S., Afolayan, A. J., & Okoh, A. I. (2008). Assessment techniques of antimicrobial properties of natural compounds of plant origin: Current methods and future trends. *African Journal of Biotechnology, 7*(12), 1797–1806. https://doi.org/10.5897/AJB07.613

Sahan, Y. (2011). Effect of *prunus laurocerasus*L. (Cherry laurel) leaf extracts on growth of bread spoilage fungi. *Bulgarian Journal of Agricultural Science, 17*(1), 83–92.

Satish, S., Mohana, D. C., Raghavendra, M. P., & Raveesha, K. A. (2007). Antifungal activity of some plant extracts against important seed borne pathogens of *Aspergillus* sp. *Journal of Agricultural Technology*

Chapter 5
Botanical Pesticides for an Eco-Friendly and Sustainable Agriculture: New Challenges and Prospects

Muzafar Riyaz, Pratheesh Mathew, S. M. Zuber, and Gulzar Ahmed Rather

Abstract The global food demand has been rapidly increasing due to expansion in the worldwide populace resulting in the waning of natural resources. The developed and emerging nations are tapping all means to feed the global population. In a run of these measures, many things brought ecological catastrophes and devastation to many organisms. In our farmlands, most of the crops are affected by a specific class of insects called pests. These pests are feeding on our crops, resulting in the collapsing of our agricultural produce. To save these crops from pests, we manufactured chemicals called pesticides which turned out to be very useful in eradicating the pests. Still, in the meantime, the excessive use of these pesticides brought massive devastation in many ways. Notably, most of the crops are dependent on cross-pollination, carried out by various types of insects. Around 80% of the crops worldwide are pollinated by insects (entomophily), especially the bees and other insect species of different families. But, with the frequent utilization of chemical pesticides, these pollinators and other beneficial insects are severely affected, resulting in a threat to their populations. The utilization of chemical pesticides affects insects, but their negative impact is also noticed in humans, aquatic organisms, birds, soil, water and the environment. So, the phytochemicals in botanical extracts are proven to be very much effective in preventing these dreadful crises due to a positive response by non-target organisms and low impact on our habitats and human health.

M. Riyaz (✉)
Division of Taxonomy & Biodiversity, Entomology Research Institute, Loyola College, Chennai, Tamil Nadu, India

P. Mathew
Department of Zoology, St. Thomas College, Palai, Kottayam, Kerala, India

S. M. Zuber
Department of Zoology, Govertment Degree College, Bijbehara, Anantnag, Kashmir, Jammu and Kashmir, India

G. A. Rather
Department of Biomedical Engineering, Satyabhama Institute of Science and Technology, Deemed to be University, Chennai, Tamil Nadu, India

S. A. Bandh (ed.), *Sustainable Agriculture*,
https://doi.org/10.1007/978-3-030-83066-3_5

Keywords Pesticides · Environment · Human health · Biodiversity · Plant extracts · Sustainable agriculture

5.1 Introduction

Since time immemorial, man has been cultivating and harvesting crops. Agriculture has consistently assumed a key position in boosting the economy of any country (Loizou et al., 2019). In the present world, agriculture is the chief fount of revenue in the nations engaged with agriculture and other farming sectors (Tang-Péronard et al., 2011). The upsurge in the global population has raised the alarm of food demand rising parallel to the worldwide population. The developed and emerging nations are tapping all means to feed the billions of people across the globe. In pest management practices, the advancement in technology has proven to be accompanying farmers to achieving higher amounts of crop yields. However, in our farmlands, most of the yields are being influenced by a particular class of insects called pests. These pests are feeding on our crops, resulting in the collapsing of most of our agricultural produce. The damage prompted to the crops by insect pests varies on the insect pests' feeding habit (Table 5.1).

Table 5.1 List of some common insect pests and their feeding habits

Common name	Order	Feeding habit
Aphids, mealybugs, whiteflies and scale insects (coccids)	Hemiptera	Plant sap
Hoppers	Hemiptera	Foliage and shoots
Caterpillars	Lepidoptera	Leaves and needles
Grasshoppers and locusts	Orthoptera	Leaves, grains, seed pods and fruits
Borers	Coleoptera, Lepidoptera	Roots, stem, shoots
Weevils	Coleoptera	Stored grains
Thrips	Thysanoptera	Fruit, leaves, shots, sap
Beetles	Coleoptera	Leaves, stem, petals, fruits
Pod bugs	Hemiptera	Seed pods
Stink bugs	Hemiptera	Leaves, fruits, stems, seed pods
Termites	Blattodea	Timber, furniture, branches
Cockroaches	Blattodea	Food, fabrics, fruits, books
Fruit flies	Diptera	Fruits, leaves
Gall midges	Diptera	Shoots, plant tissue
Saw flies, gall wasps	Hymenoptera	Plant foliage
Grubs	Coleoptera	Roots
Silverfish	Zygentoma	Books, clothes, food items

Though crops are affected by abiotic stresses, a significant portion of the crops and the harvest are influenced by the biotic stress of the insect pests. The insect pests can damage an entire or an enormous portion of our crop (Sharma et al., 2017). Around 70% of the crop can be lost to the pests if preventive measures are not taken since several species from different taxa include the natural insect predators and parasitoids that control the population of quite a few pest species. However, with the utilization of chemical pesticides, the natural enemies of the pests are becoming vulnerable. These chemical pesticides assume a significant position in agricultural and horticultural productivity (Carvalho, 2017). Pesticides have been assisting farmers by slashing the time and efforts to expel weeds and pests in farm fields for ages physically. However, due to the growing food demand, the utilization of chemical pesticides has risen enormously. Several environmental contaminations have also emerged with the considerable utilization of chemical pesticides. The soil, water and air quality got widely disrupted by the residues of these chemical pesticides. Life in aquatic ecosystems, beneficial insects and other vegetation became affected by the toxicity of these chemical pesticides (Riyaz et al., 2020).

Thus, to reduce the chemical pesticide contamination, carbon outputs, habitat destruction and fragmentation, sustainable agriculture is an effective alternative. Sustainable agricultural practices are way forward to maintain the ecological equilibrium by the eco-friendly techniques to reverse the damage done by large-scale agriculture and allied farming sectors (Slätmo et al., 2017). With a setup of a green environment and the cultivation of the crops, a lot of eco-friendly practices in sustainable farming can be utilized. These involve permaculture (Bhandari & Bista, 2019), aquaponics and hydroponics (AlShrouf, 2017), using renewable energy resources (Dudin, 2018), crop rotation and polycultures (Weißhuhn et al., 2017) and integrated pest management (Dara, 2019) (Fig. 5.1). With these practices, natural resource exploitation can be curbed. Further, diversification in the crops by crop rotation and polycultures can reduce fertilizers and pesticides. Chemical pesticides can be replaced by botanical pesticides, which are safer to handle and assure a low impact on the species of different taxa, their habitats, different ecosystems and human health (Nawaz, Juma, & Hongxia, 2016).

By introducing sustainable agricultural practices, the innovative technologies have progressed well to conserve the environment, beneficial insect diversity and human health. For eliminating the pests from farmlands, plant extracts are a creative and safe approach. The extracts can be obtained from dried or ground plant materials or crude plant. These plant extracts have been proven to be the best alternative to chemical pesticides as they can remove the pests from the farm fields while improving the quality of soil, water and air by their low impact. The plant extracts used as insecticides have a remarkable place among sustainable agriculture practices as they are safer than synthetic pesticides. In this chapter, sustainability of agriculture, botanical pesticides, their sources, usage, new challenges, prospects and effects of chemical pesticides on beneficial insect diversity, human health and aquatic ecosystems have been addressed in a detailed manner (Isman & Grieneisen, 2014).

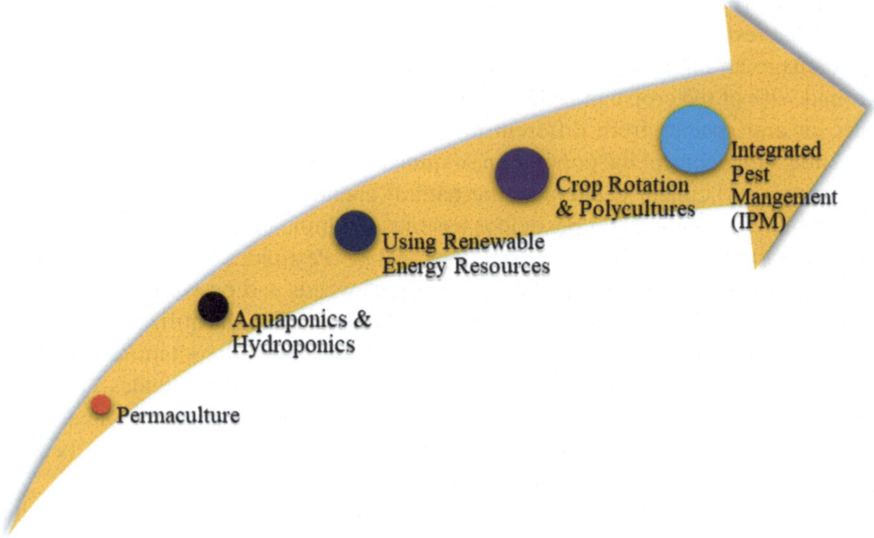

Fig. 5.1 Eco-friendly and sustainable agriculture practices

5.2 Sustainable Agriculture: A Promise to the Future

Back in the 1970s till now, there has been an enormous rise in environmental devastation brought about by the widespread agricultural activities (Majeed & Mazhar, 2019). Around 12% of the global greenhouse gas emissions are contributed by the activities such as industrial agriculture and other environmental devastations such as deforestation, habitat destruction, pesticide toxification and pollution, and intense carbon outputs caused the large-scale agribusinesses (Yue et al., 2017). Soil erosion can also be triggered by higher demands of agriculture on natural ecosystems (Nearing et al., 2017). There has been a rise in agricultural practices for a higher food demand as well, and for achieving a good result, the crops have been fertilized and sprayed with pesticides to save them from any pest damages. Safeguarding of these crops is only possible when we frequently splash them with pesticides, thereby achieving a good harvest. However, it has such severe implications on the environment and human health. For a good harvest to accomplish without menacing the soil, water, human wellbeing and the surrounding ecosystems, there is a need for eco-friendly and sustainable agriculture.

According to the Food and Agriculture Organization (FAO) of the United Nations, sustainability in agricultural development can be defined as 'the management and conservation of the natural resource base, and the orientation of technological change in such a manner, to ensure the attainment of continued satisfaction of human needs for present and future generations' (FAO, 2014). Sustainable farming becomes more substantial among the developed and developing nations

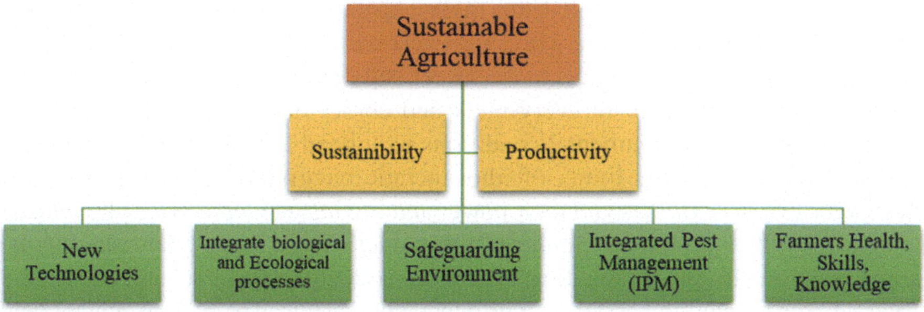

Fig. 5.2 The illustration shows the working of sustainable practices in agriculture

engaged in the agriculture sectors (Roberts & Mattoo, 2018). The deployment of sustainable agricultural practices into the farming sectors has proven to an innovative approach towards conserving natural resources like aquatic and terrestrial ecosystems, safeguarding the environment, human health and beneficial insect diversity, including pollinators and natural predators (Saunders, 2018). Sustainable agricultural practices aim to achieve higher crop yields, thus obtaining higher economic profitability (Fig. 5.2). With the advancement in technology, modern techniques can be employed in sustainable agriculture by which eco-friendly practices can be carried out so that there is the least wastage of harvest and natural resources. In sustainable agriculture, not only can we conserve our environment and natural resources but also train and exercise optimization of the usage of pesticides and fertilizers. The farmers implementing sustainable agriculture practices can achieve a higher crop yield and conserve their surrounding ecosystems contaminated by pesticide residues. Dealing with the pests in an agroecosystem and implementing integrated pest management (IPM) in sustainable agriculture are ways forward for dealing with the pests and eventually safeguarding human and environmental health. Integrated pest management (IPM) has emerged as the most ecosystem-based strategy for protecting crop and vegetable cultivations. With IPM, an aggregation of techniques can be employed such as biological, cultural, manual and chemical by implementing resistant varieties of crops and habitat management through which economic, health and environmental risk can be reduced (Peterson et al., 2018).

5.3 The Growing Pest Emergence, Problem and Utilization of Chemical Pesticides

The lower Devonian period marked the dawn of insect evolution. Because of their capability to withstand a wide range of climatic conditions, these species became the dominant creatures the planet has ever witnessed. The insects are the primary animals on the earth that adapted the flight capability, back around 400 million years ago, and dominated all the world's ecosystems (Riyaz et al., 2018). The flight ability provided immense support to the insect body for grabbing an edge over others and to get acclimatized in every nook and corner of the earth. The ability to flourish in different environmental conditions with flexible body parts helped these animals conquer even the limits of idiosyncratic environmental conditions. The arthropodic origin of insects and their power and life inside invertebrate phyla made them profoundly successful creatures of this planet. While insects flourished in the animal kingdom as dominating creatures, these creatures deliver several ecosystem services to the humankind and their world in a unique style. The pollens got shipped through cross-pollination by flying cargos of insects, and around 80% of crops across the globe are depending on the insects for transportation of their pollen from one flower to the other (McGregor, 1976). The insects have marked a natural establishment across all global ecosystems. Besides rendering the services like nutrient cycling, seed dispersal, fertility and structure of the soil, they are also proven to be a significant food source for other taxa. With all these characteristic roles they play in an ecosystem, a portion of insects turns out to be pests of many crops around the world. Saving the crop yield from the nuisances, there has been a progression in controlling these pests from time to time.

Along with the rise of agriculture in the ancient world, the eradication of the pests came on track back in 3000 BC, when ancient Egyptians employed trained cats and mongooses for controlling the pests of stored grains such as rodents and back in 500 AD in Europe when Ferrets were trained as mousers (Sherman, 2007; Taylor, 2011). Since it was simple to pulverize weeds in the farm fields by either blazing them or by tilling them out. However, with time and safeguarding crops, the ancient Sumerians utilized pesticides before 2000 BC (Pflanzenschutz-Nachrichten, 1973). Essential sulphur dusting was a conspicuously known pesticide and a key component utilized in ancient occasions around 4500 years back before Mesopotamia (Ranga Rao et al., 2007). Till the fifteenth century, toxic synthetic compounds, such as arsenic, mercury and lead, were sprinkled on yields for removing the nuisances. During the seventeenth century, nicotine sulphate from tobacco leaves was removed and utilized as a bug spray (Miller, 2002). With the disclosure of the insecticidal properties of DDT by Paul Muller in 1939, synthetic pesticides began to advance in the market. In 1948, he conceded the Nobel Prize to discover pesticide properties of DDT (Peshin et al., 2009). Eventually, in the 1960s, problems like the resistance of pests to chemicals, threat to biodiversity, aquatic and terrestrial ecosystems, climate and environment began to rise.

5.4 Erroneous Effects of Chemical Pesticides in Agriculture: Hazards to Human Health, Insect Biodiversity and Aquatic Ecosystem

'For the first time in the history of the world, every human being is now subjected to contact with dangerous chemicals, from the moment of conception until death' (Rachel Carson, 1962).

Pesticides are the chemical compounds developed for eliminating pests from agricultural fields, storage warehouses, homes, etc. Since time immemorial, man has been utilizing pesticides because their utilization has brought relief to farmers by expelling the pests from the farmlands. However, the large-scale usage of chemical pesticides has proven to be incompatible with the environment. Pesticides are generally used to remove insect pests (insecticides), fungi (fungicides), rodents (rodenticides), unwanted plants/weeds (herbicides/weedicides), nematodes (nematicides) and bacteria (bactericides) (Fig. 5.3). The impact of chemical pesticides on various life forms and different ecosystems has been reported across different world places. The nations engaged in different agricultural and allied sectors are mostly affected by it. With the rise in global population and food requirements, there has been a parallel growth in the large-scale cultivation of high-yielding monocrops. Since crop loss by the pests was controlled by the pesticides, their long-lasting adverse impact on various life forms and the natural environment is a significant challenge to be taken care of. On the contrary, the health of farmers has also declined due to their exposure to the toxicity levels of chemical pesticides. The chemical

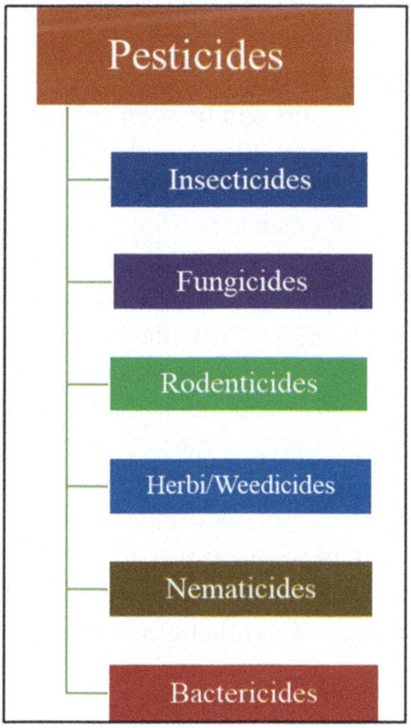

Fig. 5.3 Types of pesticides

Fig. 5.4 Pesticide sprayed in an apple orchid. (Photo Muzafar Riyaz)

pesticides have been most devastative on beneficial insect diversity, including pollinators such as honey bees, dipteran pollinators, predators, parasitoids and other useful insects that deliver several ecosystem services. In an agricultural field, several insects can be seen collecting nectar and pollinating the flowers, such as bees, wasps, hoverflies, moths and butterflies and some species checking the populations of insect pests such as parasitic wasps, hornets, beetles, lacewings, etc. While spraying the pesticides (Fig. 5.4) on crops infested with pests, 15 to 40% of an estimated fraction of pesticides are scattered into the atmosphere by either volatilization or spray drift processes (Socorro et al., 2016). After spraying, the pesticides in the atmospheric particulate phase remain in the air for about 7 to 12 days and thoroughly orbit many geographical locations worldwide. The circling of pesticides in the atmosphere alters the air quality and adds more events to climate change (Miller & Spoolman, 1996). The pesticide runoff from the agricultural lands into streams and lakes significantly impacts aquatic life and water contamination. Though runoff can transport pesticides into the aquatic ecosystem, the atmospheric dispersal of pesticides can travel to other places like grazing fields and human settlements, potentially affecting other living organisms and human wellbeing (Fig. 5.5). The impact of synthetic agrochemicals on insect diversity has been well documented

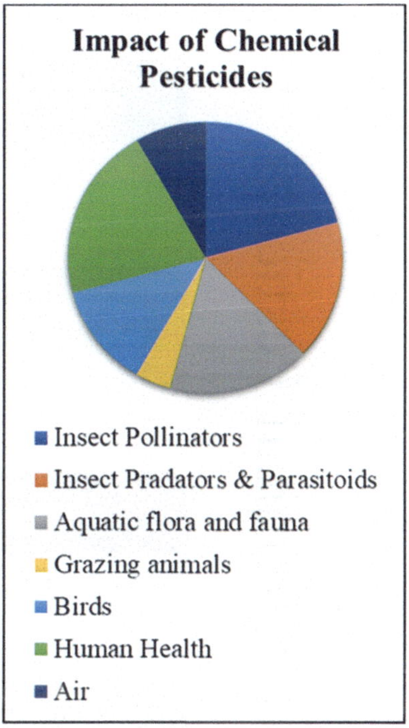

Fig. 5.5 The chart shows the impact of synthetic pesticides on different life forms

across the globe, and there has been a massive decline in insect pollinators from the past few decades due to the large-scale pesticide utilization (; Dudley & Alexander, 2017; Sánchez-Bayo & Wyckhuys, 2019). Some studies have shown that compounds of organophosphates and other pesticides can have poisonous or lethal effects resulting in the disruptions of cellular metabolism that often lead to embryonic changes and mutagenesis (Maurya et al., 2019) on fish species and birds (Tesfahunegny, 2016). Besides the impact of pesticides on the environment (Mahmood et al., 2016), soil (Joko et al., 2017) and water (de Souza et al., 2020; Hallberg, 1987), reports of pesticides influencing wildlife (Moriarty, 1972; Rattner, 2009), amphibians (Islam & Malik, 2018; McCoy & Peralta, 2018), earthworms (Yasmin & D'Souza, 2010), non-target plants (Mitra & Raghu, 1998; Saladin & Clément, 2005) and grazing animals (Choudhary et al., 2018) have also been well documented in the recent past (Fig. 5.6). Many studies worldwide have reported several health-related issues such as brain cancers, breast cancers, testis and ovarian cancers, leukaemias and lymphomas affecting people. A detailed list of health issues and diseases of humans caused by the exposure and poisoning of synthetic pesticides and their classification is given in Table 5.2.

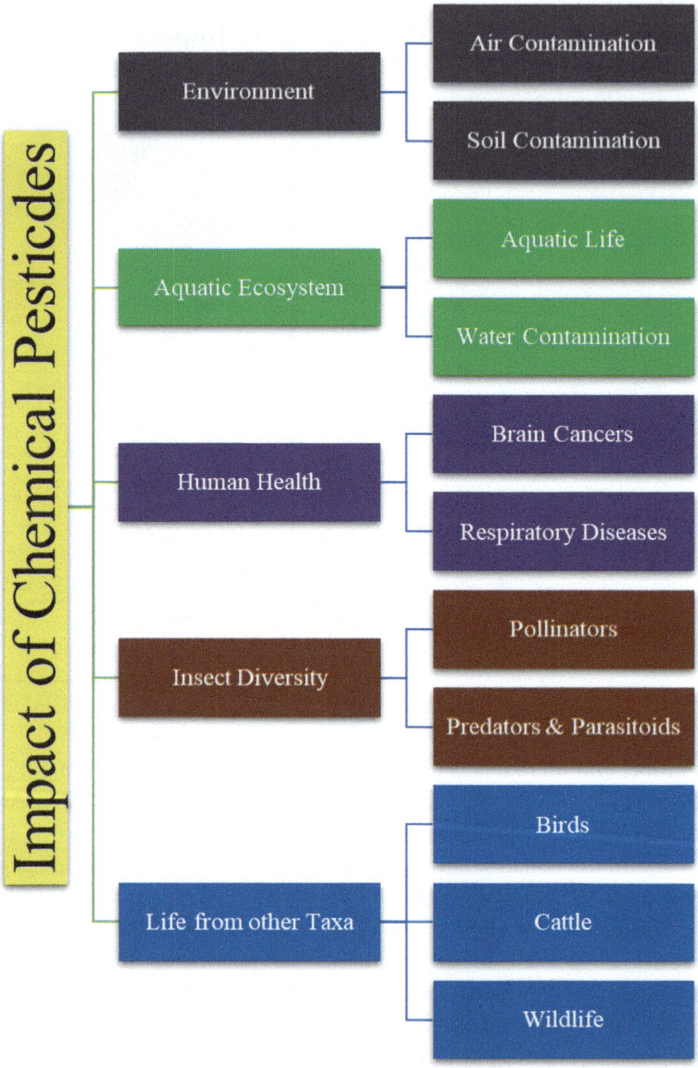

Fig. 5.6 The illustration shows the impact of synthetic pesticides on ecosystems, the environment and life from different taxa

5.5 Botanical Pesticides: A Natural Alternative for Chemical Pesticides

Synthetic pesticides are utilized as a swift remedy to the threat caused by pests in all stages of crop production (Ekeh et al., 2018). They include a wide range of chemicals that are non-biodegradable and persistent, polluting water, air and soil, leading

Table 5.2 Synthetic pesticides/insecticides: types and effects on human health

S. no.	Name (trade name)	Chemical formula	Antagonistic effects on human health	References
1	Chlorinated hydrocarbons			
2	Dichlorodiphenyltrichloroethane (DDT)	$C_{14}H_9Cl_5$	Cancer, nervous system disorders, respiratory damage, reproductive organs, immune system and endocrine disruptions, congenital disabilities	Thuy (2015); Cohn et al. (2015); Kim et al. (2017); Byard et al. (2015)
3	Methoxychlor	$C_6H_{15}Cl_3O_2$	Cancer, central nervous depression, diarrhoea, damage to the liver, kidney and heart	Chen (2014)
4	Dichlorodiphenyl ethanol	$C_{14}H_{12}Cl_2O$	Digestive tract infections, asthma, depression and morbidity, T-lymphocyte dysfunction, cancer, DNA damage	Igbinosa et al. (2013)
5	Chlorobenzilate	$C_{16}H_{14}Cl_2O_3$	Carcinogenic, genotoxic, eye damage	Lewis et al. (2016)
6	Benzene hexachloride (BHC) (lindane) (gamma-BHC or g-BHC)	$C_6H_6Cl_6$	Highly carcinogenic, dermatitis, psoriasis, burning, rashes	Loomis et al. (2015)
7	Toxaphene	$C_{10}H_{10}Cl_8$	Carcinogenic, immune system failure, reproductive organ damage, DNA damage	Wallace (2014)
8	Chlordane	$C_{10}H_6Cl_8$	Carcinogenic, type 2 diabetes, lymphoma, prostrate cancers, obesity, brain and breast cancer	Thakur and Pathania (2020); Evangelou et al. (2016); Lim et al. (2015); Tang-Péronard et al. (2011); Cook et al. (2011); Khanjani et al. (2007)

(continued)

Table 5.2 (continued)

S. no.	Name (trade name)	Chemical formula	Antagonistic effects on human health	References
9	Heptachlor	$C_{10}H_5Cl_2$	Hepatotoxicity, neurotoxicity and developmental toxicity, immune system damage, carcinogenic	Reed and Koshlukova (2014a), b)
10	Aldrin	$C_{12}H_8Cl_6$	Systemic, neurological, reproductive/ developmental, immunological, genotoxic and tumorigenic	US-EPA (2003)
11	Dieldrin	$C_{12}H_8Cl_6O$	Carcinogenic, neurological, reproductive/ developmental, immunological and genotoxic.	US-EPA (2003); Bates et al. (2008)
12	Endrin	$C_6H_8Cl_6O$	Central nervous system, headache, dizziness, nausea, vomiting, convulsions, fertility issues	Honeycutt and Jones (2014)
13	Chlordecone	$C_{10}Cl_{10}O$	Carcinogenic, body tremors, low sperm cell counts, recent memory loss, liver enlargement, oculomotor dysfunctions, ataxia	Multigner et al. (2016)
14	Endosulfan	$C_9H_6Cl_6O_2$ S	Cancer, acute and chronic toxicity, respiratory failure, endocrine disruption, reproductive failure, DNA damage	Singh et al. (2014); Sebastian and Raghavan (2016)
II	Organophosphates			
15	Tetraethyl pyrophosphate (TEPP)	C_8H_20 O_7P_2	Eye pain, blurred vision, lacrimation, rhinorrhoea	O'Neil et al. (2001)
16	Dichlorvos	$C_4H_7O_4Cl_2P_2$	Cancer, cell damage, neurotoxic, headache, sweating, nausea and vomiting	Koutros et al. (2008)

(continued)

Table 5.2 (continued)

S. no.	Name (trade name)	Chemical formula	Antagonistic effects on human health	References
17	Chlorfenvinphos	$C_{12}H_{14}O_4Cl_3P$	Developmental, reproductive and immunologic effects	Koshlukova and Reed (2014)
18	Phosphamidon	$C_{10}H_{19}O_5NClP$	Neurological disorders, cell damage	Naqvi and Hasan (1992)
19	Monocrotophos	$C_7H_{14}O_5NP$	Respiratory paresis, muscular weakness, cranial nerve palsies	Gupta and Milatovic (2014)
20	Dicrotophos	$C_8H_{16}O_5PN$	Blurred vision, pinpoint pupils, vomiting, headache, dizziness, abdominal pain, muscle spasms, diarrhoea, hypotension, respiratory paralysis	Pohanish (2015)
21	Trichlorfon	$C_4H_8Cl_3O_4P$	Nervous system disruption, nausea, respiratory paralysis, dizziness and sometimes death	Timoroğlu et al. (2014)
22	Methyl parathion	$C_8N_{10}NO_5PS$	Headaches, nausea, night-waking, diarrhoea, difficulty breathing, mental confusion, nervous system, cardiovascular and reproductive system	Edwards and Tchounwou (2005)
23	Fenthion	$C_{10}N_{15}O_3PS_2$	Neurotoxic, headache, sweating, nausea and vomiting, diarrhoea, muscle twitching and death	Moser (2014)
24	Diazinon	$C_{12}H_{21}N_2O_3PS$	Cancer, reproductive system, acute and chronic toxicity, respiratory failure, endocrine disruption	Beane Freeman et al. (2005); Harchegani et al. (2018)
25	Ethion	$C_9H_{22}O_4P_2S_4$	Clinical toxicity, abdominal pain, diarrhoea, vomiting, respiratory problems and undue secretions	Dewan et al. (2008)

(continued)

Table 5.2 (continued)

S. no.	Name (trade name)	Chemical formula	Antagonistic effects on human health	References
26	Phorate	$C_7H_{17}O_2PS_3$	DNA damage, nausea, dizziness, confusion, respiratory paralysis and death	Saquib et al. (2019)
27	Disulfoton	$C_8H_{19}O_2PS_3$	Nervous system disruption, respiratory disruptions, vomiting, diarrhoea, drooling, tremors, convulsions and sometimes even death	Fent (2014)
28	Dimethoate	$C_5H_{12}O_3PS_2N$	Cell damage, vomiting, abdominal pain, faecal incontinence, diarrhoea	Mirajkar (2014)
29	Malathion	$C_{10}H_{19}O_6PS_2$	Liver, kidney, testis, ovaries, lung, pancreas, blood, genotoxic and carcinogenic	Badr (2020)
III	Carbamates			
30	Carbaryl	$C_{12}H_{11}NO_2$	Neurological, reproductive, immunological disorders, possible carcinogen	Koshlukova and Reed (2014)
31	Aminocarb	$C_{11}H_{16}O_2N_2$	Cholinesterase inhibition, effects on the nervous system, sometimes death	Rodgers et al. (1986)
32	Carbofuran	$C_7H_{15}NO_3$	Body weakness, abdominal pain, blurred vision, nausea, sweating, muscle shuddering, coordination dysfunctions, respiratory and nervous system disorders	Gupta (1994)
33	Aldicarb	$C_7H_{14}N_2O_2\ S$	Headache, nausea, sweating, diarrhoea, coordination system disruptions and sometimes death	Baron and Merriam (1988)

(continued)

Table 5.2 (continued)

S. no.	Name (trade name)	Chemical formula	Antagonistic effects on human health	References
IV	Pyrethroids			
34	Cypermethrin	$C_{22}H_{19}Cl_2NO_3$	Neurotoxic, hepatotoxic, effects on behaviour, molecular level and reproductive system	Sharma et al. (2018)
35	Deltamethrin	$C_{22}H_{19}Br_2NO_3$	Paranaesthesia, unwanted sensations, burning and partial numbness, 'pins and needles', skin problems	Doi et al. (2006)

to unintentional hazards to humans, non-target species and the environment, including depletion of the ozone layer (Damalas & Koutroubas, 2015; Lengai et al., 2020; Wimalawansa & Wimalawansa, 2014). Uncontrolled and continuous use of synthetic pesticides can also induce pesticide resistance among pest populations and pest resurgence, yet another disastrous factor in pest management (Shabana et al., 2017). These erroneous human health issues and drastic effects on nature and biodiversity invoked the thought for an alternative (Mahmood et al., 2016). Botanical extracts are biochemical compounds extracted from different plants, biodegradable with lesser shelf life, making them nature-friendly. Plant extracts were used in various fields by human life since time immemorial in many civilizations throughout the history in China, Egypt, Greece and India (Dougoud et al., 2019). The pesticide properties of botanical extracts have shown promising results, making them suitable candidates for integrated pest management (b; Ali et al., 2014; Isman, 2017a; Isman & Grieneisen, 2014; Mkenda et al., 2015; Stevenson et al., 2017). Due to their special attributes like lower toxicity, biodegradability, diverse modes of action, efficacy and obtainability of source materials, botanical pesticides are of greater importance from planting to harvesting and storing crops (Neeraj et al., 2017).

5.5.1 Source of Botanical Pesticides

Botanical pesticides are extracted from plant sources that can kill or control pests (Chengala & Singh, 2017). Every plant in nature has developed certain natural mechanisms in evolution to adapt to various environmental conditions with their pesticide property as one among them. A worldwide estimate of more than 2500 species of plants from 235 families has been noted to possess biochemical with pesticide or deterrent or growth-regulating properties (Das, 2014; Roy et al., 2016). Major plant families with active biomolecules against pests include Apiaceae,

Apocynaceae, Asteraceae, Cupressaceae, Caesalpinaceae, Lamiaceae, Lauraceae, Liliaceae, Myrtaceae, Piperaceae, Poaceae, Rutaceae, Sapotaceae, Solanaceae, Zingiberaceae, etc. (Ahmad et al., 2017; Wanzala et al., 2016).

A wide range of common and locally available plants have also been reported to possess some pesticidal biochemical compounds like *A. indica* (neem), *Tanacetum cinerariifolium* (pyrethrum), *Allium sativum* (garlic), *Curcuma longa* (turmeric), *Rosmarinus officinalis* (rosemary), *Zingiber officinale* (ginger) and *Thymus vulgaris* (thyme) (Castillo-Sánchez, Jiménez-Osornio, Delgado-Herrera, Candelaria-Martínez, & Sandoval-Gío, Castillo-Sánchez et al., 2015). Compounds like azadirachtin from neem and pyrethrin from pyrethrum are common examples of isolated botanicals that have been commercialized due to their efficient results (Kumar et al., 2015). Selected examples of botanical pesticides against different pest groups are shown in Table 5.3.

The plant part used for extraction depends on the bioactive compound of interest and its concentration in a particular plant part, including root, rhizome, stem, bark, leaves, flower, fruit, seeds and cloves (Lengai et al., Lengai et al., 2020). The extraction and production of these botanical pesticides are economical and straightforward compared to synthetic pesticides, emitting large amounts of toxic pollutants as by-products or wastes. The process generally involves grinding of dried plant parts followed by extraction using organic solvents that maximize the extraction of target compounds (Chougule & Andoji, 2016). The extract is then concentrated, formulated and tested for evaluation in the lab and field trials (Zarubova et al., 2014).

5.5.2 Benefits of Botanical Pesticides over Synthetic Pesticides

The vast availability of source plants, diverse uses, less toxicity to non-specific targets like pollinators and fish, cheaper costs, effectiveness and reliability are the attributes responsible for the acceptability of the botanical pesticides in sustainable crop production (Castillo-Sánchez et al., 2015; Srijita, 2015). Botanical pesticides have been demonstrated to possess insecticidal properties even in their crude forms (Ali et al., 2014). Target specificity of compounds in plant extracts and essential oils ensures safeguarding non-target species and, more importantly, beneficial species like pollinators and natural predators (Nawaz, Mabubu, & Hua, 2016). The pesticide-pest interaction of botanical pesticides is biochemical, thereby decreasing the probability of pesticide resistance (Lengai et al., 2020). The efficacy of botanical pesticides can be influenced by parameters like a source of plant species, the raw material (fresh or dried) used for extraction, extraction methodology and solvents utilized for extraction (Arafat et al., 2015; Sarkar & Kshirsagar, 2014). Compared to synthetic pesticides, botanical pesticides show diverse modes of action like toxicity, repellence, growth regulation and structural modifications on target species, making them the best fit for integrated pest management (Laxmishree & Nandita, 2017). Botanical extracts, especially metabolites, can interfere with insect behaviour, morphology, metabolic pathways, biochemical processes and physiological

Table 5.3 Plants having pesticide effect on pests of different crops

Source plant	Pest	Host and disease/ damage	References
I	**Virus**		
Gossypium herbaceum	*Southern rice black streaked dwarf virus* *Tobacco mosaic virus* *Rice stripe virus*	Tobacco/tobacco mosaic Rice/leaf stripe infection	Zhao et al. (2015)
Thuja orientalis	*Watermelon mosaic virus*	Watermelon/WMV infection	Elbeshehy et al. (2015)
Cynanchum komarovii ***Celosia cristata***	*Tobacco mosaic virus*	Tobacco/TMV infection	Todorov et al. (2015)
II	**Bacteria**		
***Origanum* spp.**	*Bacillus* spp. *Serratia marcescens*	Wheat/white stripe Cucurbits/yellow wine disease	Jnaid et al. (2016); Sharoba et al. (2015)
Lantana camara	*Klebsiella pneumoniae* *Escherichia coli*		
Allium sativum	*Pseudomonas syringae*	Citrus/black pit	Mougou and Boughalleb-M'hamdi (2018)
III	**Fungi**		
Aloe vera ***Allium sativum*** ***Glycyrrhiza glabra***	*Fusarium guttiforme* *Chalara paradoxa*	Pineapple/ fusariosis	Sales et al. (2016)
A. indica ***Ocimum sanctum***	*Fusarium oxysporum*	Tomato/wilt	Chougule and Andoji (2016)
Allium sativum ***Curcuma longa*** ***Citrus limon*** ***Zingiber officinale***	*Bemisia tabaci* *Caliothrips fasciatus* *Uromyces appendiculatus* *Phaeoisariopsis griseola* *Colletotrichum lindemuthianum*	Snap bean/whitefly damage Snap bean/thrips damage Snap bean/rust Snap bean/angular leaf spot Snap bean/ anthracnose	Muthomi et al. (2017)
Hydnocarpus anthelminthicus	*Phytophthora palmivora* *Pyricularia oryzae* *Rhizoctonia solani*	Rice/fungal infection	Jantasorn et al. (2016)
IV	**Nematode**		
A. indica ***Brassica napus*** ***Lantana camara*** ***Tagetes erecta***	*Meloidogyne incognita*	Tomato/root knot	Kepenekçi and Erdo (2016)
Tagetes erecta ***Chromolaena odorata***	*Meloidogyne incognita* *Helicotylenchus* spp. *Dolichodorus* spp.	*Amaranthus* Fluted pumpkin	Ogundele et al. (2016)

Table 5.3 (continued)

Source plant	Pest	Host and disease/ damage	References
Thymus citriodorus	*Meloidogyne incognita* *Meloidogyne javanica*	Tomato/root knot	Ntalli et al. (2020)
V	**Insect**		
Cinnamomum cassia ***Cinnamomum zeylanicum*** ***Piper nigrum***	*Megalurothrips sjostedti*	Cabbage/flower damage	Abtew et al. (2015)
Aglaia odorata ***Annona squamosa*** ***Piper retrofractum***	*Crocidolomia pavonana* *Plutella xylostella*	Cabbage/foliar damage	Abtew et al. (2015)
Allium cepa ***A. sativum*** ***A. indica*** ***Curcuma zedoaria*** ***Calotropis procera*** ***Ocimum canum*** ***Phyllanthus emblica***	*Helicoverpa armigera*	Tomato/fruit damage	Sumitra et al. (2014)

activities; blocking of glucose in chemosensory receptor cells in the mouth of lepidopterans by terpenes and chemosterilant activity of some essential oils are a few examples (Lengai et al., 2020).

Utilization of botanical pesticides in pest control guarantees added benefits to farmers like food security, lowering pest intensities and enhanced superiority of harvest, drawing greater demands and higher rates in the market (Nefzi et al., 2016). Organically produced food products are in greater demand in the lucrative market, where consumers are ready to buy these foods at higher rates creating greater market openings for botanical pesticides (Misra, 2014).

5.5.3 Biodegradability of Botanical Pesticides

The botanical pesticides are quickly degraded, with their biological origin preventing their accumulation in the environment and thereby eliminating the chances of air, water and soil pollution (Sokovi' c, 2010). Exposure to sunlight, high temperature and humidity could break down their constituents depending on their nature, e.g. azadirachtin, isolated from neem (*A. indica*), has a half-life between 1 day over crops and 2 days in soil, whereas thymol, a compound extracted from *Piper nigrum*, *Satureja hortensis*, *Thymus vulgaris* and *Zataria multiflora* under sunlight, is proved to survive up to 28 hours to degrade and in soil with a duration of 8 days (Liu et al., 2017; Yan et al., 2017; Yang et al., 2017). The biodegradation process is accelerated by detoxification enzymes secreted by microorganisms present in abundance in natural conditions through oxidative metabolism (Mpumi et al., 2016).

Carboxylesterase enzyme-mediated hydrolysis of ester bonds is a common pathway of biodegradation exhibited by microorganisms such as *Bacillus cereus* and *Aspergillus niger* (Cycoń & Piotrowska-Seget, 2016). A wide variety of bacterial species are reported to degrade carbamates, organophosphates, organochlorine pesticides and pyrethroids (Cycoń & Piotrowska-Seget, 2016; Porto et al., 2011).

In soil, enzymes produced by microorganisms modify the botanical pesticides into less toxic groups that are breakable, rendering them biologically unavailable and making them non-toxic (Ortiz-Hernández et al., 2013). Microbial degradation is further influenced by physical factors and interaction with pesticides and environmental conditions (Cycoń & Piotrowska-Seget, 2016).

5.5.4 Botanical Pesticides for Integrated Pest Management

Integrated pest management (IPM) aims to achieve sustainable pest management through environment-friendly strategies for reducing pests and attaining highly profitable yields (Alam et al., 2016). Botanical pesticides are eco-friendly natural compounds effective against different pests like viruses, bacteria, fungi, nematodes and insects with varied modes of action (Feyisa et al., 2015; Todorov et al., 2015). Alongside crop, security approaches like host plant resistance or tolerance, the introduction of natural enemies like parasitoids and predators, improved agricultural practices, use of microbial pesticides and reduced use of chemical pesticides, application of botanical pesticides also serve as a critical component in IPM (Muthomi et al., 2017; Wegulo et al., 2015). Compounds extracted from plant sources are effective against a major group of pests like viruses, bacteria, fungi, nematodes and insects (Elbeshehy et al., 2015; Ingle et al., 2017; Neeraj et al., 2017; Sales et al., 2016; Salhi et al., 2017). This wide variety of botanical pesticides and their results using crude form, extractions and formulations opens a wide scope for complete replacement of synthetic pesticide with these eco-friendly pesticides, which immensely contributes to integrated pest management and sustainable agriculture for a healthy future (Fig. 5.7).

5.6 Prospects of Botanical Pesticides: Discussion and Conclusion

The large-scale utilization of chemical pesticides has affected the pollinators and other beneficial insects, but its negative impacts have also been noticed on human health, aquatic organisms, birds, wildlife, grazing animals, earthworms, soil, air, water and the environment. The phytochemicals in botanical extracts are proven to be much effective in preventing these dreadful crises due to a positive response by non-target organisms and low impact on the environment and human health. Even

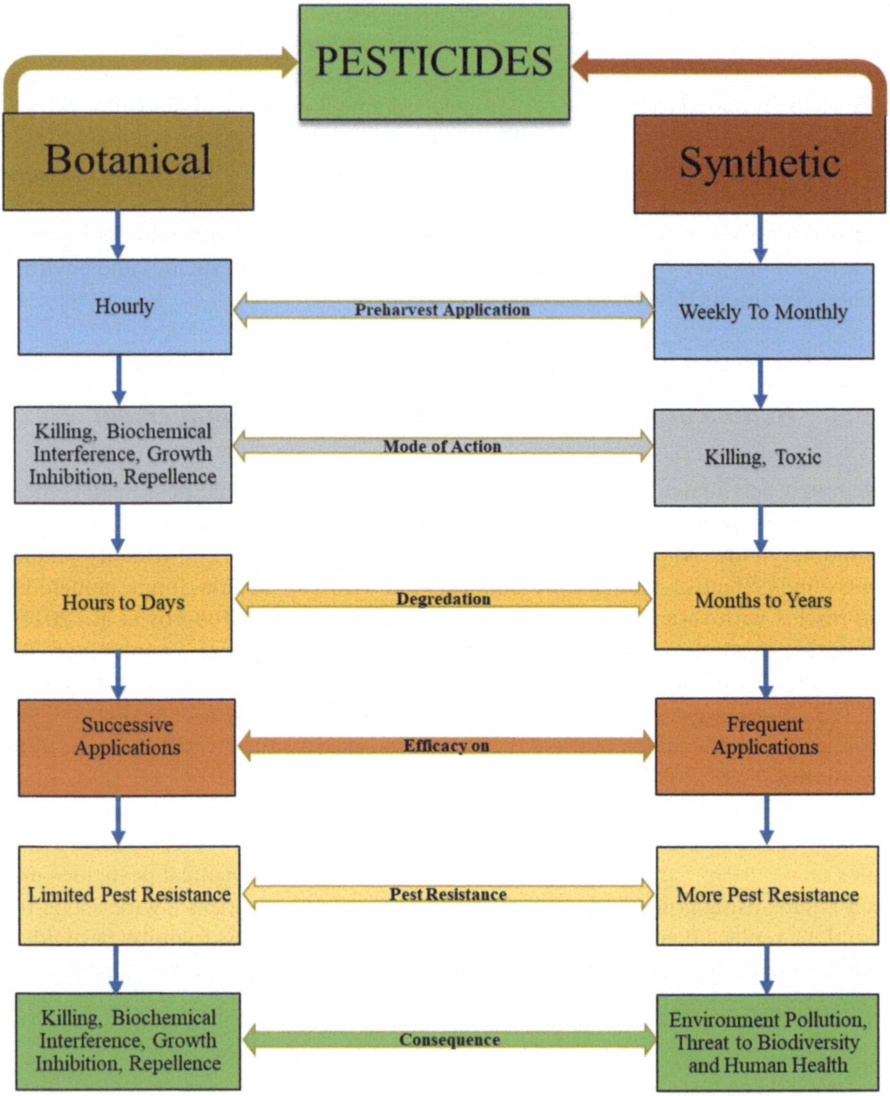

Fig. 5.7 An illustration comparing synthetic and botanical pesticides based on different parameters

with great scope, botanical pesticides are not much represented in the pesticide market (Kekuda et al., 2016). Plants that are being used for food are less preferred for pesticide production by farmers due to a greater demand for food security (Srijita, 2015). Farmers have shown interest in synthetic pesticides over botanicals due to their unrestricted availability and established production facilities, leading to cheaper rates in the market, longer shelf life and simpler application methods (Lengai et al., 2020). Little awareness among farmers, complex regulatory procedures for production, chances of biodegradability with physical factors, reported rare side effects on non-target species, dependency on extraction methodology for promising results, etc. diminish the presence of botanical pesticides in agricultural fields, which needs to be addressed with comprehensive future research (Ekpo et al., 2017; Okunlola & Akinrinnola, 2014; Sales et al., 2016; Stevenson & Belmain, 2017).

In light of these facts, the governments by implementing agricultural laws utilizing natural pesticides in economic ways for farmers could bring around eco-friendly farming practices. Considering the drastic effects of synthetic pesticides and the benefits of botanical pesticides, it is an environmental emergency to replace synthetic pesticides with botanicals and other organic pesticides. With technological advancements and the exploration of more plants with pesticide effects, botanical pesticides could eventually replace synthetic pesticides for safer, environment-friendly and sustainable agriculture soon. This could also bring about the use of these plants and plant products as a source of income for many societies, especially the rural and tribal, which will direct towards eradicating unemployment and sustainable utilization of available natural resources contributing to the development of the country and humanity.

References

Abtew, A., Subramanian, S., Cheseto, X., Kreiter, S., Garzia, G. T., & Martin, T. (2015). Repellency of plant extracts against the legume flower thrips Megalurothrips sjostedti (Thysanoptera: Thripidae). *Insects, 6*(3), 608–625. https://doi.org/10.3390/insects6030608

Agriculture Organization (Ed.). (2014). Opportunities and challenges. Food and agriculture org. *State of World Fisheries and Aquaculture, 2014*.

Ahmad, W., Shilpa, S., & Sanjay, K. (2017). Phytochemical screening and antimicrobial study of Euphorbia hirta extracts. *Journal of Medicinal Plants Studies, 2*, 183–186.

Alam, M. Z., Haque, M. M., Islam, M. S., Hossain, E., Hasan, S. B., Hasan, S. B., & Hossain, M. S. (2016). Comparative study of integrated pest management and farmers practices on sustainable environment in the rice ecosystem. *International Journal of Zoology, 2016*, 1–12. https://doi.org/10.1155/2016/7286040

Ali, S., Muhammad, S. M. H., Muneer, A., Faisal, H., Muhammad, F., Dilbar, H., … Abdul, G. (2014). Insecticidal activity of turmeric (Curcuma longa) and garlic (Allium sativum) extracts against red flour beetle, Tribolium castaneum: A safe alternative to insecticides in stored commodities. *Journal of Entomology and Zoology Studies, 3*, 201–205.

AlShrouf, A. (2017). Hydroponics, aeroponic and aquaponic as compared with conventional farming. *American Scientific Research Journal for Engineering, Technology and Sciences (ASRJETS), 27*(1), 247–255.

Arafat, Y., Shahida, K., Wenxiong, L., Changxun, F., Sehrish, S., Niaz, A., & Saadia, A. (2015). Allelopathic evaluation of selected plants extract against broad and narrow leaves weeds and their associated crops, Acad. *Journal of Agricultural Research, 10*, 226–234.

Badr, A. M. (2020). Organophosphate toxicity: Updates of malathion potential toxic effects in mammals and potential treatments. *Environmental Science and Pollution Research International, 27*(21), 26036–26057. https://doi.org/10.1007/s11356-020-08937-4

Baron, R. L., & Merriam, T. L. (1988). Toxicology of aldicarb. In *Reviews of environmental contamination and toxicology* (Vol. 105, pp. 1–70). Springer. https://doi.org/10.1007/978-1-4612-3876-8_1

Bates, L., Clifford, H., Coyle, R., Ertz, S., McClure, V., McKenzie, A., . . . Butler, C. D. (2008). Dieldrin and breast cancer: A literature [review], 1–22.

Beane Freeman, L. E., Bonner, M. R., Blair, A., Hoppin, J. A., Sandler, D. P., Lubin, J. H., … Alavanja, M. C. (2005). Cancer incidence among male pesticide applicators in the agricultural health study cohort exposed to diazinon. *American Journal of Epidemiology, 162*(11), 1070–1079. https://doi.org/10.1093/aje/kwi321

Bhandari, D., & Bista, B. B. (2019). Permaculture: A key driver for sustainable agriculture in Nepal. *International Journal of Applied Sciences and Biotechnology, 7*(2), 167–173. https://doi.org/10.3126/ijasbt.v7i2.24647

Byard, J. L., Paulsen, S. C., Tjeerdema, R. S., & Chiavelli, D. (2015). DDT, chlordane, toxaphene and PCB residues in Newport Bay and watershed: Assessment of hazard to wildlife and human health. In *Reviews of environmental contamination and toxicology* (Vol. 235, pp. 49–168). Springer. https://doi.org/10.1007/978-3-319-10861-2_3

Carson, R. (1962). *Silent spring* (1st ed.). Houghton Mifflin Harcourt.

Carvalho, F. P. (2017). Pesticides, environment, and food safety. *Food and Energy Security, 6*(2), 48–60. https://doi.org/10.1002/fes3.108

Castillo-Sánchez, L. E., Jiménez-Osornio, J. J., Delgado-Herrera, M. A., Candelaria-Martínez, B., & Sandoval-Gío, J. J. (2015). Effects of the hexanic extract of neem Azadirachta indica against adult whitefly Bemisia tabaci. *Journal of Entomology and Zoology Studies, 5*, 95–99.

Chen, G. (2014). Methoxychlor. In P. Wexler (Ed.), *Encyclopedia of toxicology* (3rd ed., pp. 254–255). Academic Press.

Chengala, L., & Singh, N. (2017). Botanical pesticides—A major alternative to chemical pesticides: A review. *International Journal of Life Sciences, 5*(4), 722–729.

Choudhary, S., Yamini, N. R., Yadav, S. K., Kamboj, M., & Sharma, A. (2018). A review: Pesticide residue: Cause of many animal health problems. *Journal of Entomology and Zoology Studies, 6*(3), 330–333.

Chougule, P. M., & Andoji, Y. S. (2016). Antifungal activity of some common medicinal plant extracts against soil borne phytopathogenic fungi Fusarium oxysporum causing wilt of tomato. *International Journal of Development Research, 6*(3), 7030–7033.

Cohn, B. A., La Merrill, M., Krigbaum, N. Y., Yeh, G., Park, J. S., Zimmermann, L., & Cirillo, P. M. (2015). DDT exposure in utero and breast cancer. *Journal of Clinical Endocrinology and Metabolism, 100*(8), 2865–2872. https://doi.org/10.1210/jc.2015-1841

Cook, M. B., Trabert, B., & McGlynn, K. A. (2011). Organochlorine compounds and testicular dysgenesis syndrome: Human data. *International Journal of Andrology, 34*(4 pt. 2), e68–e84; discussion e84. https://doi.org/10.1111/j.1365-2605.2011.01171.x

Cycoń, M., & Piotrowska-Seget, Z. (2016). Pyrethroid-degrading microorganisms and their potential for the bioremediation of contaminated soils: A review. *Frontiers in Microbiology, 7*, 1463. https://doi.org/10.3389/fmicb.2016.01463

Damalas, C. A., & Koutroubas, S. D. (2015). Farmers' exposure to pesticides: Toxicity types and ways of prevention. *Toxics, 1*, 1–10.

Dara, S. K. (2019). The new integrated pest management paradigm for the modern age. *Journal of Integrated Pest Management, 10*(1), 12. https://doi.org/10.1093/jipm/pmz010

Das, S. K. (2014). Recent development and future of botanical pesticides in India. *Popular Kheti, 2*(2), 93–99.

de Souza, R. M., Seibert, D., Quesada, H. B., de Jesus Bassetti, F., Fagundes-Klen, M. R., & Bergamasco, R. (2020). Occurrence, impacts and general aspects of pesticides in surface water: A review. *Process Safety and Environmental Protection, 135*, 22–37. https://doi.org/10.1016/j.psep.2019.12.035

Dewan, A., Patel, A. B., Pal, R. R., Jani, U. J., Singel, V. C., & Panchal, M. D. (2008). Mass ethion poisoning with high mortality. *Clinical Toxicology, 46*(1), 85–88. https://doi.org/10.1080/15563650701517251

Doi, H., Kikuchi, H., Murai, H., Kawano, Y., Shigeto, H., Ohyagi, Y., & Kira, J. (2006). Motor neuron disorder simulating ALS induced by chronic inhalation of pyrethroid insecticides. *Neurology, 67*(10), 1894–1895. https://doi.org/10.1212/01.wnl.0000244489.65670.9f

Dougoud, J., Toepfer, S., Bateman, M., & Jenner, W. H. (2019). Efficacy of homemade botanical insecticides based on traditional knowledge. A review. *Agronomy for Sustainable Development, 39*(4), 37. https://doi.org/10.1007/s13593-019-0583-1

Dudin, M. (2018). Renewable energy sources as an instrument to support the competitiveness of agro-industrial enterprises and reduce their costs.

Dudley, N., Attwood, S. J., Goulson, D., Jarvis, D., Bharucha, Z. P., & Pretty, J. (2017). How should conservationists respond to pesticides as a driver of biodiversity loss in agroecosystems? *Biological Conservation, 209*, 449–453. https://doi.org/10.1016/j.biocon.2017.03.012

Edwards, F. L., & Tchounwou, P. B. (2005). Environmental toxicology and health effects associated with methyl parathion exposure–a scientific review. *International Journal of Environmental Research and Public Health, 2*(3–4), 430–441. https://doi.org/10.3390/ijerph2005030007

Ekpo, P. B., Uno, U. U., Effiong, E. C., & Etta, S. E. (2017). Acute toxicity of *Tephrosia vogelli* on the early life stages of farmed clariid. *(Clarias gariepinus) Asian J. Adv. Agricultural Research, 3*(2), 1–5.

Elbeshehy, E. K. F., Metwali, E. M. R., & Almaghrabi, O. A. (2015). Antiviral activity of Thuja orientalis extracts against watermelon mosaic virus (WMV) on Citrullus lanatus. *Saudi Journal of Biological Sciences, 22*(2), 211–219. https://doi.org/10.1016/j.sjbs.2014.09.012

Evangelou, E., Ntritsos, G., Chondrogiorgi, M., Kavvoura, F. K., Hernández, A. F., Ntzani, E. E., & Tzoulaki, I. (2016). Exposure to pesticides and diabetes: A systematic review and meta-analysis. *Environment International, 91*, 60–68. https://doi.org/10.1016/j.envint.2016.02.013

Fent, G. M. (2014). Disulfoton. In P. Wexler (Ed.), *Encyclopedia of toxicology* (3rd ed., pp. 210–211). Academic Press.

Feyisa, B., Lencho, A., Selvaraj, T., & Getaneh, G. (2015). Evaluation of some botanicals and *Trichoderma harzianum* for the management of tomato root-knot nematode (meloidogyne incognita (Kofoid and white) chit wood). *Advances in Crop Science and Technology, 1*, 1–10.

Gupta, R. C., & Milatovic, D. (2014). Insecticides. In R. C. Gupta (Ed.), *Biomarkers in toxicology* (pp. 389–407). Academic Press.

Gupta, R. C. (1994). Carbofuran toxicity. *Journal of Toxicology and Environmental Health, 43*(4), 383–418. https://doi.org/10.1080/15287399409531931

Hallberg, G. R. (1987). The impacts of agricultural chemicals on ground water quality. *GeoJournal, 15*(3), 283–295. https://doi.org/10.1007/BF00213456

Harchegani, A. B., Rahmani, A., Tahmasbpour, E., Kabootaraki, H. B., Rostami, H., & Shahriary, A. (2018). Mechanisms of diazinon effects on impaired spermatogenesis and male infertility. *Toxicology and Industrial Health, 34*(9), 653–664. https://doi.org/10.1177/0748233718778665

Honeycutt, M., & Jones, L. (2014). Endrin. In P. Wexler (Ed.), *Encyclopedia of toxicology* (3rd ed., pp. 344–347). Academic Press.

Igbinosa, E. O., Odjadjare, E. E., Chigor, V. N., Igbinosa, I. H., Emoghene, A. O., Ekhaise, F. O., ... Idemudia, O. G. (2013). Toxicological profile of chlorophenols and their derivatives in the environment: The public health perspective. *The Scientific World Journal, 2013*, 460215. https://doi.org/10.1155/2013/460215

Ingle, K. P., Deshmukh, A. G., Padole, D. A., Dudhare, M. S., Moharil, M. P., & Khelurkar, V. C. (2017). Bioefficacy of crude extracts from Jatropha curcas against Spodoptera litura. *Journal of Entomology and Zoology Studies, 1*, 36–38.

Islam, A., & Malik, M. F. (2018). Impact of pesticides on amphibians: A review. *Journal of Analytical Toxicology, 1*(2), 3.

Isman, M. B. (2017a). Bridging the gap: Moving botanical insecticides from the laboratory to the farm. *Industrial Crops and Products, 110*, 10–14. https://doi.org/10.1016/j.indcrop.2017.07.012

Isman, M. B., & Grieneisen, M. L. (2014). Botanical insecticide research: Many publications, limited useful data. *Trends in Plant Science, 19*(3), 140–145. https://doi.org/10.1016/j.tplants.2013.11.005

Isman, M. B. (2017b). Bridging the gap: Moving botanical insecticides from the laboratory to the farm. *Industrial Crops and Products, 110*, 10–14. https://doi.org/10.1016/j.indcrop.2017.07.012

Jantasorn, A., Boontida, M., & Tida, D. (2016). In vitro antifungal activity evaluation of five plant extracts against five plant pathogenic fungi causing rice and economic crop diseases. *Journal of Biopesticides, 1*, 1–7.

Jnaid, Y., Yacoub, R., & Al-Biski, F. (2016). Antioxidant and antimicrobial activities of *Origanum vulgare* essential oil. *International Food Research Journal, 4*, 1706–1710.

Joko, T., Anggoro, S., Sunoko, H. R., & Rachmawati, S. (2017). Pesticides usage in the soil quality degradation potential in wanasari subdistrict, Brebes, Indonesia. *Applied and Environmental Soil Science, 2017*, 1–7. https://doi.org/10.1155/2017/5896191

Kekuda, P. T. R., Akarsh, S., Nawaz, S. A. N., Ranjitha, M. C., Darshini, S. M., & Vidya, P. (2016). *In vitro* antifungal activity of some plants against *Bipolaris sarokiniana* (Sacc.) Shoem. *International Journal of Current Microbiology and Applied Sciences, 6*, 331–337.

Kepenekçi, I., & Erdo. (2016). ˘gu¸s D, Erdo ˘gan P. effects of some plant extracts on root-knot nematodes in vitro and in vivo conditions, Turk. *Journal of Entomology, 40*(1), 3–14.

Khanjani, N., Hoving, J. L., Forbes, A. B., & Sim, M. R. (2007). Systematic review and meta-analysis of cyclodiene insecticides and breast cancer. *Journal of Environmental Science and Health. Part C, Environmental Carcinogenesis and Ecotoxicology Reviews, 25*(1), 23–52. https://doi.org/10.1080/10590500701201711

Kim, K. H., Kabir, E., & Jahan, S. A. (2017). Exposure to pesticides and the associated human health effects. *Science of the Total Environment, 575*, 525–535. https://doi.org/10.1016/j.scitotenv.2016.09.009

Koshlukova, S., & Reed, N. (2014). Carbaryl. In: Richardson RJ (ed) Encyclopedia of toxicology, 3rd. Elsevier, Amsterdam, pp 668–672.

Koutros, S., Mahajan, R., Zheng, T., Hoppin, J. A., Ma, X., Lynch, C. F., … Alavanja, M. C. (2008). Dichlorvos exposure and human cancer risk: Results from the agricultural health study. *Cancer Causes and Control, 19*(1), 59–65. https://doi.org/10.1007/s10552-007-9070-0

Kumar, M. M., Kumar, S., Prasad, C. S., & Kumar, P. (2015). Management of gram pod borer, Helicoverpa armigera (Hubner) in chickpea with botanical and chemical insecticide. *Journal of Experimental Zoology India, 18*(2), 741.

Laxmishree, C., & Nandita, S. (2017). Botanical pesticides –a major alternative to chemical pesticides: A review. *International Journal of Life Sciences, 4*, 722–729.

Lengai, G. M. W., Muthomi, J. W., & Mbega, E. R. (2020). Phytochemical activity and role of botanical pesticides in pest management for sustainable agricultural crop production. *Scientific African, 7*, e00239. https://doi.org/10.1016/j.sciaf.2019.e00239

Lewis, K. A., Tzilivakis, J., Warner, D. J., & Green, A. (2016). An international database for pesticide risk assessments and management. *Human and Ecological Risk Assessment: An International Journal, 22*(4), 1050–1064. https://doi.org/10.1080/10807039.2015.1133242

Lim, J. E., Park, S. H., Jee, S. H., & Park, H. (2015). Body concentrations of persistent organic pollutants and prostate cancer: A meta-analysis. *Environmental Science and Pollution Research International, 22*(15), 11275–11284. https://doi.org/10.1007/s11356-015-4315-z

Liu, B., Chen, B., Zhang, J., Wang, P., & Feng, G. (2017). The environmental fate of thymol, a novel botanical pesticide, in tropical agricultural soil and water. *Toxicological and Environmental Chemistry, 99*(2), 223–232. https://doi.org/10.1080/02772248.2016.1198907

Loizou, E., Karelakis, C., Galanopoulos, K., & Mattas, K. (2019). The role of agriculture as a development tool for a regional economy. *Agricultural Systems, 173*(173), 482–490. https://doi.org/10.1016/j.agsy.2019.04.002

Loomis, D., Guyton, K., Grosse, Y., El Ghissasi, F., Bouvard, V., Benbrahim-Tallaa, L., … International Arctic Research Center, Monograph Working Group. Carcinogenicity of lindane. (2015). Carcinogenicity of lindane, DDT, and 2,4-dichlorophenoxyacetic acid. *Lancet Oncology, 16*(8), 891–892. https://doi.org/10.1016/S1470-2045(15)00081-9

Mahmood, I., Imadi, S. R., Shazadi, K., Gul, A., & Hakeem, K. R. (2016). Effects of pesticides on environment. In *Plant, soil and microbes* (pp. 253–269). Springer.

Majeed, M. T., & Mazhar, M. (2019). Environmental degradation and output volatility: A global perspective. *Pakistan Journal of Commerce and Social Sciences (PJCSS), 13*(1), 180–208.

Maurya, P. K., Malik, D. S., & Sharma, A. (2019). Impacts of pesticide application on aquatic environments and fish diversity. *Contaminants in Agriculture and Environment: Health Risks and Remediation, 1*, 111.

McCoy, K. A., & Peralta, A. L. (2018). Pesticides could alter amphibian skin microbiomes and the effects of Batrachochytrium dendrobatidis. *Frontiers in Microbiology, 9*, 748. https://doi.org/10.3389/fmicb.2018.00748

McGregor, S. E. (1976). *Insect pollination of cultivated crop plants*. Agricultural Research Service.

Miller, G. (2002). *Living in the environment* (12th ed.). Thomson Learning.

Miller, G. T., & Spoolman, S. (1996). *Living in the environment: Principles. Connections, and solutions*. Wodsworth.

Mirajkar, N. S. (2014). Dimethoate. In P. Wexler (Ed.), *Encyclopedia of toxicology* (3rd ed., pp. 55–157). Academic Press.

Misra, H. P. (2014). Role of botanicals, biopesticides and bioagents in integrated pest management, Odisha. *Rev*, 62–67.

Mitra, J., & Raghu, K. (1998). Pesticides-non target plants interactions: An overview. *Archives of Agronomy and Soil Science, 43*(6), 445–500. https://doi.org/10.1080/03650349809366059

Mkenda, P. A., Stevenson, P. C., Ndakidemi, P., Farman, D. I., & Belmain, S. R. (2015). Contact and fumigant toxicity of five pesticidal plants against Callosobruchus maculatus (Coleoptera: Chrysomelidae) in stored cowpea (Vigna unguiculata). *International Journal of Tropical Insect Science, 35*(4), 172–184. https://doi.org/10.1017/S174275841500017X

Moriarty, F. (1972 November 1). The effects of pesticides on wildlife: Exposure and residues. *Science of the Total Environment, 1*(3), 267–288. https://doi.org/10.1016/0048-9697(72)90023-x

Moser, V. C. (2014). Fenthion. In P. Wexler (Ed.), *Encyclopedia of toxicology* (3rd ed., pp. 583–585). Academic Press.

Mougou, I., & Boughalleb-M'hamdi, N. (2018). Biocontrol of *Pseudomonas syringae* pv. *Syringae* affecting citrus orchards in Tunisia by using indigenous *Bacillus* spp. and garlic extract. *Egyptian Journal of Biological Pest Control, 28*(1), 60. https://doi.org/10.1186/s41938-018-0061-0

Mpumi, N., Mtei, K., Machunda, R., & Ndakidemi, P. A. (2016). The toxicity, persistence and mode of actions of selected botanical pesticides in Africa against insect pests in common beans, P. vulgaris: A review. *American Journal of Plant Sciences, 7*, 138–151.

Multigner, L., Kadhel, P., Rouget, F., Blanchet, P., & Cordier, S. (2016). Chlordecone exposure and adverse effects in French West Indies populations. *Environmental Science and Pollution Research International, 23*(1), 3–8. https://doi.org/10.1007/s11356-015-4621-5

Muthomi, J., Fulano, A. M., Wagacha, J. M., & Mwang'ombe, A. W. (2017). Management of snap bean insect pests and diseases by use of antagonistic fungi and plant extracts. *Sustainable Agriculture Research, 6*(3), 52. https://doi.org/10.5539/sar.v6n3p52

Naqvi, S. M., & Hasan, M. A. (1992). Acetylhomocysteine thiolactone protection against phosphamidon-induced alteration of regional superoxide dismutase activity in the central nervous system and its correlation with altered lipid peroxidation. *Indian Journal of Experimental Biology, 30*(9), 850–852.

Nawaz, M., Juma, M., & Hongxia, H. (2016). Current status and advancement of biopesticides: Microbial and botanical pesticides. *Journal of Entomology and Zoology Studies, 2*, 241–246.

Nawaz, M., Mabubu, J. I., & Hua, H. (2016). Current status and advancement of biopesticides: Microbial and botanical pesticides. *Journal of Entomology and Zoology Studies, 4*(2), 241–246.

Nearing, M. A., Xie, Y., Liu, B., & Ye, Y. (2017). Natural and anthropogenic rates of soil erosion. *International Soil and Water Conservation Research, 5*(2), 77–84. https://doi.org/10.1016/j.iswcr.2017.04.001

Neeraj, G. S., Kumar, A., Ram, S., & Kumar, V. (2017). Evaluation of nematicidal activity of ethanolic extracts of medicinal plants to meloidogyne incognita (kofoid and white) Chitwood under lab conditions. *Indian Journal of Pure & Applied Biosciences, 1*, 827–831.

Nefzi, A., Abdallah, B. A. R., Jabnoun-Khiareddine, H., Saidiana-Medimagh, H. R., & Danmi-Remadi. (2016). Antifungal activity of aqueous and organic extracts from *Withania som-*

nifera L. against *Fusarium oxysporum* f. sp. *radicis-lycopersici. Journal of Microbial and Biochemical Technology, 3*, 144–150.

Ekeh, F. N., Odo, G. E., Nzei, J. I., Ohanu, C. M., Ugwu, F., Ngwu, G., & Reginald, N. (2018). Effects of aqueous and oil leaf extracts of Pterocarpus santalinoides on the maize weevil, Sitophilus zeamais, pest of stored maize grains. *African Journal of Agricultural Research, 13*(13), 617–626. https://doi.org/10.5897/AJAR2018.13014

Ntalli, N., Bratidou Parlapani, A., Tzani, K., Samara, M., Boutsis, G., Dimou, M., … Monokrousos, N. (2020 February). Thymus citriodorus (Schreb) botanical products as ecofriendly nematicides with bio-fertilizing properties. *Plants, 9*(2), 202. https://doi.org/10.3390/plants9020202

Ogundele, R. A., Oyedele, D. J., & Adekunle, O. K. (2016). Management of *Meloidogyne incognita* and other phytonematodes infecting *Amaranthus cruentus* and *Telfairia occidentalis* with African marigold (*Tagetes erecta*) and Siam weed (*Chromolaena odorata*). *Australasian Plant Pathology, 45*(5), 537–545. https://doi.org/10.1007/s13313-016-0438-z

Okunlola, A. I., & Akinrinnola, O. (2014). Effectiveness of botanical formulations in vegetable production and bio-diversity preservation in Ondo state, Nigeria. *Journal of Horticulture and Forestry, 1*, 6–13.

O'Neil, M. J., Smith, A., Heckelman, P. E., & Budavari, S. (2001). *The merck index-An encyclopedia of chemicals, drugs, and BioLogicals, 767* p. 4342. Whitehouse Station, NJ: Merck & Co., Inc.

Ortiz-Hernández, M. L., Sánchez-Salinas, E., Dantán-González, E., & Castrejón-Godínez, M. (2013). Pesticide biodegradation: Mechanisms, genetics and strategies to enhance the process. *Biodegrad. Life Sciences*, 251–287.

Peshin, R. et al. (2009). *Integrated pest management: A global overview of history, programs and adoption. Integrated pest management: Innovation-development process* (1st ed) (pp. 1–49). Dordrecht: Springer.

Peterson, R. K. D., Higley, L. G., & Pedigo, L. P. (2018). Whatever happened to IPM? *American Entomologist, 64*(3), 146–150. https://doi.org/10.1093/ae/tmy049

Pohanish, R. P.. D. (2015) R. P. Pohanish (Ed.). *Sittig's handbook of pesticides and agricultural chemicals* (2nd ed) (pp. 196–133). William Andrew Publishing.

Porto, A. L. M., Melgar, G. Z., Kasemodel, M. C., & Nitschke, M. (2011). Biodegradation of pesticides. In M. Stoytcheva (Ed.), *Pesticides in the Modern World–Pesti- cides Use and Management, 1*, Tech (pp. 407–438).

Ranga Rao, G. V., Rupela, O. P., Rao, V. R., & Reddy, Y. V. (2007). Role of biopesticides in crop protection: Present status and future prospects. *Indian Journal of Plant Protection, 35*(1), 1–9.

Rattner, B. A. (2009). History of wildlife toxicology. *Ecotoxicology, 18*(7), 773–783. https://doi.org/10.1007/s10646-009-0354-x

Reed, N. R., & Koshlukova, S. (2014a). Chlorfenvinphos. In P. Wexler (Ed.), *Encyclopedia of toxicology* (3rd ed., pp. 851–854). Academic Press.

Reed, N. R., & Koshlukova, S. (2014b). Heptachlor. In P. Wexler (Ed.), *Encyclopedia of toxicology* (3rd ed., pp. 840–844). Academic Press.

Riyaz, M., Aamir Iqbal, W. A., Sivasankaran, K., & Ignacimuthu, S. (2020). Impact on farmers' health due to the pesticide exposure in the agrarian zones of Kashmir valley: A review. *Acta Scientific Agriculture, 4*(2), 01–07. https://doi.org/10.31080/ASAG.2020.04.impact-on-farmers-health-due-to-the-pesticide-exposure-in-the-agrarian-zones-of-kashmir-valley-a-review.

Riyaz, M., Mathew, P., Paulraj, G., & Ignacimuthu, S. (2018). Entomophily of apple ecosystem in Kashmir valley, India: A review. *International Journal of Scientific Research in Biological Sciences (IJSRBS), 5*(5), 46–154.

Roberts, D. P., & Mattoo, A. K. (2018). Sustainable agriculture—Enhancing environmental benefits, food nutritional quality and building crop resilience to abiotic and biotic stresses. *Agriculture, 8*(1), 8. https://doi.org/10.3390/agriculture8010008

Rodgers, K. E., Leung, N., Imamura, T., & Devens, B. H. (1986). Rapid in vitro screening assay for immunotoxic effects of organophosphorus and carbamate insecticides on the generation of cytotoxic T-lymphocyte responses. *Pesticide Biochemistry and Physiology, 26*(3), 292–301.

Roy, S., Handique, G., Muraleedharan, N., Dashora, K., Roy, S. M., Mukhopadhyay, A., & Babu, A. (2016). Use of plant extracts for tea pest management in India. *Applied Microbiology and Biotechnology, 100*(11), 4831–4844. https://doi.org/10.1007/s00253-016-7522-8

Saladin, G., & Clément, C. (2005). Physiological side effects of pesticides on non-target plants. *Agriculture and Soil Pollution: New Research*, 53–86.

Sales, M. D. C., Costa, H. B., Patrícia, M. B. F., Jose, A. V., & Debora, D. M. (2016). Antifungal activity of plant extracts with potential to control plant pathogens in pineapple. *Asian Pacific Journal of Tropical Biomedicine, 1*, 26–31.

Salhi, N., Mohammed Saghir, S. A., Terzi, V., Brahmi, I., Ghedairi, N., & Bissati, S. (2017). Antifungal activity of aqueous extracts of some dominant Algerian medicinal plants. *BioMed Research International, 2017*, 7526291. https://doi.org/10.1155/2017/7526291

Sánchez-Bayo, F., & Wyckhuys, K. A. G. (2019). Worldwide decline of the entomofauna: A review of its drivers. *Biological Conservation, 232*, 8–27. https://doi.org/10.1016/j.biocon.2019.01.020

Saquib, Q., Faisal, M., Ansari, S. M., & Wahab, R. (2019). Phorate triggers oxidative stress and mitochondrial dysfunction to enhance micronuclei generation and DNA damage in human lymphocytes. *Saudi Journal of Biological Sciences, 26*(7), 1411–1417. https://doi.org/10.1016/j.sjbs.2019.04.008

Sarkar, M., & Kshirsagar, R. (2014). Botanical pesticides: Current challenges and reverse pharmacological approach for future discoveries. *Journal of Biofertilizers & Biopesticides, 5*(e), 125.

Saunders, M. E. (2018). Insect pollinators collect pollen from wind-pollinated plants: Implications for pollination ecology and sustainable agriculture. *Insect Conservation and Diversity, 11*(1), 13–31. https://doi.org/10.1111/icad.12243

Sebastian, R., & Raghavan, S. C. (2016). Induction of DNA damage and erroneous repair can explain genomic instability caused by endosulfan. *Carcinogenesis, 37*(10), 929–940. https://doi.org/10.1093/carcin/bgw081

Shabana, Y. M., Abdalla, M. E., Shahin, A. A., El-Sawy, M. M., Draz, I. S., & Youssif, A. W. (2017). Efficacy of plant extracts in controlling wheat leaf rust disease caused by Puccinia triticina. *Egyptian Journal of Basic and Applied Sciences, 4*(1), 67–73. https://doi.org/10.1016/j.ejbas.2016.09.002

Sharma, A., Yadav, B., Rohatgi, S., & Yadav, B. (2018). Cypermethrin toxicity: A review. *Journal of Forensic Sciences & Criminal Investigation, 9*(4) PubMed: 555767.

Sharma, S., Kooner, R., & Arora, R. (2017). Insect pests and crop losses. *Inbreeding Insect Resistant Crops for Sustainable Agriculture*, 45–66.

Sharoba, A. M., El Mansy, H. A., El Tanahy, H. H., El Waseif, K. H., & Ibrahim, M. A. (2015). Chemical composition, antioxidant and antimicrobial properties of the essential oils and extracts of some aromatic plants, Middle East. *Journal of Applied Sciences, 2*, 344–352.

Sherman, D. M. (2007 November 27). *Tending animals in the global village: A guide to international veterinary medicine*. John Wiley & Sons.

Singh, P., Volger, B., & Gordon, E. (2014). Endosulfan. In P. Wexler (Ed.), *Encyclopedia of toxicology* (3rd ed., pp. 341–343). Academic Press.

Slätmo, E., Fischer, K., & Röös, E. (2017). The framing of sustainability in sustainability assessment frameworks for agriculture. *Sociologia Ruralis, 57*(3), 378–395. https://doi.org/10.1111/soru.12156

Socorro, J., Durand, A., Temime-Roussel, B., Gligorovski, S., Wortham, H., & Quivet, E. (2016). The persistence of pesticides in atmospheric particulate phase: An emerging air quality issue. *Scientific Reports, 6*, 33456. https://doi.org/10.1038/srep33456

Srijita, D. (2015). Biopesticides: An ecofriendly approach for pest control. *World Journal of Pharmacy and Pharmaceutical Sciences (WJPPS), 4*(6), 250–265.

Stevenson, P. C., & Belmain, S. R. (2017). Tephrosia Vogelii. In *A pesticide of the future for African farming* (pp. 19–22). Boletín SEEA.

Stevenson, P. C., Isman, M. B., & Belmain, S. R. (2017). Pesticidal plants in Africa: A global vision of new biological control products from local uses. *Industrial Crops and Products, 110*, 2–9. https://doi.org/10.1016/j.indcrop.2017.08.034

Sumitra, A., Kanojia, A. K., Kumar, A., Mogha, N., & Sahu, V. (2014). Biopesticide formulation to control tomato lepidopteran pest menace. *Current Science, 7*, 1051–1057.

Tang-Péronard, J. L., Andersen, H. R., Jensen, T. K., & Heitmann, B. L. (2011). Endocrine-disrupting chemicals and obesity development in humans: A review. *Obesity Reviews, 12*(8), 622–636. https://doi.org/10.1111/j.1467-789X.2011.00871.x

Taylor, D. (2011). The complete contented cat: Your ultimate guide to feline fulfilment. Google.co.uk/books?id=Cc5BM_aPegkC&dq=pest+cat+rats&source=gbs_navlinks_s. Retrieved from https://web.archive.org/web/20150615023812/https://books. David & Charles p. 9.

Tesfahunegny, W. (2016). Impact of pesticides on birds from DDT to current fatality: A literature review. *Journal of Zoology Studies, 3*(2), 44–55.

Thakur, M., & Pathania, D. (2020 January 1). Environmental fate of organic pollutants and effect on human health. In *Inabatement of environmental pollutants* (pp. 245–262). Elsevier.

Pflanzenschutz-Nachrichten. (1973). "Ancient and medieval plant pathology". *The history of integrated pest management.* which cites Orlob, G.B., *26* (pp. 65–294). NY: Cornell University.

Thuy, T. T. (2015). Effects of ddt on environment and human health. *Journal of Education and Social Sciences, 2*, 108–114.

Timoroğlu, İ., Yüzbaşıoğlu, D., Ünal, F., Yılmaz, S., Aksoy, H., & Çelik, M. (2014). Assessment of the genotoxic effects of organophosphorus insecticides phorate and trichlorfon in human lymphocytes. *Environmental Toxicology, 29*(5), 577–587. https://doi.org/10.1002/tox.21783

Todorov, D., Shishkova, K., Dragolova, D., Hinkov, A., Kapchina-Toteva, V., & Shishkov, S. (2015). Antiviral activity of medicinal plant *Nepeta nuda. Biotechnology and Biotechnological Equipment, 1*, 39–43.

US EPA (United States Environmental Protection Agency). (2003). *Health effects support document for aldrin/dieldrin. EPA 822-R-03-001, 4304t. Office of Water.* Health and Ecological Criteria Division.

Wallace, D. R. (2014). Toxaphene. In P. Wexler (Ed.), *Encyclopedia of toxicology* (3rd ed., pp. 606–609). Academic Press.

Wanzala, W., Wagacha, J. M., Dossaji, S. F., & Gakuubi, M. M. (2016). *Bioactive properties of Tagetes minuta L. (Asteraceae) essential oils: A review.*

Wegulo, S. N., Baenziger, P. S., Hernandez Nopsa, J. H., Bockus, W. W., & Hallen-Adams, H. (2015). Management of Fusarium head blight of wheat and barley. *Crop Protection, 73*, 100–107. https://doi.org/10.1016/j.cropro.2015.02.025

Weißhuhn, P., Reckling, M., Stachow, U., & Wiggering, H. (2017). Supporting agricultural ecosystem services through the integration of perennial polycultures into crop rotations. *Sustainability, 9*(12), 2267. https://doi.org/10.3390/su9122267

Wimalawansa, S. A., & Wimalawansa, S. J. (2014). Agrochemical-related environmental pollution: Effects on human health. *Global Journal of Biology, Agriculture and Health Sciences, 3*, 72–83.

Yan, Y., Feng, C. C., & Chang, K. T. T. (2017). Towards enhancing integrated pest management based on volunteered geographic information. *ISPRS International Journal of Geo-Information, 6*(7), 224. https://doi.org/10.3390/ijgi6070224

Yang, X., Huang, Q., Jiang, T., & Xu, H. (2017). Degradation dynamics of azadirachtin in cabbage and soil. *Journal of South China Agricultural University, 38*(4), 37–40.

Yasmin, S., & D'Souza, D. (2010). Effects of pesticides on the growth and reproduction of earthworm: A review. *Applied and Environmental Soil Science, 2010*, 1–9. https://doi.org/10.1155/2010/678360

Yue, Q., Xu, X., Hillier, J., Cheng, K., & Pan, G. (2017). Mitigating greenhouse gas emissions in agriculture: From farm production to food consumption. *Journal of Cleaner Production, 149*, 1011–1019. https://doi.org/10.1016/j.jclepro.2017.02.172

Zarubova, L., Lenka, K., Pavel, N., Miloslav, Z., Ondrej, D., & Skuhrovec, J. (2014). In *Botanical pesticides and Their Human Health Safety on the Example of Citrus sinensis Essential oil and Oulema Melanopus Under Laboratory Conditions, Mendel Net* (pp. 330–336).

Zhao, L., Feng, C., Hou, C., Hu, L., Wang, Q., & Wu, Y. (2015). First discovery of acetone extract from cottonseed oil sludge as a novel antiviral agent against plant viruses. *PLoS One, 2*, 1–13.

Chapter 6
The Role of Plant-Mediated Biosynthesised Nanoparticles in Agriculture

Humeraha Nazneen, Gulzar Ahmed Rather, Aarif Ali, and Arghya Chakravorty

Abstract Different types of nanomaterials and different strategies could be used in the betterment of the overstressed agriculture. We have tried to focus on the different characterisation techniques involved in nanomaterial synthesis like UV-Vis spectroscopy, scanning electron microscopy (SEM), X-ray diffraction (XRD), transmission electron microscopy (TEM), atomic force microscopy (AFM) and Fourier transform infrared spectroscopy (FTIR). Further, the limitations of physical and chemical methods have also been discussed. We have talked about the organic strategies in detail, like microorganisms and plant-intervened biosynthesis of nanomaterials.

Keywords Nanotechnology · Green synthesis · Nanopesticides · Eco-friendly

6.1 Introduction

Nanotechnology is the link between physical and biological sciences. The plan and improvement of nanomaterials result from information on material designing and its exercises, explored by knowing natural science (Majeed et al., 2020). The

H. Nazneen (✉)
Nizam College, Autonomous with CPE Status, A Constituent College of Osmania University, Basheerbagh, Hyderabad, India

G. A. Rather
Department of Biomedical Engineering, Sathyabama Institute of Science and Technology, Deemed to be University, Chennai, India

A. Ali
Department of Clinical Biochemistry, University of Kashmir, Srinagar, Jammu & Kashmir, India

A. Chakravorty
School of Bio-Sciences & Technology, Vellore Institute of Technology, Vellore, India

S. A. Bandh (ed.), *Sustainable Agriculture*,
https://doi.org/10.1007/978-3-030-83066-3_6

rapidly growing field of nanotechnology is the interdisciplinary research and developmental field in physics, chemistry and biology. It explores the design, manufacture, assemblage and characterisation of materials that are more modest than 100 nanometres in size, just as the utilisation of scaled-down useful frameworks got from these materials (Nadaroglu et al., 2017). Nanotechnology has expected a colossal part in the agriculture industry, named nano-agribusiness, which infers that this advancement is regularly used to fabricate the yield (Duhan et al., 2017). Metal nanoparticles pulled in scientists due to their wide application and advances in various zones (Herlekar et al., 2014; Zhang et al., 2018). The main drawback of conventional methods is their environmental toxicity; therefore, the requirement for alternatives is increasing faster (Kaur et al., 2014). The nanoparticles synthesised through plants is a straightforward and eco-friendly approach to reduce toxicity, time and cost. There are various assessments subjected to plant-interceded biosynthesis of nanoparticles (Gupta et al., 2018), e.g. TiO_2 NPs impel spinach seed germination and plant improvement (Zheng et al., 2005), and ZnO nanoparticles significantly improve the transport and metabolic processes in plants (Jayarambabu et al., 2014). Similarly, Ag NPs obtained from neem (*A. indica*), dark tulsi (*Ocimum tenuiflorum*) and banana (*Musa balbisiana*) applied on mung bean (*Vigna transmit*) demonstrate a fundamental augmentation in shoot and root lengths (Banerjee et al., 2014). In this chapter, the main focus has been on collecting studies focused on plant-mediated nanomaterial synthesis and its essential function in agriculture.

6.2 Types of Different Nanoparticles (NPs)

The NPs can be categorised into different types:

(i) Inorganic-based nanomaterials.
(ii) Organic-based nanomaterials.
(iii) Carbon-based nanomaterials.
(iv) Composite-based nanomaterials.

6.2.1 *Inorganic-Based Nanomaterials*

The inorganic NMs consist of metal oxide and metal NPs. The metallic nanoparticles include Ag, Zn, Au, etc., while the metal oxide nanomaterials include TiO_2 and ZnO and semiconductors like ceramics and silicon.

6.2.2 Organic-Based Nanomaterials

These nanomaterials are made mainly from organic matter. The use of weak interactions (noncovalent) for design and molecular self-assemblage helps in turning organic NMs into structures such as liposomes, polymer, dendrimers and desired micelle NPs (Jaison et al., 2018).

6.2.3 Carbon-Based Nanomaterials

These NMs are found in different morphologies like hollow tubes, spheres or ellipsoids and contain carbon. The carbon-based NMs include MXene, carbon nanotubes (CNTs), graphene, fullerenes, carbon nanofibers and carbon black (C60). The methods for the preparation of carbon-based nanomaterials include laser ablation, chemical vapour deposition (CVD) and arc discharge (Kumar & Kumbhat, 2016; Paul et al., 2020; Syamsai & Grace, 2020).

6.2.4 Composite-Based Nanomaterials

Composite-based NMS are materials with at least one of the phases in the nanometre range. They comprise an assemblage of two materials of different types, allowing us to obtain a material of greater quality. Composite-based nanomaterials are a combination of carbon, organic or metal nanomaterials and some forms of polymer bulk, metal or ceramic materials (Jaison et al., 2018).

6.3 Techniques for the Readiness of Nanoparticles

Nanomaterials can be blended top-down and bottom-up methodologies, which are additionally partitioned into various strategies.

6.3.1 Top-Down Approach

This method involves the destruction of bulk materials into smaller molecules, which are later converted into NMs. Physical vapour deposition, milling or grinding are a few examples of the top-down approach.

6.3.2 *Bottom-Up Approach*

The bottom-up approach is a type of constructive strategy, opposite to that of the top-down approach. In this approach, NMs are obtained through simpler substances. Some examples of the bottom-up approach include sol-gel, pyrolysis and biological synthesis (Yadav et al., 2009).

6.4 Methods of Nanoparticle Production

6.4.1 *Physical Methods*

Physical methods for NM synthesis employ mechanical strain, high-energy radiations, electrical energy or thermal energy that leads materials to evaporation, condensation, abrasion or melting to produce nanoparticles. Based on physical procedures for NM preparation, they are usually divided into the following types:

6.4.1.1 Mechanical Attrition

Mechanical methods employ the technique of mechanical alloying that gained huge attention over a long time to manufacture various kinds of nanomaterials. Mechanical alloying is considered one of the novel techniques that can be carried out at room temperature. The strategy includes completed force plants, diffusive plants and vibratory industrial facilities (Dhand et al., 2015).

6.4.1.2 Condensation of Inert Gas

It is based on the application of inert gases like helium or argon and sometimes liquid nitrogen on the substrate to synthesise nanomaterials. The nanomaterials, after being evaporated, are transported along with the inert gases over the substrate, which gets condensed with liquid nitrogen. This method was first used by Ward et al. (2006) for the amalgamation of Mn nanomaterials.

6.4.1.3 Physical Vapour Deposition

The physical vapour deposition process is a group of techniques that are widely utilised for nanomaterial synthesis. They help in the formation of thin layers of nanomaterials of a few nanometres. Physical vapour deposition methods are environmentally safe and include three basic steps: vaporisation of materials, transport of vaporised materials and their nucleation to grow them into thin fibres.

6.4.2 Chemical Methods

These are the methods using certain chemical elements for the synthesis of nanomaterials. Different specialist substances like sodium borohydride, hydrazine and hydrogen are utilised for the synthesis (Egorova & Revina, 2000). In light of nanomaterials' compound union, they can also be isolated into two principle types: (i) gas-phase synthesis and (ii) liquid-phase synthesis.

6.4.3 Gas-Phase Synthesis

Gas-phase synthesis is a type of bottom-up approach of nanomaterial synthesis, and among this type of synthesis, gas pyrolysis and gas condensation are the most common types. In gas pyrolysis, the aerosol droplets resulting from metal salt are formed by flame heating. Droplets disperse in the gas, and dehydration decreases their size. Another method is gas condensation, which involves the evaporation of metal salts inside the chamber by different heat sources like laser beams, electron or radio frequencies, etc. The vapours are being pushed into the cooler chamber consisting of inert gases and after that collected from the chamber. The major drawback associated with this method is the agglomeration and amalgamates of nanomaterials (Naveed Ul Haq et al., 2017).

6.4.4 Liquid-Phase Synthesis

It is one of the precipitation methods in which inorganic alkalis act as reducing agents and is reacting with the metal salts to form an insoluble or soluble precipitated product. The product is washed and calcinated at a suitable temperature to produce a particular nanomaterial with variable morphology. With this method, the size can be tailored by optimising synthetic conditions. The liquid-phase synthesis can be divided into different types like sol-gel synthesis and colloidal, hydrothermal and solvothermal methods (Rai et al., 2013). Figure 6.1 shows the different physical and substance systems for nanomaterial synthesis.

6.5 Limitations of Chemical and Physical Methods

Though nanomaterials' physical and chemical syntheses are popular, they are also associated with a large number of risks. Using physical methods, we may obtain nanomaterials of high purity, but they typically require refined equipment, chemical materials, radiations and high energy consumption, leading to high operating costs.

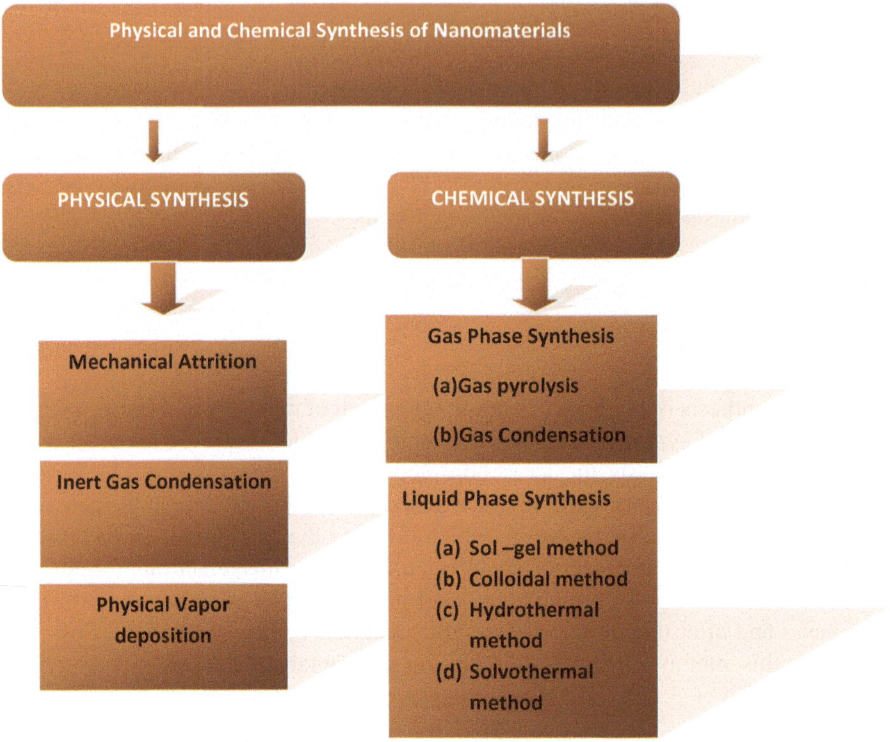

Fig. 6.1 Physical and chemical synthesis of nanomaterials

Again, the chemical synthesis generates many toxic chemicals that are non-biodegradable and harmful and can restrict the manufacturing process. In addition, certain toxic materials may contaminate the surface of nanomaterials and make them unsuitable for different applications. In this context the researcher's main focus is to formulate the alternate route for nanomaterial synthesis to defeat the restrictions of substances and actual strategies (Khandel et al., 2018).

6.6 Characterisation of Nanomaterials

Nanomaterials are characterised by different methods, including UV-Vis spectroscopy, X-ray diffraction (XRD), transmission electron microscopy (TEM), atomic force microscopy (AFM), scanning electron microscopy (SEM) and Fourier transmission infrared spectroscopy (FTIR).

6.6.1 UV-vis Spectroscopy

This is the most straightforward technique used to check the construction of nanoparticles. Different nanoparticles show different peaks, which confirm the structure of NPs in the aqueous medium. The UV-Vis spectroscopy works based on the intensity of light. It detects, analyses and investigates the nanomaterials' optical properties. This method is usually used to check the particles' distribution size (Rajasekaran & Raghavan, 2020; Velappan et al., 2020).

6.6.2 Scanning Electron Microscopy (SEM)

The SEM is one of the versatile techniques used to check the texture, morphology and size of nanoparticles. In this technique, electrons are used instead of light to scan the specimen surface to generate various signals, which give detailed information about the interaction, nature, composition and structure of materials (Raghavan et al., 2020; Sitaaraman et al., 2020).

6.6.3 X-Ray Diffraction (XRD)

The XRD is another technique for nanomaterial characterisation. It gives detailed information about the crystalline structure, crystalline size and lattice parameter of materials. However, this technique only uses the dried powder samples for characterisation. The data obtained from XRD analysis is compared with reference patterns from the Joint Committee on Powder Standards (JCPDS) (Krupa et al., 2019; Mourdikoudis et al., 2018).

6.6.4 Transmission Electron Microscopy (TEM)

Transmission electron microscopy is another quantitative technique used to characterise the morphology and homogeneity of nanomaterials. TEM gives the actual size of nanomaterials and accurate images of the nanoparticles. In TEM the uniform electron beam touches the samples and diffuses through them. The formation of images by TEM analysis is due to the interaction of samples with an electron, wherein the imaging device further magnifies the samples. We can get the maximum resolution through TEM than other characterisation techniques. The information about size, structure, shape, morphology and agglomeration is only possible through this technique (Chakravorty et al., 2020).

6.6.5 *Fourier Transmission Infrared Spectroscopy (FTIR)*

It is one of the analytical techniques used to study the different kinds of practical gatherings present in the biomolecules. During nanomaterials, the functional groups that act as capping and reducing agents are studied through this technique. It gives information about the molecular structure, nature of bonds and the functional groups involved in nanomaterial biosynthesis. It works on an electromagnetic absorption spectrum and wavelength ranging from 400 to 4000 cm^{-1} (Busó-Rogero et al., 2016).

6.6.6 *Atomic Force Microscopy*

Atomic force microscopy is one of the microscopic techniques that can produce three-dimensional pictures of the sample surfaces. The basic principle of atomic force microscopy is the interaction of forces involved between samples and fine probe. AFM gives the detailed size, shape and surface area of nanomaterials.

6.7 Biological Synthesis of Nanomaterials

The biological process of nanomaterial synthesis is the alternative to the physical and chemical methods (Fig. 6.2). It is a cheap, non-toxic, environmentally friendly option of nanomaterial synthesis compared to its physical and chemical counterparts. Nanomaterials with different sizes and shapes can be prepared through biological synthesis (Shah et al., 2015). The synthesis of nanomaterial through biological routes leads to safer, ecologically appropriate and non-toxic nanomaterial through the involvement of bacteria, fungi and plants (Nayantara & Kaur, 2018).

6.7.1 *Bacteria-Mediated Biosynthesis of Nanomaterials*

Research has been heavily based on prokaryotes as the easy and ideal ways to synthesise different nanoparticles. Because of their ability to adjust to extreme conditions and their abundance in the environment, microbes are decent contenders in nano-research. They can be handily controlled as they are fast-growing and inexpensive to cultivate in large quantities. Their growth conditions, including oxygen, incubation and temperature, can be easily monitored, and controlling such parameters can produce nanoparticles of different sizes (Pantidos & Horsfall, 2014). Different strains of bacteria like yeast, moulds and microalgae have been utilised to integrate metallic and non-metallic nanoparticles (Hulkoti & Taranath, 2014). Different types of nanomaterials like gold, silver and selenium with different

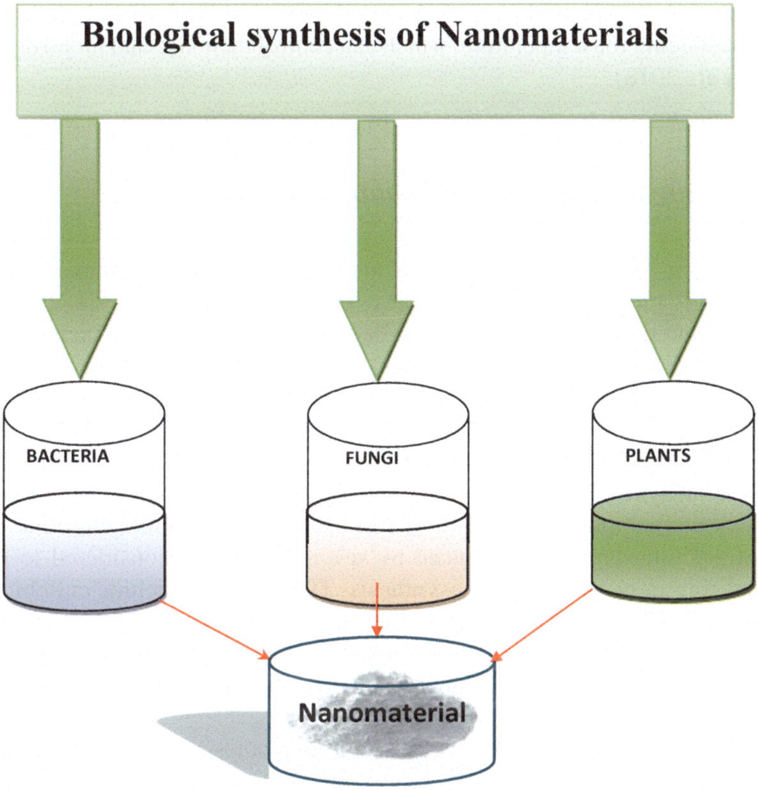

Fig. 6.2 Biological synthesis of nanomaterials

properties and different purposes like imaging, biosensors, in vitro antibacterial, anticancer, antioxidant and anticoagulant activities have been synthesised from bacteria over time (Grasso et al., 2019). *Bacillus* species have been widely considered because of their ability to bioaccumulate metals (Pantidos & Horsfall, 2014). Microbe-intervened amalgamation of nanomaterials can be classified into intracellular or extracellular by the guide of chemicals or proteins present in microorganisms which can go about as lessening specialists and convert metal salts into specific nanomaterials (Nadaroglu et al., 2017). Different bacteria like *Pseudomonas stutzeri*, *Pseudomonas aeruginosa*, *Escherichia coli* and *Vibrio cholera* have been utilised to combine diverse metallic nanoparticles through intracellular and extracellular strategies (Nayantara & Kaur, 2018; Srinath & Ravishankar Rai, 2015). Extracellular is the simplest method as it occurs outside the bacterial cells and does not involve the breakdown of the cell wall. It includes the usage of bacterial biomass, supernatant and cell-free extracts. The extracellular synthesis is preferred over intracellular synthesis as it does not involve complex downstream processes. The main challenge in microbe-based nanomaterials is the selection of choosing the right microbe, depending on its essential properties like replication, growth rate and biochemical pathways to be studied. Another aspect of microbe-interceded nanomaterials is reducing the temperature, which can control their size and mono dispersion (Ovais et al., 2018).

6.7.2 Fungal-Mediated Nanomaterials

Fungi have a great potential for the manufacture of different nanomaterials; around 6400 bioactive substances have been separated from the filamentous organisms and their connected species. Because of the substantial metal resilience and ability to disguise and bio-gather metals, organisms go about as significant balancing out and decreasing specialists. In addition, fungi can be quickly grown on a large scale and can produce size-controlled nanomaterials with definite morphologies (Guilger-Casagrande & de Lima, 2019). Fungus is a great contender for nanomaterial synthesis as it goes about as apparatus for a huge amount of proteins and quick and simple combination of nanomaterials (Alghuthaymi et al., 2015). The extracellular enzymes produced by several fungi are considered to play an essential role in nanomaterial biosynthesis. The enzymes include cellobiohydrolase D, glucosidase, acetyl xylan esterase and β-glucosidase. The enzyme nitrate reductase, released by fungi, acts as a reducing agent in nanomaterial production. Silver and gold nanoparticles have been produced from *Fusarium oxysporum* (Ovais et al., 2018).

Similarly, *Duggingyonia flagans* are used to synthesise silver nanoparticles by using insect carapaces as a source of substrate for fungi (Costa Silva et al., 2017). *A. alternata* can be used to synthesise silver nanoparticles (Ibrahim & Hassan, 2016). The microbial-assisted synthesis of nanomaterials has been found to be very easily scalable, co-friendly and consistent. Still, the production is more expensive because of extended time maintenance of cultures, and chances of contamination are very high. Moreover, microbial-based techniques require high aseptic conditions and maintenance that are not appropriate for nanomaterials' large-scale production, so plant-based nanomaterial production is preferred over microbial-based. The key advantage of bio-based methods over physical and chemical methods is that large-scale nanomaterials involve environmentally friendly, simple and one-step processes rather than chemicals, high temperature and pressure (Khandel et al., 2018).

6.7.3 Plant-Based Nanomaterials

Plants are the bio-factories for many active compounds like flavonoids, terpenes, alkaloids, enzymes and proteins, acting as capping and balancing out specialists for the nanomaterial synthesis (Fig. 6.3). The mechanism of nanoparticle synthesis

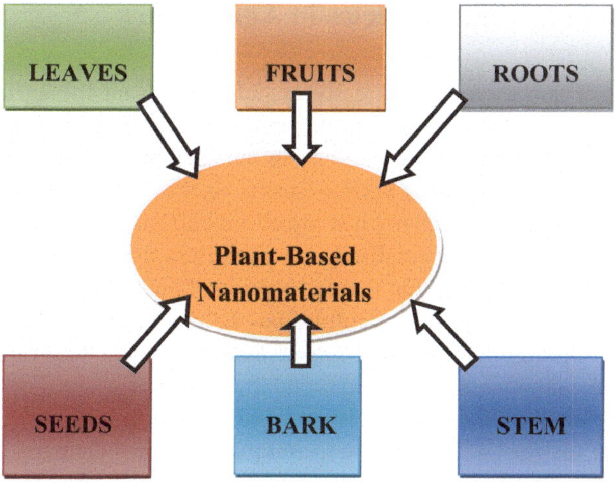

Fig. 6.3 Different parts of plants for nanomaterial biosynthesis

from plants is the same as that of microbial synthesis, but it's cheap, less costly and environmentally friendly (Khandel et al., 2018). Different parts of plants like leaves, roots, stem, bark, shoots, latex, seeds, peels, oils, natural products and so on can be utilised for the nanomaterial combination, as they can go about as great wellsprings of phytochemicals (Dauthal & Mukhopadhyay, 2016). Diverse metal oxide nanoparticles have been set up through a green approach. Zinc oxide nanoparticles use different plant extracts like *Cassia alata*, *Bauhinia tomentosa* and *Catharanthus roseus* (Happy et al., 2019; Gupta et al., 2018; Sharmila et al., 2018). Spherical-shaped nanoparticles were synthesised from various plant parts like *A. calamus* roots and *A. dentata* leaves (Kumar et al., 2014; Nakkala et al., 2014). Different types of plants like clove buds, cardamom, black pepper and saffron have been used for various types of nanomaterial synthesis (Chakravorty et al., 2020).

Plant-based selenium nanoparticles (SeNPS) have been carried out using different plant extracts like *Citrus reticulate*, *Catharanthus roseus*, *Leucas lavandulifolia*, *Allium sativum*, *Aloe vera* and *Asteriscus graveolens* (Anu et al., 2017; Deepa & Ganesan, 2015; Fardsadegh & Jafarizadeh, 2019; Kirupagaran et al., 2016; Sasidharan et al., 2014; Zeebaree et al., 2020). *Euphrasia officinalis* leaf extract mediated biosynthesis of gold nanoparticles (AuNPs) and silver nanoparticles (AgNPs) (Singh et al., 2018), and *Ziziphus* leaf extract mediated gold nanoparticles (Aljabali et al., 2018). *Indigofera tinctoria* leaf extract mediated silver (AgNPs), and gold nanoparticles (AuNPs) are a major highlight of the plant-based nanomaterials (Vijayan et al., 2018).

6.8 The Role of Nanoparticles in Agriculture

Human beings obtain their food directly or indirectly from the agriculture sector and keep in view the overgrowing world population. It is imperative to use new technologies like bio- and nanotechnology in the agricultural industry. In developing countries, the development of agribusiness is seen as a context for development. The field of nanotechnology has not just improved current horticultural practices by making them more secure, specialised and powerless yet, in addition, raised the nature of farming items by making them exceptionally nutritious and infection safe. The use of nanotechnology in horticulture has helped create work openings, new rural items, stockpiling/bundling techniques and the longer timeframe of realistic usability and, in a manner, has also improved the nature of water. The field of nanotechnology can improve the production and quality of food. A report published by Wheeler (2005) suggested that modern techniques can meet growing food demands and boost health, economic and environmental sectors as well. Lately, the significance of nanotechnology in the farming area has been acknowledged, although its examination started some 50 years back (Mukhopadhyay, 2014). In developing countries, more than 60% of the people earn their livelihood directly or indirectly from agriculture, thus acting as a backbone of their economy (b; Brock et al., 2011; Qamar et al., 2014; Rai & Ingle, 2012a). In the rural area, nanotechnology has arisen as one of the best basic instruments, and soon, it might turn into an anticipated thrust. To improve crop productivity, nanotechnology employs different approaches that involve the use of novel delivery systems and chemical agents posing a lesser threat to the welfare of living beings. Nanotechnology offers answers for the current issues in farming regions and gives trust in improving yield efficiency by better administration and protection programmes. Due to the extraordinary physicochemical properties of nanoparticles, nanotechnology offers incredible breadth to fulfil the food needs of the rising total populace. These nanoparticles control a broad scope of utilisations, essentially their utilisation in treating human sicknesses and in the agricultural area. In the rural area, nanoparticles have different applications as depicted in Fig. 6.4.

However, the most important aim of nanomaterials that is of a greater significance is to improve crop productivity and plant protection as discussed below.

6.8.1 Crop Productivity

Nanomaterials have been utilised to improve crop profitability and effectiveness. In the agricultural sector, a new strategy based on nanoparticle use has been commonly employed to address crop yield and efficiency problems. For supportable farming, nanotechnology can expand world food production, improve the nature of foods, screen plant development, distinguish sicknesses of plants/animals and give insurance to plants and capacity to decrease squanders (Biswal et al., 2012; Ditta, 2012;

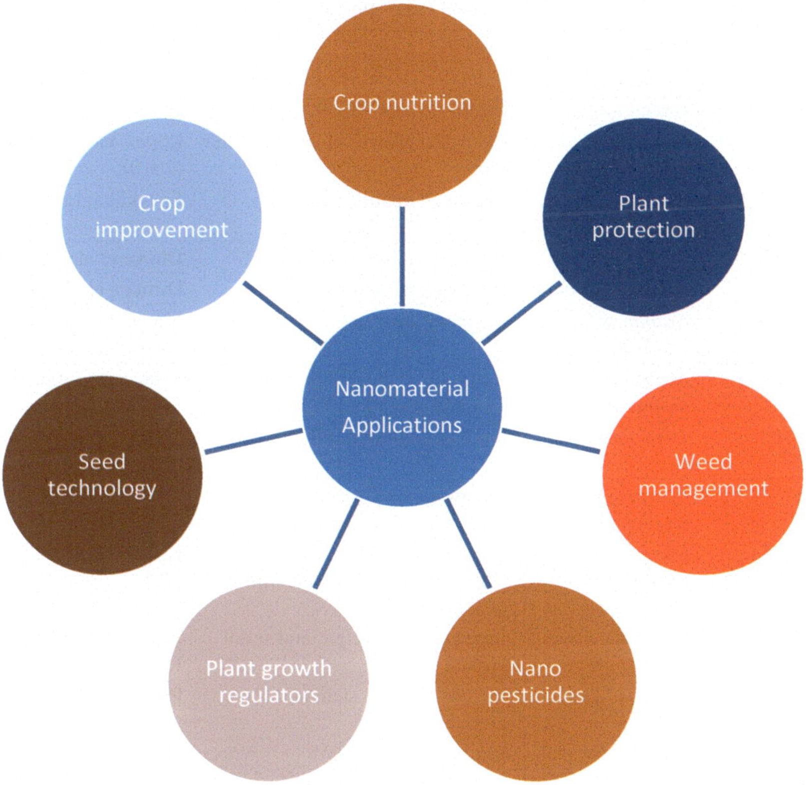

Fig. 6.4 Applications of nanomaterials in agriculture

Frewer et al., 2011; Gruere et al., 2011; Perez-de-Luque & Hermosín, 2013; Prasad et al., 2014; Sonkaria et al., 2012). The plants hereditarily incited by nanomaterial-based substances can assume a vital part in expanding agrarian efficiency (Kuzma, 2006; Scott, 2007). In plants and animals at cellular/molecular levels, the induction of molecules by gene delivery, site-specific drug delivery and nano-array-based gene modifications have been used (Maysinger, 2007). The factors that determine nanoparticle efficiency include size, chemical composition, reactivity and surface area. On plant development and improvement, nanoparticles may display both positive and negative impacts. A study on tomato seeds (Khodakovskaya et al., 2009) reported that the inserted carbon nanotubes (CNTs) increased their germination efficiency multiple times because CNTs improved the capacity of water take-up (Khodakovskaya et al., 2009). The growth of spinach increased with the use of TiO_2 nanoparticles, which enhanced the activity of the Rubisco enzyme and improved the absorbance of light (Hong et al., 2005; Yang et al., 2006). It was found that Tio_2 nanoparticles enhanced nitrogen metabolism, which ultimately improved spinach growth (Yang et al., 2007). A study reported by DeRosa et al. (2010) found that in corn and ryegrass, seed germination was inhibited by ZnO nanoparticles. The use of

Table 6.1 Significant nanoparticles for plant development and advancement

Nanoparticles	Plant	Effects	References
TiO_2	*Spinacia oleracea*	Induction of enzyme activity	Yang et al. (2006)
Alumina NPs	*Lemna minor*	Increased root length	Juhel et al. (2011)
Cerium oxide NPs	*Corn, soybean, alfalfa*	Increased growth of the stem and root	López-Moreno et al. (2010)
Iron oxide NPs	*Glycine max*	Improved quality and yield	Sheykhbaglou et al. (2010)
Iron oxide NPs	*Vigna radiate*	Biomass enhancement	Dhoke et al. (2013)
CuO NPs	*Triticum aestivum*	Increased biomass	Dimkpa et al. (2012)
CeO_2 NPs	*Arabidopsis thaliana*	Increased biomass	Ma et al. (2013)
G NPs	*Arabidopsis thaliana*	Early flowering and increased root and shoot length	Kumar et al. (2013)
TiO_2 NPs	*Triticum aestivum*	Increased chlorophyll content	Mahmoodzadeh et al. (2013)
CNTs	*Lycopersicum esculentum*	Enhanced seed germination and growth	Morla et al. (2011)
MWCNTs	*Lycopersicum esculantum*	Improved height of the plant along with an increased number of flowers	Khodakovskaya et al. (2013)
ZnO NPs	*Cicer arietinum* L.	Increased dry weight and shoot growth	Burman, Saini and Kumar et al. (2013)
ZnO NPs	*Arachis hypogea*	Increased yield, stem and root growth	Prasad et al. (2012)
AI NPs	Radish	Improved root growth	Lin and Xing (2007)
Au NPs	Lettuce, cucumber	Increased germination index	Barrena et al. (2009)

ZnO-based nanomaterial left some porous spaces in the roots of these plants, thus creating a potential route for nutrient delivery systems.

Other nanoparticles that have been discovered to be significant for plant development and advancement are listed in Table 6.1.

6.8.2 Plant Protection

In addition to enhancing crop productivity, nanoparticles are also known to protect plants from various diseases. Several approaches have been used to manage crop diseases, particularly genetic breeding, sanitation schemes, new pesticides and integrated pest management. New insights have been provided by nanotechnology for improving and modifying present crop management methods. Techniques such as spraying and broadcasting are conventionally used for applying plant protection chemicals and nutrients. However, the minimum required amounts of chemicals/nutrients do not reach the target site because of leaching, hydrolysis and microbial degradation. The conventional methods of crop protection generally involved the use of large-scale herbicides, insecticides and fungicides. Over 90% of the pesticides used for pest control were either lost in the environment or were unable to reach the target sites (Nuruzzaman et al., 2016). The utilisation of pesticides expanded the cost expenses and caused degradation of the general climate. In this regard, a better initiative that was needed in the agricultural sector to protect plants from microbial diseases was the development of nanoformulations or encapsulation of pesticides. These nanoformulations contain a small number of tiny particles with pesticides as active ingredients. Nanoparticles of carbon, silver, silica and alumina silicates have been used to control plant diseases caused by various phytopathogens. The epitomised nanoformulations encourage the controlled arrival of dynamic fixings into the objective zones of plants and hence give better outcomes. The pesticides of conventional origin have various limitations like limited solubility, increased resistance and nanoformulations; these problems are decreased (Dwivedi et al., 2016). Therefore, to accomplish higher harvest creation, the criticalness of nanotechnology has expanded dramatically. A study reported by Petosa et al. (2017) showed that the pesticide nanoformulations boosted crop yield by enhancing the efficacy of pesticides by regulating their transport potential. Their study combined polymeric nanocapsules with pyrethroid bifenthrin (Ncap-BIF), which ended up being a promising conveyance vehicle for plant security. The catalytic activity of trypsin, known as a viral protease, was reduced by fabricated bioactive AuNPs, thereby proving effective in controlling insects. This change in catalytic activity was believed to be due to the interaction of proteins with metallic nanoparticles (Patil et al., 2016). For instance, to control the growth of *Penicillium expansum*, *Alternaria alternate*, *Rhizopus stolonifer*, *A. flavus*, *Fusarium graminearum* and pathogenic bacteria, ZnO nanoparticles have proven to be effective (Dwivedi et al., 2016; Vanathi et al., 2016). Further, Si and TiO_2 have shown a promise in suppressing crop diseases through its antimicrobial activity. In sustainable agriculture development, nanomaterial-encapsulated pesticides, herbicides and fungicides have shown a tremendous scope.

6.9 Conclusions

Nanotechnology has emerged as the most innovative science with widespread applications. It has solved many agriculture-related issues like nutrient uptake efficiency, insect pest control and crop production. The synthetic pesticides available in the markets have negatively impacted the environment due to their toxic and persistent nature. To avoid the limitations of these synthetic pesticides, pest management could be done by involving nanotechnology-based nanopesticides. The nanopesticides are non-toxic and safe to use. To beat the impediments of biological or green synthesis methods, nanoparticles are favoured as eco-friendly, less toxic and healthy alternatives. In this regard, we will understand the role of green synthesised nanoparticles in agriculture management in this chapter.

References

Alghuthaymi, M. A., Almoammar, H., Rai, M., Said-Galiev, E., & Abd-Elsalam, K. A. (2015). Myconanoparticles: synthesis and their role in phytopathogens management. *Biotechnology & Biotechnological Equipment, 29*(2), 221–236.

Aljabali, A. A. A., Akkam, Y., Al Zoubi, M. S., Al-Batayneh, K. M., Al-Trad, B., Abo Alrob, O., … Evans, D. J. (2018 March). Synthesis of gold nanoparticles using leaf extract of Ziziphus zizyphus and their antimicrobial activity. *Nanomaterials, 8*(3), 174. https://doi.org/10.3390/nano8030174

Anu, K., Singaravelu, G., Murugan, K., & Benelli, G. (2017 January 1). Green-synthesis of selenium nanoparticles using garlic cloves (Allium sativum): Biophysical characterisation and cytotoxicity on vero cells. *Journal of Cluster Science, 28*(1), 551–563. https://doi.org/10.1007/s10876-016-1123-7

Banerjee, P., Satapathy, M., Mukhopahayay, A., & Das, P. (2014 December 1). Leaf extract mediated green synthesis of silver nanoparticles from widely available Indian plants: Synthesis, characterisation, antimicrobial property and toxicity analysis. *Bioresources and Bioprocessing, 1*(1), 3. https://doi.org/10.1186/s40643-014-0003-y

Barrena, R., Casals, E., Colón, J., Font, X., Sánchez, A., & Puntes, V. (2009). Evaluation of the ecotoxicity of model nanoparticles. *Chemosphere, 75*(7), 850–857. https://doi.org/10.1016/j.chemosphere.2009.01.078

Biswal, S. K., Nayak, A. K., Parida, U. K., & Nayak, P. L. (2012). Applications of nanotechnology in agriculture and food sciences. *IJSID, 2*(1), 21–36.

Brock, D. A., Douglas, T. E., Queller, D. C., & Strassmann, J. E. (2011). Primitive agriculture in a social amoeba. *Nature, 469*(7330), 393–396. https://doi.org/10.1038/nature09668

Busó-Rogero, C., Brimaud, S., Solla-Gullon, J., Vidal-Iglesias, F. J., Herrero, E., Behm, R. J., & Feliu, J. M. (2016 February 15). Ethanol oxidation on shape-controlled platinum nanoparticles at different pHs: A combined in situ IR spectroscopy and online mass spectrometry study. *Journal of Electroanalytical Chemistry, 763*, 116–124. https://doi.org/10.1016/j.jelechem.2015.12.034

Chakravorty, A., Rather, G. A., Ali, A., Bhat, B. A., Sana, S. S., Abhishek, N., & Nanda, A. (2020). Nano approach: Indian spices as antimicrobial agents. In *Advances in medical diagnosis, treatment, and care* (pp. 205–241). IGI Global. https://doi.org/10.4018/978-1-7998-2524-1.ch016

Costa Silva, L. P., Oliveira, J. P., Keijok, W. J., da Silva, A. R., Aguiar, A. R., Guimarães, M. C. C., … Braga, F. R. (2017). Extracellular biosynthesis of silver nanoparticles using the cell-free filtrate of nematophagous fungus *Duddingtonia flagrans*. *International Journal of Nanomedicine, 12*, 6373–6381. https://doi.org/10.2147/IJN.S137703

Dauthal, P., & Mukhopadhyay, M. (2016 September 14). Noble metal nanoparticles: Plant-mediated synthesis, mechanistic aspects of synthesis, and applications. *Industrial and Engineering Chemistry Research, 55*(36), 9557–9577. https://doi.org/10.1021/acs.iecr.6b00861

Deepa, B., & Ganesan, V. (2015). Bioinspiredsynthesis of selenium nanoparticles using flowers of Catharanthus roseus (L.) G. Don. And Peltophorum pterocarpum (DC.) backer ex Heyne–a comparison. *Int. J. Chem. Technol. Res., 7*, 725–733.

DeRosa, M. C., Monreal, C., Schnitzer, M., Walsh, R., & Sultan, Y. (2010). Nanotechnology in fertilisers. *Nature Nanotechnology, 5*(2), 91–91. https://doi.org/10.1038/nnano.2010.2

Dhand, C., Dwivedi, N., Loh, X. J., Jie Ying, A. N., Verma, N. K., Beuerman, R. W., … Ramakrishna, S. (2015). Methods and strategies for the synthesis of diverse nanoparticles and their applications: A comprehensive overview. *RSC Advances, 5*(127), 105003–105037. https://doi.org/10.1039/C5RA19388E. PubMed: 105003

Dhoke, S. K., Mahajan, P., Kamble, R., & Khanna, A. (2013). Effect of nanoparticles suspension on the growth of mung (Vigna radiata) seedlings by foliar spray method. *Nanotechnology Development, 3*(1). https://doi.org/10.4081/nd.2013.e1

Dimkpa, C. O., McLean, J. E., Latta, D. E., Manangón, E., Britt, D. W., Johnson, W. P., … Anderson, A. J. (2012). CuO and ZnO nanoparticles: Phytotoxicity, metal speciation, and induction of oxidative stress in sand-grown wheat. *Journal of Nanoparticle Research, 14*(9), 1–15. https://doi.org/10.1007/s11051-012-1125-9

Ditta, A. (2012). How helpful is nanotechnology in agriculture? *Advances in Natural Sciences: Nanoscience and Nanotechnology, 3*(3). https://doi.org/10.1088/2043-6262/3/3/033002. PubMed: 033002.

Duhan, J. S., Kumar, R., Kumar, N., Kaur, P., Nehra, K., & Duhan, S. (2017 September 1). Nanotechnology: The new perspective in precision agriculture. *Biotechnology Reports, 15*, 11–23. https://doi.org/10.1016/j.btre.2017.03.002

Dwivedi, S., Saquib, Q., Al-Khedhairy, A. A., & Musarrat, J. (2016). Understanding the role of nanomaterials in agriculture. In D. P. Singh, H. B. Singh, & R. Prabha (Eds.), *Microbial inoculants in sustainable agricultural productivity* (pp. 271–288). Springer.

Egorova, E. M., & Revina, A. A. (2000 July 31). Synthesis of metallic nanoparticles in reverse micelles in the presence of quercetin. *Colloids and Surfaces A: Physicochemical and Engineering Aspects, 168*(1), 87–96. https://doi.org/10.1016/S0927-7757(99)00513-0

Fardsadegh, B., & Jafarizadeh-Malmiri, H. (2019). Aloe vera leaf extract mediated green synthesis of selenium nanoparticles and assessment of their in vitro antimicrobial activity against spoilage fungi and pathogenic bacteria strains. *Green Processing and Synthesis, 8*(1), 399–407.

Frewer, L. J., Norde, W., Fischer, A. R. H., & Kampers, F. W. H. (2011). *Nanotechnology in the Agri-food sector: Implications for the future*. Wiley-VCH Press.

Grasso, G., Zane, D., & Dragone, R. (2019 January). Microbial nanotechnology: Challenges and prospects for green biocatalytic synthesis of nanoscale materials for sensoristic and biomedical applications. *Nanomaterials, 10*(1), 11. https://doi.org/10.3390/nano10010011

Gruere, G., Narrod, C., & Abbott, L. (2011). *Agriculture, food, and water nanotechnologies for the poor: Opportunities and constraints. Policy brief 19*. International Food Policy Research Institute.

Guilger-Casagrande, M., & de Lima, R. (2019). Synthesis of silver nanoparticles mediated by fungi: A review. *Frontiers in Bioengineering and Biotechnology, 7*, 287. https://doi.org/10.3389/fbioe.2019.00287

Gupta, M., Tomar, R. S., Kaushik, S., Mishra, R. K., & Sharma, D. (2018 September 3). Effective antimicrobial activity of green ZnO Nano particles of Catharanthus roseus. *Frontiers in Microbiology, 9*, 2030. https://doi.org/10.3389/fmicb.2018.02030

Happy, A., Soumya, M., Kumar, S. V., Rajeshkumar, S., Sheba, R. D., Lakshmi, T., & Nallaswamy, V. D. (2019). Phyto-assisted synthesis of zinc oxide nanoparticles using Cassia alata and its antibacterial activity against Escherichia coli. *Biochemistry and Biophysics Reports, 17*, 208–211.

Herlekar, M., Barve, S., & Kumar, R. (2014). Plant-mediated green synthesis of iron nanoparticles. *Journal of Nanoparticles, 2014*, 1–9. https://doi.org/10.1155/2014/140614

Hong, F., Zhou, J., Liu, C., Yang, F., Wu, C., Zheng, L., & Yang, P. (2005). Effect of Nano-TiO2 on photochemical reaction of chloroplasts of spinach. *Biological Trace Element Research, 105*(1–3), 269–279. https://doi.org/10.1385/BTER:105:1-3:269

Hulkoti, N. I., & Taranath, T. C. (2014 September 1). Biosynthesis of nanoparticles using microbes—A review. *Colloids and Surfaces B, Biointerfaces, 121*, 474–483. https://doi.org/10.1016/j.colsurfb.2014.05.027

Ibrahim, H. M. M., & Hassan, M. S. (2016 October 20). Characterization and antimicrobial properties of cotton fabric loaded with green synthesised silver nanoparticles. *Carbohydrate Polymers, 151*, 841–850. https://doi.org/10.1016/j.carbpol.2016.05.041

Jaison, J., Barhoum, A., Chan, Y. S., Dufresne, A., & Danquah, M. K. (2018). Review on nanoparticles and nanostructured materials: History, sources, toxicity and regulations. *Beilstein Journal of Nanotechnology*.

Jayarambabu, N., Kumari, B. S., Rao, K. V., & Prabhu, Y. T. (2014). Germination and growth characteristics of mungbean seeds (Vigna radiata L.) affected by synthesised zinc oxide nanoparticles. *International Journal of Current Engineering and Technology, 5161*(4(5)), 2347.

Juhel, G., Batisse, E., Hugues, Q., Daly, D., van Pelt, F. N., O'Halloran, J., & Jansen, M. A. (2011). Alumina nanoparticles enhance growth of Lemna minor. *Aquatic Toxicology, 105*(3–4), 328–336. https://doi.org/10.1016/j.aquatox.2011.06.019

Kaur, P., Jain, P., Kumar, A., & Thakur, R. (2014 June 1). Biogenesis of PbS nanocrystals by using rhizosphere fungus ie, Aspergillus sp. isolated from the rhizosphere of chickpea. *BioNanoScience, 4*(2), 189–194. https://doi.org/10.1007/s12668-014-0135-8

Khandel, P., Yadaw, R. K., Soni, D. K., Kanwar, L., & Shahi, S. K. (2018). Biogenesis of metal nanoparticles and their pharmacological applications: Present status and application prospects. *Journal of Nanostructure in Chemistry, 8*(3), 217–254. https://doi.org/10.1007/s40097-018-0267-4

Khodakovskaya, M., Dervishi, E., Mahmood, M., Xu, Y., Li, Z., Watanabe, F., & Biris, A. S. (2009). Carbon nanotubes are able to penetrate plant seed coat and dramatically affect seed germination and plant growth. *ACS Nano, 3*(10), 3221–3227. https://doi.org/10.1021/nn900887m

Khodakovskaya, M. V., Kim, B. S., Kim, J. N., Alimohammadi, M., Dervishi, E., Mustafa, T., & Cernigla, C. E. (2013). Carbon nanotubes as plant growth regulators: Effects on tomato growth, reproductive system, and soil microbial community. *Small, 9*(1), 115–123. https://doi.org/10.1002/smll.201201225

Kirupagaran, R., Saritha, A., & Bhuvaneswari, S. (2016 December 31). Green synthesis of selenium nanoparticles from leaf and stem extract of leucas lavandulifolia sm. And their application. *NanoScience and Technology*, 224–226.

Krupa, N. D., Grace, A. N., & Raghavan, V. (2019 February 13). Process optimisation for green synthesis of ZnO nanoparticles and evaluation of its antimacrofouling activity. *IET Nanobiotechnology, 13*(5), 510–514. https://doi.org/10.1049/iet-nbt.2018.5396

Kumar, V., Guleria, P., Kumar, V., & Yadav, S. K. (2013). Gold nanoparticle exposure induces growth and yield enhancement in Arabidopsis thaliana. *Science of the Total Environment, 461–462*, 462–468. https://doi.org/10.1016/j.scitotenv.2013.05.018

Kumar, N., & Kumbhat, S. (2016). *Essentials in nanoscience and nanotechnology* (pp. 189–236). John Wiley & Sons,. Nanomaterials, C.-B.

Kumar, D. A., Palanichamy, V., & Roopan, S. M. (2014 June 5). Green synthesis of silver nanoparticles using Alternanthera dentata leaf extract at room temperature and their antimicrobial activity. *Spectrochimica Acta. Part A, Molecular and Biomolecular Spectroscopy, 127*, 168–171. https://doi.org/10.1016/j.saa.2014.02.058

Kuzma, J. (2006). Moving forward responsibly: Oversight for the nanotechnology-biology interface. *Journal of Nanoparticle Research, 9*(1), 165–182. https://doi.org/10.1007/s11051-006-9151-0

Lin, D., & Xing, B. (2007). Phytotoxicity of nanoparticles: Inhibition of seed germination and root growth. *Environmental Pollution, 150*(2), 243–250. https://doi.org/10.1016/j.envpol.2007.01.016

López-Moreno, M. L., De La Rosa, G., Hernández-Viezcas, J. A., Castillo-Michel, H., Botez, C. E., Peralta-Videa, J. R., & Gardea-Torresdey, J. L. (2010). Evidence of the differential biotransformation and genotoxicity of ZnO and CeO_2 nanoparticles on soybean (Glycine max) plants. *Environmental Science and Technology, 44*(19), 7315–7320. https://doi.org/10.1021/es903891g

Ma, C., Chhikara, S., Xing, B., Musante, C., White, J. C., & Dhankher, O. P. (2013). Physiological andmolecular response of Arabidopsis thaliana (L.) to nanoparticle cerium and indium oxide exposure. *ACS Sustainable Chemistry and Engineering, 1*(7), 768–778. https://doi.org/10.1021/sc400098h

Mahmoodzadeh, H., Nabavi, M., & Kashefi, H. (2013). Effect of nanoscale titanium dioxide particles on the germination and growth of canola (Brassica napus). *Journal of Ornamental Horticulture Plants, 3*, 25–32.

Majeed, S., Danish, M., Ibrahim, M. N., Sekeri, S. H., Ansari, M. T., Nanda, A., & Ahmad, G. (2020 September 4). Bacteria mediated synthesis of iron oxide nanoparticles and their antibacterial, antioxidant, cytocompatibility properties. *Journal of Cluster Science*, 1–2.

Maysinger, D. (2007). Nanoparticles and cells: Good companions and doomed partnerships. *Organic and Biomolecular Chemistry, 5*(15), 2335–2342. https://doi.org/10.1039/b704275b

Morla, S., Ramachandra Rao, C. S. V., & Chakrapani, R. (2011). Factors affecting seed germination and seedling growth of tomato plants cultured in vitro conditions, J Chem bio. *Physiological Sciences, B1*, 328–334.

Mourdikoudis, S., Pallares, R. M., & Thanh, N. T. K. (2018). Characterisation techniques for nanoparticles: Comparison and complementarity upon studying nanoparticle properties. *Nanoscale, 10*(27), 12871–12934. https://doi.org/10.1039/c8nr02278j

Mukhopadhyay, S. S. (2014). Nanotechnology in agriculture prospects and constraints. *Nanotechnology, Science and Applications, 7*, 63–71. https://doi.org/10.2147/NSA.S39409

Nadaroglu, H., Güngör, A. A., & Selvi, İ. N. (2017 August). Synthesis of nanoparticles by green synthesis method. *International Journal of Innovative Research and Reviews, 1*(1), 6–9.

Nakkala, J. R., Mata, R., Gupta, A. K., & Sadras, S. R. (2014 October 6). Biological activities of green silver nanoparticles synthesised with Acorous calamus rhizome extract. *European Journal of Medicinal Chemistry, 85*, 784–794. https://doi.org/10.1016/j.ejmech.2014.08.024

Naveed Ul Haq, A., Nadhman, A., Ullah, I., Mustafa, G., Yasinzai, M., & Khan, I. (2017 April 18). Synthesis approaches of zinc oxide nanoparticles: The dilemma of ecotoxicity. *Journal of Nanomaterials, 2017*, 1–14. https://doi.org/10.1155/2017/8510342

Nayantara, P., & Kaur, P. (2018 January 1). Biosynthesis of nanoparticles using eco-friendly factories and their role in plant pathogenicity: A review. *Biotechnology Research and Innovation, 2*(1), 63–73. https://doi.org/10.1016/j.biori.2018.09.003

Nuruzzaman, M., Rahman, M. M., Liu, Y. J., & Naidu, R. (2016). Nanoencapsulation, nano-guard for pesticides: A new window for safe application. *Journal of Agricultural and Food Chemistry, 64*(7), 1447–1483. https://doi.org/10.1021/acs.jafc.5b05214

Ovais, M., Khalil, A. T., Ayaz, M., Ahmad, I., Nethi, S. K., & Mukherjee, S. (2018 December). Biosynthesis of metal nanoparticles via microbial enzymes: A mechanistic approach. *International Journal of Molecular Sciences, 19*(12), 4100. https://doi.org/10.3390/ijms19124100

Pantidos, N., & Horsfall, L. E. (2014 September 1). Biological synthesis of metallic nanoparticles by bacteria, fungi and plants. *Journal of Nanomedicine and Nanotechnology, 5*(5), 1.

Patil, C. D., Borase, H. P., Suryawanshi, R. K., & Patil, S. V. (2016). Trypsin inactivation by latex fabricated gold nanoparticles: A new strategy towards insect control. *Enzyme and Microbial Technology, 92*, 18–25. https://doi.org/10.1016/j.enzmictec.2016.06.005

Paul, A. M., Sajeev, A., Nivetha, R., Gothandapani, K., Bhardwaj, P., K, G., … Grace, A. N. (2020 April 28). Cuprous oxide (Cu2O)/graphitic carbon nitride (g-C3N4) nanocomposites for electrocatalytic hydrogen evolution reaction. *Diamond and Related Materials, 107*. https://doi.org/10.1016/j.diamond.2020.107899, PubMed: 107899.

Perez-de-Luque, A., & Hermosín, M. C. (2013). Nanotechnology and its use in agriculture. In D. Bagchi, M. Bagchi, H. Moriyama, & F. Shahidi (Eds.), *Bio-nanotechnology: A revolution in food, Bomedical and health sciences* (pp. 299–405). Wiley-Blackwell, West.

Petosa, A. R., Rajput, F., Selvam, O., Öhl, C., & Tufenkji, N. (2017). Assessing the transport potential of polymeric nanocapsules developed for crop protection. *Water Research, 111*, 10–17.

Prasad, R., Kumar, V., & Prasad, K. S. (2014). Nanotechnology in sustainable agriculture: Present concerns and future aspects. *African Journal of Biotechnology, 6*, 13705–13713.

Prasad, T. N. V. K. V., Sudhakar, P., Sreenivasulu, Y., Latha, P., Munaswamy, V., Reddy, K. R., … Pradeep, T. (2012). Effect of nanoscale zinc oxide particles on the germination, growth and yield of peanut. *Journal of Plant Nutrition, 35*(6), 905–927. https://doi.org/10.1080/01904167.2012.663443

Qamar, Z., Nasir, I. A., & Husnain, T. (2014). In-vitro development of cauliflower synthetic seeds and conversion to plantlets. *Advances in Life Sciences, 1*(2), 34–41.

Raghavan, V., Deb, A., & Grace, A. N. (2020 June 19). Honokiol-camptothecin loaded graphene oxide nanoparticle towards combinatorial anticancer drug delivery. *IET Nanobiotechnology.*

Rai, M., & Ingle, A. (2012a). Role of nanotechnology in agriculture with special reference to management of insect pests. *Applied Microbiology and Biotechnology, 94*(2), 287–293. https://doi.org/10.1007/s00253-012-3969-4

Rai, M., & Ingle, A. (2012b). Role of nanotechnology in agriculture with special reference to management of insect pests. *Applied Microbiology and Biotechnology, 94*(2), 287–293. https://doi.org/10.1007/s00253-012-3969-4

Rai, P., Kwak, W. K., & Yu, Y. T. (2013 April 24). Solvothermal synthesis of ZnO nanostructures and their morphology-dependent gas-sensing properties. *ACS Applied Materials and Interfaces, 5*(8), 3026–3032. https://doi.org/10.1021/am302811h

Rajasekaran, S. J., & Raghavan, V. (2020 November 1). Facile synthesis of activated carbon derived from Eucalyptus globulus seed as efficient electrode material for supercapacitors. *Diamond and Related Materials, 109*. https://doi.org/10.1016/j.diamond.2020.108038. PubMed: 108038.

Sasidharan, S., Sowmiya, R., & Balakrishnaraja, R. (2014). Biosynthesis of selenium nanoparticles using citrus reticulata peel extract. *World Journal of Pharmaceutical Research, 4*, 1322–1330.

Scott, N. R. (2007). Nanoscience in veterinary medicine. *Veterinary Research Communications, 31*(Suppl. 1), 139–144. https://doi.org/10.1007/s11259-007-0083-7

Shah, M., Fawcett, D., Sharma, S., Tripathy, S. K., & Poinern, G. E. J. (2015 November). Green synthesis of metallic nanoparticles via biological entities. *Materials, 8*(11), 7278–7308. https://doi.org/10.3390/ma8115377

Sharmila, G., Muthukumaran, C., Sandiya, K., Santhiya, S., Pradeep, R. S., Kumar, N. M., … Thirumarimurugan, M. (2018 September 1). Biosynthesis, characterisation, and antibacterial activity of zinc oxide nanoparticles derived from Bauhinia tomentosa leaf extract. *Journal of Nanostructure in Chemistry, 8*(3), 293–299. https://doi.org/10.1007/s40097-018-0271-8

Sheykhbaglou, R., Sedghi, M., Shishevan, M. T., & Sharifi, R. S. (2010). Effects of nano-iron oxide particles on agronomic traits of soybean. *Notulae Scientia Biologicae, 2*(2), 112–113. https://doi.org/10.15835/nsb224667

Singh, H., Du, J., Singh, P., & Yi, T. H. (2018 August 18). Ecofriendly synthesis of silver and gold nanoparticles by Euphrasia officinalis leaf extract and its biomedical applications. *Artificial Cells, Nanomedicine, and Biotechnology, 46*(6), 1163–1170. https://doi.org/10.1080/21691401.2017.1362417

Sitaaraman, S. R., Santhosh, R., Kollu, P., Jeong, S. K., Sellappan, R., Raghavan, V., … Grace, A. N. (2020 October 1). Role of graphene in NiSe2/graphene composites-Synthesis and testing for electrochemical supercapacitors. *Diamond and Related Materials, 108*. PubMed: 107983.

Sonkaria, S., Ahn, S. H., & Khare, V. (2012). Nanotechnology and its impact on food and nutrition: A review. *Recent Patents on Food, Nutrition and Agriculture, 4*(1), 8–18. https://doi.org/10.2174/2212798411204010008

Srinath, B. S., & Ravishankar Rai, V. R. (2015 October 1). Biosynthesis of highly monodispersed, spherical gold nanoparticles of size 4–10 nm from spent cultures of Klebsiella pneumoniae. *3 Biotech, 5*(5), 671–676. https://doi.org/10.1007/s13205-014-0265-2

Syamsai, R., & Grace, A. N. (2020 March 1). Synthesis, properties and performance evaluation of vanadium carbide MXene as supercapacitor electrodes. *Ceramics International, 46*(4), 5323–5330. https://doi.org/10.1016/j.ceramint.2019.10.283

Vanathi, P., Rajiv, P., & Sivaraj, R. (2016). Synthesis and characterisation of Eichhornia-mediated copper oxide nanoparticles and assessing their antifungal activity against plant pathogens. *Bulletin of Materials Science, 39*(5), 1165–1170. https://doi.org/10.1007/s12034-016-1276-x

Velappan, S., Nivedhita, P., Vimala, R., & Raja, S. (2020 November 1). Role of Nano titania on the thermomechanical properties of silicon carbide refractories. *Ceramics International, 46*(16), 25921–25926. https://doi.org/10.1016/j.ceramint.2020.07.077

Vijayan, R., Joseph, S., & Mathew, B. (2018 May 19). Indigofera tinctoria leaf extract mediated green synthesis of silver and gold nanoparticles and assessment of their anticancer, antimicrobial, antioxidant and catalytic properties. *Artificial Cells, Nanomedicine, and Biotechnology, 46*(4), 861–871. https://doi.org/10.1080/21691401.2017.1345930

Ward, M. B., Brydson, R., & Cochrane, R. F. (2006) (Vol. 26, No. 1, p. 296). Mn nanoparticles produced by inert gas condensation. In *Journal of Physics: Conference Series*. IOP Publishing, 296–299. https://doi.org/10.1088/1742-6596/26/1/071.

Wheeler, S. (2005). *Factors influencing agricultural professionals' attitudes toward organic agriculture and biotechnology. Center for Regulation and market analysis*. University of South Australia.

Yadav, B. C., Srivastava, R., & Yadav, A. (2009 January 1). Nanostructured zinc oxide synthesised via hydroxide route as liquid petroleum gas sensor. *Sensors and Materials, 21*, 87–94.

Yang, F., Hong, F., You, W., Liu, C., Gao, F., Wu, C., & Yang, P. (2006). Influence of nano-anatase TiO2 on the nitrogen metabolism of growing spinach. *Biological Trace Element Research, 110*(2), 179–190. https://doi.org/10.1385/bter:110:2:179

Yang, F., Liu, C., Gao, F., Su, M., Wu, X., Zheng, L., … Yang, P. (2007). The improvement of spinach growth by nano-anatase TiO2 treatment is related to nitrogen photoreduction. *Biological Trace Element Research, 119*(1), 77–88. https://doi.org/10.1007/s12011-007-0046-4

Zeebaree, S. Y. S., Zeebaree, A. Y. S., & Zebari, O. I. H. (2020 March 1). Diagnosis of the multiple effect of selenium nanoparticles decorated by Asteriscus graveolens components in inhibiting HepG2 cell proliferation. *Sustainable Chemistry and Pharmacy, 15*. https://doi.org/10.1016/j.scp.2019.100210. PubMed: 100210.

Zhang, K., Lv, S., Lin, Z., Li, M., & Tang, D. (2018 March 15). Bio-bar-code-based photo electrochemical immunoassay for sensitive detection of prostate-specific antigen using rolling circle amplification and enzymatic biocatalytic precipitation. *Biosensors and Bioelectronics, 101*, 159–166. https://doi.org/10.1016/j.bios.2017.10.031

Zheng, L., Hong, F., Lu, S., & Liu, C. (2005 April 1). Effect of Nano-TiO 2 on strength of naturally aged seeds and growth of spinach. *Biological Trace Element Research, 104*(1), 83–92. https://doi.org/10.1385/BTER:104:1:083

Chapter 7
The Role of Green Synthesised Zinc Oxide Nanoparticles in Agriculture

Gulzar Ahmed Rather, Saima Hamid, Muzafar Riyaz, Musheerul Hassan, Mohmmad Ashaq Sofi, Ifrah Manzoor, and Anima Nanda

Abstract Nanotechnology is the most innovative field of the twenty-first century, as detailed research is underway to develop nanoproducts worldwide. Because of their unique properties, nanoparticles have grown considerably. These nanomaterials are used in photovoltaic systems, fuel cells and biomedical fields, with zinc oxide as an actual example. Concerning their synthesis, ZnO-NPs may be synthesised by numerous chemical techniques, including vapour transport, precipitation and hydrothermal processes. The green synthesis of ZnO-NPs is also popular nowadays because it is facile, safe, non-toxic and environmentally friendly. The green synthesis includes bacteria, fungi, yeast, algae, plants and their parts like seeds, fruits, leaves, stem and pulp. Among green synthesis, using plant extracts is the most popular as it is a single-step process. In this paper, we will discuss the role of plant-mediated synthesis of ZnO-NPs in crop production.

Keywords Nanotechnology · ZnO-NPs · Crop production · Green synthesis · Plant extracts

---

G. A. Rather · M. A. Sofi · A. Nanda (✉)
Department of Biomedical Engineering, Sathyabama Institute of Science and Technology, Deemed to be University, Chennai, Tamil Nadu, India

S. Hamid
Centre of research for development, Department of Environmental Sciences, University of Kashmir, Srinagar, Jammu & Kashmir, India

M. Riyaz
Entomology Research Institute, Loyola College, Chennai, India

M. Hassan
Department of Life Sciences, Pacific University, Udaipur, Rajasthan, India

I. Manzoor
Department of Biochemistry, University of Kashmir, Srinagar, Jammu & Kashmir, India

S. A. Bandh (ed.), *Sustainable Agriculture*,
https://doi.org/10.1007/978-3-030-83066-3_7

7.1 Introduction

In modern days, due to their distinct and attractive applications, ZnO-NPs have received tremendous attention in many fields like physics, chemistry, biology, medicine, electronics, etc. These characteristics can be endowed with their large surface area, reduced scale and surface-specific binding position, availability and catalytic, electronic and thermal properties. There are different routes for the biosynthesis of ZnO-NPs using physical and chemical methods that involve different set-up, high-temperature and high-pressure conditions (Agarwal et al., 2018). Biosynthesis of nanoparticles via the green route is natural, environmentally friendly, cost-effective, safe and biocompatible (Abdul Salam et al., 2014). The green synthesis method involves the use of bacteria, fungi, algae, yeast, plant extracts, etc. They permit the mass production of ZnO-NPs without additional impurities (Yuvakkumar et al., 2014). The utilisation of nanotechnology in agriculture is one of the potential sectors that might enhance sustainable crop management. For example, the use of nano-based pesticides to release chemicals has been effectively deployed in a regulated and specifically tailored manner that makes for a cleaner and simpler pest control system. Nanoparticles that are essential for improving agricultural production are generally considered to improve productivity and sustainability. Zinc (Zn) is a significant micronutrient for the proper growth and development of plants and animals. In this regard, numerous studies have been carried out on the role of zinc in the growth and metabolism of plants. For crop nutrition, it is crucial since it is needed in numerous enzyme processes, metabolism and oxidation-reduction processes. Zinc is important for the proper functioning of many enzymes such as isomerases, dehydrogenases, transphosphorylases, aldolases and DNA and RNA polymerases used for a wide variety of essential physiological and metabolic processes. Zinc also takes part in tryptophan synthesis, cell division, photosynthetic maintenance and membrane structure. Besides, it also acts as an essential cofactor in controlling the biosynthesis of proteins (Lacerda et al., 2018). Appropriate Zn fertilisation, therefore, helps enhance cereal, vegetable and food production. Zinc deficiency is characterised by reduced leaf size, interveinal necrosis and ribbed leaf margins.

Under the extreme conditions of zinc shortage, SPAD values, low leaf area values and total N and NO3 are observed (Castillo et al., 2019). With the deteriorating deficiency, catalase, superoxide dismutase and glutathione peroxide activities often increase. Declines in cross-sectional area, yield and kernel percentages are also found with severe Zn shortage. Increased activity of the enzyme superoxide dismutase, catalase and peroxides is related to reactive oxygen detoxification (Ali et al., 2020).

7.2 Zinc Oxide Nanoparticles (ZnO-NPs)

In numerous cutting-edge applications such as communications, environmental biosafety, biology, medicine, cosmetics and electronics, ZnO-NPs have been used within the broad family of metal oxide nanoparticles. In addition, ZnO-NPs have immense potential in biomedical applications such as gene transfer, bio-labelling, nanomedicine and biological sensing (Bala et al., 2015). Zinc oxide is an n-type semiconductor metal oxide that has been extensively used for numerous applications like the manufacture of rubber and separation of arsenic and sulphur from water. It has an excellent property of protein adsorption and is also used in dental applications. It has been registered as GRAS (generally recognised as safe) among the other metal oxides by the US FDA. From the past years, ZnO-NPs gained tremendous attention in the area of research due to its wide bandgap and large excitation-binding energy (Anbuvannan et al., 2015; Jamdagni, Khatri, et al., 2018; Taranath & Patil, 2016; Pulit-Prociak et al., 2016; Sundrarajan et al., 2015). Due to its wide bandgap and high binding energy ZnoNPS have shown excellent antibacterial, antifungal, wound healing, antioxidant and optical properties. There are different methods available to synthesise ZnO-NPs like physical, chemical and biological or green synthesis. Due to the limitation of the physical and chemical methods, biological or green synthesis of ZnO-NPs is usually preferred. The physical and chemical methods are costly, time-consuming, employed on high pressure and temperature and generating large quantities of secondary waste products and toxic chemicals into the environment. As a result of these limitations, biogenic or green synthesis is the ideal method for nanoparticle synthesis as it is safer, cheaper and less toxic (Sharmila et al., 2019).

7.3 Nanoparticles Synthesis

For the development of nanomaterials of definite form and scale, two methods called top-down and bottom-up approaches (Fig. 7.1) are commonly used. The principle of synthesis for both approaches is different, but they produce nanomaterials with desired characteristics. In top-down approaches, the bulk material is crushed into small pieces leading to the formation of nanomaterials. The nanomaterial produced in this manner utilises the photolithographic techniques, sputtering, grinding and milling. The top-down approach is a relatively feasible nanoparticle production method that results in a large mass of nanomaterials being generated. The limitations associated with the top-down approach are surface imperfection of nanomaterials. Another approach is the bottom-down approach, in which assembly of atoms by atoms, molecules by molecules and clusters by clusters is done to produce a wide range of nanomaterials. The techniques employed for the development of nanomaterials through the bottom-up method are chemical or electrochemical nanostructural precipitation, chemical vapour deposition, laser pyrolysis, and bio-assisted

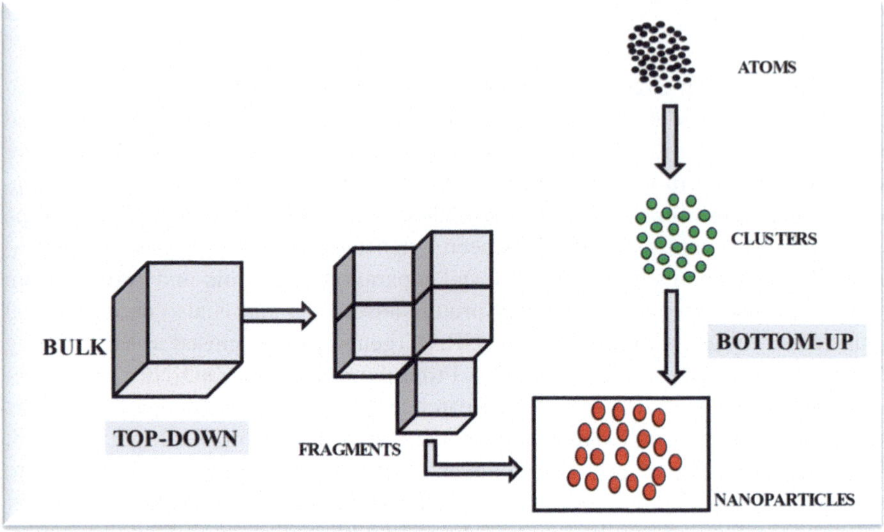

Fig. 7.1 Top-Down and Bottom-Up Approaches of Nanomaterial Synthesis

synthesis (Dhand et al., 2015; Gwo et al., 2016; Patil & Chandrasekaran, 2020) (Dhand et al., 2015; Gwo et al., 2016; Patil & Chandrasekaran, 2020).

7.4 Methods of Nonmaterial Synthesis

7.4.1 Physical Synthesis

The techniques like ball milling, sputtering and deposition are included in the physical synthesis of nanomaterials. The rate of production of nanomaterials through these techniques is meagre. In ball milling, the yield of nanomaterial synthesis is only 50%. In the case of laser ablation and plasma techniques, high consumption of energy is required. For most physical technologies that cannot be adopted for actual commercial applications, vast size distribution, slow production rate, waste by-products and significant energy consumption make it exceedingly costly (Seetharaman et al., 2018). (Seetharaman et al., 2018).

7.4.2 Chemical Synthesis

The chemical synthesis of nanomaterials includes various methods like sol-gel deposition, hydrothermal deposition, microemulsion and chemical vapour deposition (Król et al., 2017). Based on solid phases, wet chemical synthesis is one of the

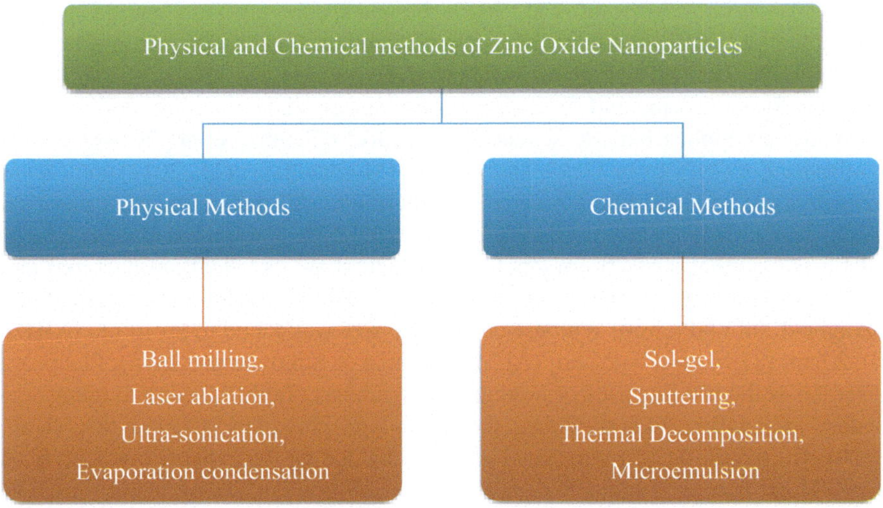

Fig. 7.2 Physical and chemical methods for nonmaterial synthesis

essential tools for the production of nanomaterials (Król et al., 2017; Malfatti et al., 2015). In industrial wet chemical synthesis, capping agents and stabilising agents are typically used to control the particles' size and prevent agglomeration due to toxicity. Triethylamine (TEA), thioglycerol, oleic acid and polyethylene glycol are significant capping or stabilising agents, even though they are apoptotic, immunogenic and necrotic (Naveed Ul Haq et al., 2017). In the microemulsion method, stable fluid droplets from immiscible hydrocarbons and waters are formed in a thermodynamically stable manner. A method to monitor the size (˂200 nm), structure (hexagonal crystal) and shape of ZnO-NPs based on minimum emulsion using TEA has been documented by Fricke et al. (2015). The chemical vapour deposition has also been used for the synthesis of ZnO-NPs, efficiently. Figure 7.2 shows the physical and chemical synthesis of nanomaterials.

7.4.3 *Biological Synthesis*

As biological methods are environmentally friendly, they provide exciting possibilities than physical and chemical synthesis (Agarwal et al., 2018). The biosynthesis of ZnO-NPs was systematically examined with microorganisms, proteins, DNA and plant extracts (Ishwarya et al., 2017; Raja et al., 2018). However, biological synthesis is not fully understood as the mechanical method for producing ZnO-NPs. By going through enzymatic and biological pathways, ZnO-NPs can be produced. A synthesis of ZnO-NPs (10–95 nm; rod, cubic, multiform, triangle, acicular) was performed with bacteria, including *Halomonas elongate* IBRC-M 10214 (Taran et al., 2018), *Sphingobacterium thalpophilum* (Rajabairavi et al., 2017) and

Staphylococcus aureus (Rauf et al., 2017). Fungal species such as Candida albicans (Mashrai et al., 2017) and Aspergillus niger (Kalpana et al., 2018) can also be used to synthesise ZnO-NPs. Fungal-mediated ZnO-NPs with a spherical and quasi-spherical form of 61 nm and 25 nm were used for steroidal pyrazoline synthesis and antimicrobial applications. As a yeast system, JA2 (Chauhan et al., 2014) can also synthesise ZnO-NPs from *Pichia kudriavzevii* (Moghaddam et al., 2017) and *Pichia fermentans*. The algae like *Chlamydomonas reinhardtii* and *Sargassum muticum* (Azizi et al., 2014) are also used to synthesise ZnO-NPs. Gelatin has also been helpful for the synthesis of 20 nm ZnO-NPs, which show excellent antibacterial and anti-angiogenic activity (Divya et al., 2018). By using egg albumin, spherical and hexagonal ZnO-NPs were synthesised, and the size was found to be 16 nm (XRD), 10–20 nm (TEM) and 8–22 nm (AFM) (Ambika & Sundrarajan, 2015). Plants are regarded as the bio-factories of nonmaterial synthesis (Table 7.1) due to many secondary metabolites like alkaloids, flavonoids and phenolics (Chakravorty et al., 2020). The synthesis of nonmaterial is started by adding the extracts obtained from the different plant components to the aqueous solution containing metal ions. The secondary metabolites present in the plant extract functions as reducing and capping agents to reduce metal ions into nanoparticles (Rather et al., 2020).

Nanoparticles, which have proven to be one of the critical needs for growth and development in the era of nanotechnology (Keat et al., 2015), can be used for a broad range of applications. Therefore, the study is going on to increase the efficiency of nanoparticles to make them more application-specific. However, the methods used to synthesise nanoparticles should be cost-effective and eco-friendly. This suggests that the method of NP synthesis should avoid the use of toxic chemicals, and there should be a negligible generation of any harmful by-products.

Table 7.1 Different Parts of Plants Used for Nanoparticle Synthesis

S. no.	Name of plant	Part of plant used	UV absorption peak (nm)	Shape of nanoparticles	References
1.	*Musa acuminate*	Peel	344	Triangular	Abdullah et al. (2020)
2.	*Melia azedarach*	Leaf	372	Spherical	Dhandapani et al. (2020)
3.	*Cassia alata*	Leaf	320	Spherical	Happy et al. (2019)
4.	*Coccinia abyssinica*	Tuber	365	Hexagonal	Safawo et al. (2018)
5.	*Azadirachta indica*	Leaf	375	Spherical	Singh et al. (2019)
6.	*Artemisia annua*	Bark	330	Spherical	
7.	*Ziziphus jujube*	Fruit	376	Spherical	Golmohammadi et al. (2020)
8.	*Berberis aristata*	Leaf	343	Needle	Chandra et al. (2019)
9.	*Codonopsis lanceolata*	Root	356	Flower	Lu et al. (2019)

The method of green synthesis (biosynthesis) uses basic methods, readily available raw materials and the atmosphere for the synthesis process, where the precursors used are healthy, with a small possibility of developing harmful by-products (Dizaj et al., 2014 (Hossain et al., 2019; Ogunyemi et al., 2019)). Biosynthesised NPs are observed and characterised using various methods, such as XRD, AFM, FT-IR, DLS, SEM, TEM, UV-Vis spectroscopy, zeta potential analysis, etc. (Salem & Fouda, 2021).

Plant sections including leaves, roots, flowers and stem have been utilised to synthesise zinc oxide nanoparticles (Azizi et al., 2016; Lingaraju et al., 2016; Raj, 2015; Suresh et al., 2018). They contain many phytochemicals that help stabilise and reduce zinc to zinc nanoparticles (Iqbal et al., 2019). The simple method used for synthesising zinc nanoparticles is that the plant part is collected and washed under tap water or double-distilled water. Then the plant part is shade dried for few days and finally ground to powder using mortar and pestle or electric grinder. Milli-Q water is used for the preparation of plant extract and the synthesis of nanoparticles. Some volume of plant extract is added to the zinc acetate, zinc oxide solution. The mixture is boiled at the desired temperature to ensure optimal mixing. Some optimisation such as pH, temperature and extract concentration in the process of synthesis may be required. The mixture is centrifuged at 7000–8000 rpm and oven-dried at 6–80 °C. The oven-dried material is calcined in a muffle furnace at 400–450 °C for 2 h to obtain a white powder containing zinc oxide nanoparticles (Alamdari et al., 2020; Demissie et al., 2020; Selim et al., 2020). Several studies have reported the synthesis of zinc nanoparticles from plants (Lakshmi, 2017; Selim et al., 2020).

Happy et al. (2019) reported the plant-mediated synthesis of Zn nanoparticles from *Cassia alata* leaf. Aqueous extract of *Eucalyptus* has also been exploited for the synthesis of zinc oxide nanoparticles of size ranging between 52 and 70 nm, confirmed by SEM and TEM analysis (Ahmad et al., 2020). Zinc oxide nanostructures have also been synthesised using the seed extract of *Nigella saliva*. Chemingui et al. (2019) reported the synthesis of zinc oxide nanoparticles using *Laurus nobilis* plant extract. Synthesised nanoparticles were of wurtzite hexagonal structure, and the average size of the nanocrystals was between 20 and 35 nm.

The synthesis of metal nanoparticles using microorganisms has recently been investigated and now accepted as an effective way to exploit microorganisms as cost-effective nanofactories. The biological synthesis of nanoparticles using microbes provides a benefit over plants because microbes are readily replicated. Nonetheless, several drawbacks like careful monitoring are required, culturing bacteria, and the media used for bacterial culture is also quite costly (Ahmed et al., 2017). The involvement of different enzymes, proteins and other biomolecules from microbes such as bacteria, fungi and yeasts plays a crucial role in the reduction process (Ali et al., 2018). These organic products secreted in suspension or growth medium result in multisite mono- and polydispersed nanoparticles and serve as capping agents to stabilise nanoparticles (Gahlawat & Choudhury, 2019).

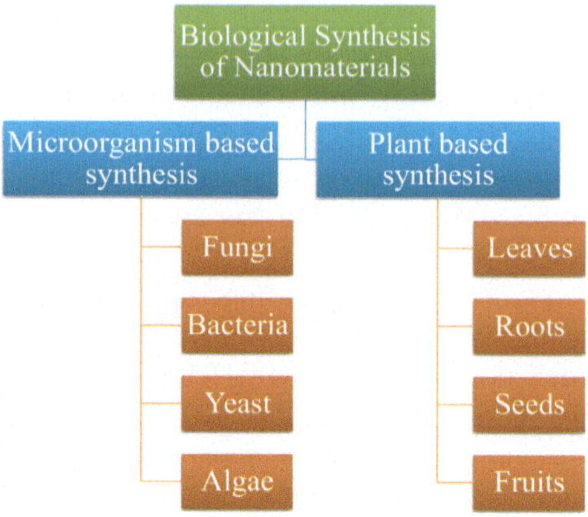

Fig. 7.3 Biological synthesis of nanomaterials

Lactic acid bacteria have raised interest in the wide-scale manufacturing of metal oxide nanoparticles because of their non-pathogenic features and the enormous synthesis of enzymes (Yusof et al., 2019). In addition, LAB has been acknowledged to have beneficial effects on human health. Moog et al. (2020) synthesised monodispersed zinc oxide nanoparticles utilising zinc-tolerant probiotic *Lactobacillus plantarum* strain TA4 with an average particle size of 124.2 nm, as validated by DLS analysis.

Fungi and yeasts have also been widely used to synthesise nanoparticles because of their high binding capacity, tolerance and better bioaccumulation ability (Boroumand Moghaddam et al., 2015a, 2015b; Pati et al., 2014). As study in which synthesis of ZnOnps was successfully conducted by using culture filtrates of Aspergillus niger have shown the size of ZnOnps were in the range of 84–91 nm as confirmed by SEM analysis. Extracellular mycosynthesis of zinc oxide nanoparticles using *A. alternata*, *A. tenuissima* (an endophytic fungi) and *Pichia kudriavzevii* (a yeast strain) has been reported in the scientific literature (Abdelhakim et al., 2020; Moghaddam et al., 2017; Sarkar et al., 2014), and the same have been confirmed by the SEM and TEM studies. Figure 7.3 shows biological methods of zinc oxide nanomaterials.

7.5 Limitations of Conventional Methods for ZnO Nanoparticle Synthesis

The physical methods for the synthesis of ZnO nanoparticles hold their limitations. The method microemulsion involves impacting the crystallisation process, due to which it is difficult to achieve a uniform size of nanocrystalline oxides. While in the sol-gel preparation method, the drawback is to purify the final products from residues of the solvents and polymers used. Meanwhile, the magnetron sputtering method holds a disadvantage involving the high cost of equipment relative complexity and technical implementation. The green synthesis, which provides a comparatively pollution-free mechanism, optimises reaction conditions to achieve improved performance, desirable features and instability. The stability of biosynthesised ZnO is unpredictable to calculate and could otherwise cause a significant harm to the biological systems. Thus, reaction mechanisms and technologies both for chemical and green syntheses must be properly built and optimised. In addition, most ZnO synthetic chemicals are classified as harmful chemical compounds, and the legal exhaust cap in most countries is well established. To assess the threat presented by ZnO, there is a clear need for environmental risk assessment. In addition, replication models should be built to parallel the growing usefulncss, ZnO synthetisation quantity and contaminants expired during these operations.

7.6 Characterisation of ZnO Nanoparticles

In the past years, nanotechnology has gained immense popularity in science, be it medicine, electronics, agriculture and engineering. Nanomaterials can be characterised through different techniques like XRD, TEM, SEM, EDAX, FT-IR, DLS, AFM and UV-Vis spectroscopy (Fig. 7.4). Through these techniques, nanomaterial properties like morphology, composition, structure, form, physical and chemical properties can be studied. There is no single method that satisfies to analyse the properties of nonmaterial (Hasan, 2015; Nivetha et al., 2020).

7.6.1 UV-Visible Spectroscopy

UV-visible spectroscopy is one of the basic analytical techniques used to analyse nanomaterials. In UV-visible spectroscopy, light intensity is measured, and by this technique, the optical properties of nanomaterials can be studied. The nanomaterials like Au and Cu show normal UV-visible excitation spectrum due to the presence of a signal in the range of the visible portion of the spectrum (Hendel et al., 2014). With the help of UV-visible spectroscopy, the molar concentration of nanomaterials can be calculated (Paramelle et al., 2014).

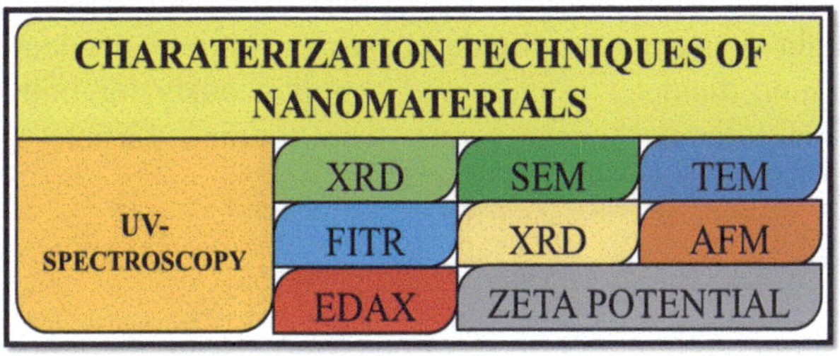

Fig. 7.4 Characterisation techniques of nanomaterial synthesis

7.6.2 Transmission Electron Microscopy

Transmission microscopy is one of the advanced methods used to characterise nanomaterials. By this technique, the actual size and images of the nanoparticles could be captured. During this technique, a uniform beam of electrons touches the specimen and spreads out through it. After the interaction between the electron beam and the specimen, an image is formed that is magnified and focused with the help of an imaging tool. The contact between the sample and beam depends on the density, elemental composition of the material, size and morphology. Compared to any other characterisation technique, TEM has significantly higher image resolution and can visualise extremely small atoms. TEM is one of the versatile techniques used to assess nanomaterials' in vitro absorption, size, shape, morphology and aggregation. With the help of TEM, it is possible to determine the degree of nanoparticles' penetration, their position inside the cell and the mechanism of cell death in cancerous cells. The dispersion of nanoparticles can also be analysed through TEM (Laborda et al., 2016; Yue et al., 2017).

7.6.3 Scanning Electron Microscopy

Scanning electron microscopy is one of the widely used characterisation method to measure the morphology of nanomaterials. Due to its high resolution, it is used to capture high-quality images of nanomaterials. The principle optical microscope and SEM are the same, but in SEM, the electron dispersion is quantified through electrical potential instead of photons. In this technology, the electrons are employed to make pictures, which further interacts and scans the surface of samples to create distinct signals to offer information on the type of interaction, composition and the arrangement of samples (Arun et al., 2020; Machado et al., 2015; Rajeev et al., 2019).

7.6.4 Dynamic Light Scattering

DLS is the most powerful technique employed to analyse the size of nanomaterials in colloidal solutions. The nanomaterials are in continuous Brownian motion in the aqueous form and produce both positive and destructive interfaces. By DLS, the light scattering amplitude causes the time-dependent oscillations, and hydrodynamic diameter is measured simultaneously for both nanomaterials and solvent molecule at the same rate. The average particle size can also be estimated (Chakravorty et al., 2020).

7.6.5 Energy-Dispersive X-Ray Spectroscopy

The energy dispersive x-ray spectroscopy (EDX) is used to evaluate nanomaterials' chemical characterisation and elemental composition. The composition of elemental analysis is achieved, as every element has its unique structure, and it can produce distinctive peaks on the x-ray spectrum. EDX is an essential tool for the examination and the extent of purity in ZnO-NPs. The plant extract acts as a source of reducing agents and elements like oxygen and carbon on EDX (Taziwa et al., 2017).

7.6.6 X-Ray Diffraction

X-ray diffraction (XRD) is one of the widely used techniques for the characterisation of nanomaterials. It is an advantageous and non-destructive method for the characterisation of nanomaterials. XRD gives information about the crystalline phase, structure, shape and pressures of nanomaterials. In this technique, the powdered samples are used, and by calculating the position and amplitude, the composition of nanomaterials can be analysed. The peaks are obtained by using the monochromatic beam of x-rays.

7.6.7 Fourier Transforms Infrared Spectroscopy

Fourier transform infrared spectroscopy (FTIR) is another flexible method used to analyse the composition of nanomaterials, their molecular structure, type of bonding and the existence of related functional groups. The principle of FTIR is based on the absorption spectrum of electromagnetic wavelength in the range of 4000–400 cm^{-1} (Blanco Andujar, 2014). With the help of the FTIR technique, characterisation of ZnO-NPs from plant extracts to determine the number of functional groups such as -OH, C=O, C=C and C-N present in the ZnO-NPs can be done (Shao et al., 2018).

7.6.8 Atomic Force Microscopy (AFM)

AFM, a type of microscopy, provides high-resolution 3D images. The main principle of this approach involves the force of attraction between the specimen and a fine probe. Because of attraction or repulsive forces, when AFM scans the sample between the tip and the sample surface, the information collected from the laser gap and the final result is decided by combining forces. When characterising the nanoparticles, the latter is the most common one. By this technique, topological characterisation of small nanoparticles (≤6 nm), such as ion-doped Y_2O_3, has been achieved without any special treatment (Krupa & Vimala, 2017).

7.7 The Role of Green Synthesised Zinc Oxide Nanoparticles (ZnO-NPs) in Agriculture

Agriculture is the primary sector or source of Third World economies. Still, unfortunately, it is facing numerous global challenges like drastic climate changes, urbanisation, environmental issues like runoff and the accumulation of pesticides and fertilisers. Metal oxide nanoparticles have shown excellent effects on the agriculture sector, including effective growth, increased yield and nutritional quality. Besides, they are acting as antipathogenic agents in plant protection. Being biologically important, Zn has played an essential role in plant systems, including its role in metal protein complexes. Zinc is required in very small amounts and serves a variety of activities in plants, including cell membrane stabilisation, protein synthesis, plant protection, cell elongation, and resistance to environmental stress (Kolenčík et al., 2019; Sabir et al., 2014). The positive effects of ZnO-NPs on the germination and development of pearl millet are also reported (Nandhini et al., 2019).

Similarly, the beneficial effects of ZnO-NPs on the quantitative and physiological parameters of *Zea mays* and *Triticum aestivum* L. have also been described (Singh et al., 2019). In corn and wheat, the chlorophyll formation is positively enhanced by ZnO-NPs (Rizwan et al., 2019). In *Solanum tuberosum* L., the usage of 100, 350 and 500 mgL-1 of ZnO-NPs has increased the common starch content (Raigond et al., 2017). The ZnO-NPs at lower concentrations increase seed germination (Ramesh et al., 2014). The foliar application of ZnO-NPs on coffee plants (*Coffea arabica* L.) demonstrated improved growth and physiology compared to Zn (ZnSO4) salt due to its greater leaf penetration (Rossi et al., 2019). The antioxidant and phenolic compounds in *Capsicum annum* L. have been boosted during seed germination by ZnO-NPs (García-López et al., 2018). In nursery phases with foliar spraying of ZnO-NPs, a significant increase in growth was observed in karanj, milk wood-pine and meem seedlings (Chaudhuri & Malodia, 2017). ZnO-NPs also promote embryogenesis, plantlet regeneration and certain enzymes in MS media that are allowed to survive in biotic tension (Helaly et al., 2014). The study conducted on *Fagopyrum esculentum* showed the presence of ZnO-NPs that had strengthened the antioxidant properties, photosynthetic efficacy and increased proline accumulation, providing plant stabilisation (Faizan et al., 2018). ZnO-NPs serve as new fertiliser for crop yield and food quality enhancement with unique physicochemical properties (Yusefi-Tanha et al., 2020). Studies performed on *Nicotiana benthamiana* have shown that it deactivates TMV (*Tomato mosaic virus*) and activates its immunity through the use of ZnO-NPs (Cai et al., 2019). Studies on ZnO-NPs have shown increased plant resistance to a wide range of microbes, increased crop production and decreased disease severity (Tripathi et al., 2017). Zinc oxide nanoparticles (ZnO-NPs) have shown excellent pesticide activity against the *Artemia salina* larva (Singh et al., 2018). In the growth of *Arachis hypogea* plant, the fertility study of ZnO-NPs was carried out. The results showed an increase in seed germination, rapid shoot growth, increased seedling vigour, enhanced root growth and rapid flowering and yield. In *Solanum lycopersicum* the fertility efficacy of ZnO-NPs was conducted, and the results showed increased seed germination and high protein content (Singh et al., 2016).

There was a significant rise in root growth and dry weight in onions after adding ZnO-NPs (Laware & Raskar, 2014). The photosynthetic pigment levels in millet were increased by applying ZnO-NPs (Tarafdar et al., 2014). The mechanism and behaviour of ZnO-NPs in plants are not well defined. Figure 7.5 shows the role of zinc oxide nanomaterials in agriculture.

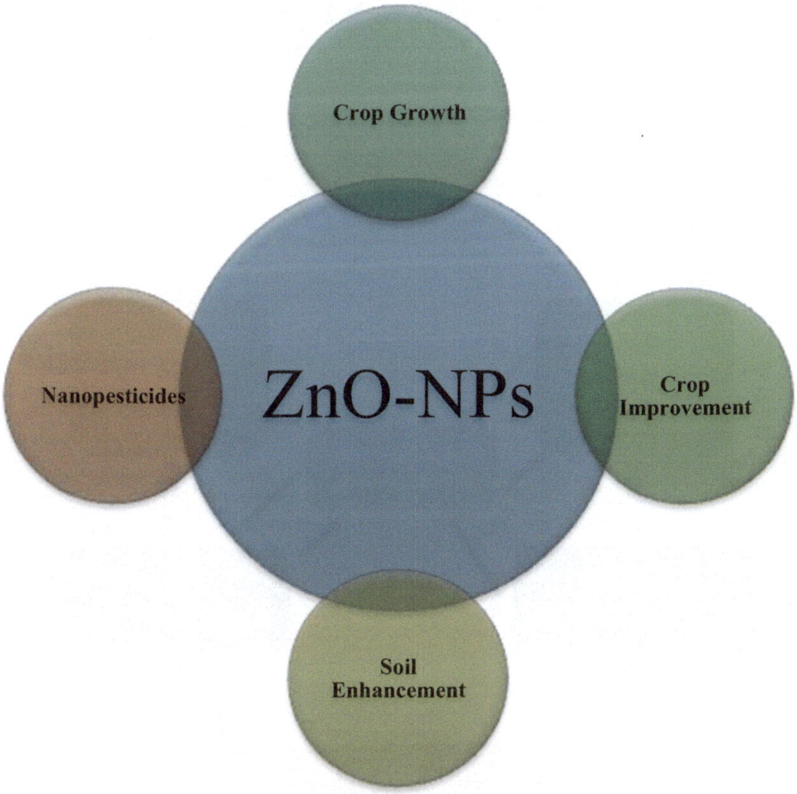

Fig. 7.5 The role of green synthesised zinc oxide nanoparticles in agriculture

7.8 The Role of ZnO-NPs under Abiotic Stress

The application of ZnO-NPs mitigates the damaging effect of ROS on cells. The ZnO-NPs trigger antioxidant enzymes, free amino acids and nutrients, which play a major role in providing plant protection from different stresses (Fig. 7.6). It has shown successfully that the application of ZnO-NPs at reduced doses was very efficient in relieving various abiotic stresses and promoting plant growth and development (Venkatachalam et al., 2017; Wang et al., 2018).

The applications of nanoparticles have proven to be promising among the several other strategies implemented to combat drought-inducing damage to plants (Table 7.2). The application of ZnO-NPs to the wheat crop has minimised the adverse effects of drought stress and increased crop yield (Taran et al., 2017). According to Venkatachalam et al. (2017), the toxicity in *Leucaena leucocephala* seedlings by Cd and Pb has been reduced by applying ZnO-NPs. The foliar applications of ZnO-NPs have reduced salinity's adverse effects on the growth of sunflower

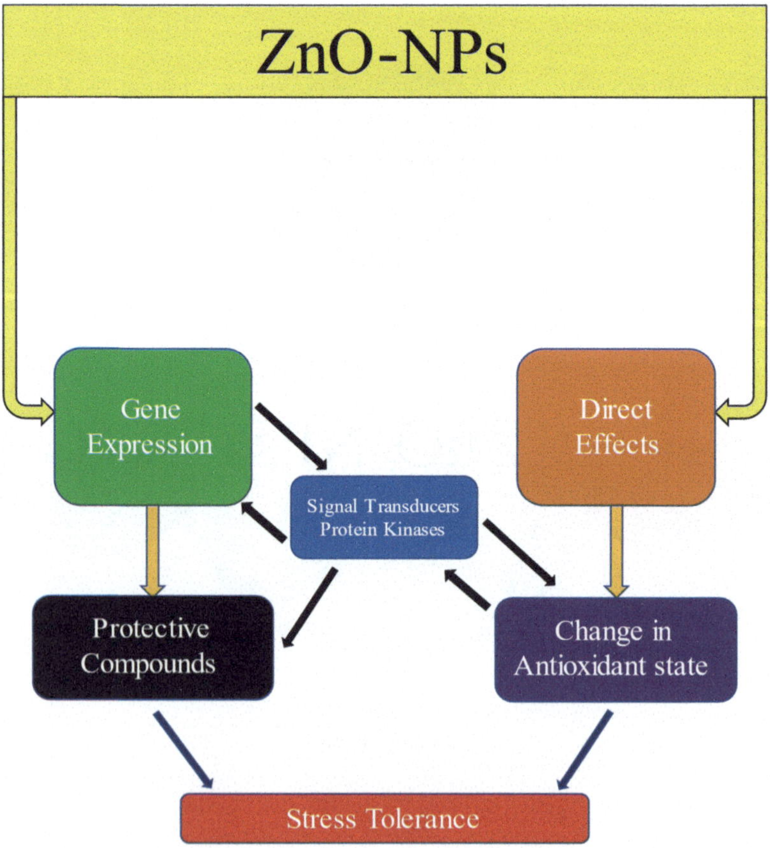

Fig. 7.6 The role of zinc oxide nanoparticles under abiotic stress

Table 7.2 Applications of green synthesised ZnO-NPs in the field of agriculture

S. no.	Type of nanoparticles	Plant	Source of nanoparticles	Size in nm	Application in agriculture	References
1.	Zinc oxide nanoparticles	*Aloe barbadensis*	Leaves	35 nm	Seedling (enhancing growth)	Singh et al. (2019)
2.	Zinc oxide nanoparticles	*Curcuma longa*	Tuber	20 nm	Seedling (root and shoot elongation)	Jayarambabu et al. (2015)
3.	Zinc oxide nanoparticles	*Nyctanthes arbor-tristis*	Flower	12–32 nm	Plant (antifungal activity)	Jamdagni, Khatri and Rana (2018, Jamdagni, Rana, et al., (2018)
4.	Zinc oxide nanoparticles	*Solanum lycopersicum*	Leaves	48.5 nm	Plant (antibacterial activity)	Ogunyemi et al. (2019)
5.	Zinc oxide nanoparticles	*Eucalyptus globules*	Leaves	52–70 nm	Plant (antifungal activity)	Ahmad et al. (2020)
6.	Zinc oxide nanoparticles	*Nigella sativa*	Leaves	20 nm	Plant (increases branching, growth)	Alaghemand et al. (2018)
7.	Zinc oxide nanoparticles	*Citrus limon*	Fruit extract	388 nm	Plant root (antibacterial effect)	Hossain et al. (2019)
8.	Zinc oxide nanoparticles	*Scadoxus multiflorus*	Leaves	31 nm	Plant (antilarval)	Al-Dhabi and Valan Arasu (2018)
9.	Zinc oxide nanoparticles	*Calotropis gigantea*	Leaf extract	1.5–8 nm	Seedling (acts as nanofertiliser and improves growth and development)	Chaudhuri and Malodia (2017)
10.	Zinc oxide nanoparticles	*Aloe barbadensis*	Leaf extract	35 nm	Enhances plant growth	Singh et al. (2019)
11.	Zinc oxide nanoparticles	*Elaeagnusan-gustifolia*	Flower extract	40 nm	Enhances growth and germination of seedling	Singh, Kim, et al. (2016); Singh, Singh, et al. (2016)

(continued)

Table 7.2 (continued)

S. no.	Type of nanoparticles	Plant	Source of nanoparticles	Size in nm	Application in agriculture	References
12.	Zinc oxide nanoparticles	*Nigella sativa*	Seed extract	24 nm	Enhances growth of the shoot and root and increases leaf chlorophyll content	Awan et al. (2020)
13.	Zinc oxide nanoparticles	*Nyctanthes arbor-tristis*	Flower extract	76 nm	Acts as fungicidal	Jamdagni, Khatri, & Rana (2018); Jamdagni, Rana, et al. (2018)
14.	Zinc oxide nanoparticles	*Solanum lycopersicum*	Plant extract	500 nm	Acts as strong antibacterial agent	Abdallah et al. (2020)
15.	Zinc oxide nanoparticles	*Azadirachta indica*	Leaf extract	27.8 nm	Plant (antimicrobial activity)	Mankad et al. (2016)
16.	Zinc oxide nanoparticles	*Mentha spicata*	Leaf extract	11–88 nm	Acts as antiviral agent against TMV	Abdelkhalek and Al-Askar (2020)
17.	Zinc oxide nanoparticles	*Psidium guajava*	Leaf extract	12 nm	Shows antibacterial activity and enhances fabric properties	Saha et al. (2018)

(Torabian et al., 2018). ZnO-NPs have led to the alleviation of soil salinity. About 60 mg/L of ZnO-NPs has shown to reduce the salt stress in *Moringa peregrina* plants (Soliman, 2015). Under stressful and stress-free conditions, ZnO-NPs act as essential materials for plant growth. ZnO-NPs have shown an increase in plant production as well as plant growth. Under stressful conditions like lipid peroxidation, ZnO-NPs have played an essential role in modulating the physiological parameters in plants. Due to their small size, ZnO-NPs have facilitated their easy penetration inside the plant cells and controlled the water channels that assisted plant growth and seed germination in plants (Faizan et al., 2018).

References

Abdallah, Y., Liu, M., Ogunyemi, S. O., Ahmed, T., Fouad, H., Abdelazez, A., … Li, B. (2020 January). Bioinspired green synthesis of chitosan and zinc oxide nanoparticles with strong antibacterial activity against rice pathogen Xanthomonas oryzae pv. oryzae. *Molecules, 25*(20), 4795. https://doi.org/10.3390/molecules25204795

Abdelhakim, H. K., El-Sayed, E. R., & Rashidi, F. B. (2020 June). Biosynthesis of zinc oxide nanoparticles with antimicrobial, anticancer, antioxidant and photocatalytic activities by the endophytic Alternaria tenuissima. *Journal of Applied Microbiology, 128*(6), 1634–1646. https://doi.org/10.1111/jam.14581

Abdelkhalek, A., & Al-Askar, A. A. (2020 January). Green synthesised ZnO nanoparticles mediated by Mentha spicata extract induce plant systemic resistance against tobacco mosaic virus. *Applied Sciences, 10*(15), 5054. https://doi.org/10.3390/app10155054

Abdul Salam, H. A., Sivaraj, R., & Venckatesh, R. (2014 September 15). Green synthesis and characterisation of zinc oxide nanoparticles from Ocimum basilicum L. var. purpurascens Benth.-Lamiaceae leaf extract. *Materials Letters, 131*, 16–18. https://doi.org/10.1016/j.matlet.2014.05.033

Abdullah, F. H., Abu Bakar, N. H. H., & Abu Bakar, M. (2020 March 1). Low temperature biosynthesis of crystalline zinc oxide nanoparticles from Musa acuminata peel extract for visible-light degradation of methylene blue. *Optik, 206*. https://doi.org/10.1016/j.ijleo.2020.164279. PubMed: 164279.

Agarwal, H., Menon, S., Venkat-Kumar, S., et al. (2018). Mechanistic study of the antibacterial action of zinc oxide nanoparticles synthesised using green route. *Chemico-biol Inter, 286*, 60–70.

Ahmad, H., Venugopal, K., Rajagopal, K., De Britto, S., Nandini, B., Pushpalatha, H. G., … Jogaiah, S. (2020 March). Green synthesis and characterisation of zinc oxide nanoparticles using Eucalyptus globules and their fungicidal ability against pathogenic fungi of apple orchards. *Biomolecules, 10*(3), 425. https://doi.org/10.3390/biom10030425

Ahmed, S., Annu, C., & S. A., & Ikram, S. (2017 January 1). A review on biogenic synthesis of ZnO nanoparticles using plant extracts and microbes: A prospect towards green chemistry. *Journal of Photochemistry and Photobiology B, Biology, 166*, 272–284. https://doi.org/10.1016/j.jphotobiol.2016.12.011

Alaghemand, A., Khaghani, S., Bihamta, M. R., Gomarian, M., & Ghorbanpour, M. (2018 January 1). Green synthesis of zinc oxide nanoparticles using Nigella sativa L. extract: The effect on the height and number of branches. *Journal of Nanostructures, 8*(1), 82–88.

Alamdari, S., Sasani Ghamsari, M., Lee, C., Han, W., Park, H. H., Tafreshi, M. J., … Ara, M. H. M. (2020 January). Preparation and characterisation of zinc oxide nanoparticles using leaf extract of Sambucus ebulus. *Applied Sciences, 10*(10), 3620. https://doi.org/10.3390/app10103620

Al-Dhabi, N. A., & Valan Arasu, M. (2018 July). Environmentally friendly green approach for the production of zinc oxide nanoparticles and their anti-fungal, ovicidal, and larvicidal properties. *Nanomaterials, 8*(7), 500. https://doi.org/10.3390/nano8070500

Ali, A., Bhat, B. A., Rather, G. A., Malla, B. A., & Ganie, S. A. (2020). Proteomic studies of micronutrient deficiency and toxicity. In *Inplant micronutrients* (pp. 257–284). Springer.

Ali, J., Irshad, R., Li, B., Tahir, K., Ahmad, A., Shakeel, M., … Khan, Z. U. H. (2018 June 1). Synthesis and characterisation of phytochemical fabricated zinc oxide nanoparticles with enhanced antibacterial and catalytic applications. *Journal of Photochemistry and Photobiology B, 183*, 349–356. https://doi.org/10.1016/j.jphotobiol.2018.05.006

Ambika, S., & Sundrarajan, M. (2015). Green biosynthesis of ZnO nanoparticles using Vitex negundo L. extract:Spectroscopic investigation of interaction between ZnO nanoparticles and human serum albumin. *Journal of Photochemistry and Photobiology B, Biology, 149*, 143–148. https://doi.org/10.1016/j.jphotobiol.2015.05.004

Anbuvannan, M., Ramesh, M., Viruthagiri, G., Shanmugam, N., & Kannadasan, N. (2015). Synthesis, characterisation and photocatalytic activity of ZnO nanoparticles prepared by biological method. *Spectrochimica Acta. Part A: Molecular and Biomolecular Spectroscopy, 143*, 304–308. https://doi.org/10.1016/j.saa.2015.01.124

Arun, D., Vimala, R., & Devendranath Ramkumar, K. D. (2020 May 6). Investigating the microbial-influenced corrosion of UNS S32750 stainless-steel base alloy and weld seams by biofilm-forming marine bacterium Macrococcus equipercicus. *Bioelectrochemistry, 135*, 107546. https://doi.org/10.1016/j.bioelechem.2020.107546

Awan, S., Shahzadi, K., Javad, S., Tariq, A., Ahmad, A., & Ilyas, S. (2020 October 16). A preliminary study of influence of zinc oxide nanoparticles on growth parameters of Brassica oleracea var italic. *Journal of the Saudi Society of Agricultural Sciences*.

Azizi, S., Ahmad, M. B., Namvar, F., & Mohamad, R. (2014). Green biosynthesis and characterisation of zinc oxide nanoparticles using brown marine macroalga Sargassum muticum aqueous extract. *Materials Letters, 116*, 275–277. https://doi.org/10.1016/j.matlet.2013.11.038

Azizi, S., Mohamad, R., Bahadoran, A., Bayat, S., Rahim, R. A., Ariff, A., & Saad, W. Z. (2016 August 1). Effect of annealing temperature on antimicrobial and structural properties of bio-synthesised zinc oxide nanoparticles using flower extract of Anchusa italica. *Journal of Photochemistry and Photobiology B, 161*, 441–449. https://doi.org/10.1016/j.jphotobiol.2016.06.007

Bala, N., Saha, S., Chakraborty, M., Maiti, M., Das, S., Basu, R., & Nandy, P. (2015). Green synthesis of zinc oxide nanoparticles using Hibiscus subdariffa leaf extract: Effect of temperature on synthesis, antibacterial activity and anti-diabetic activity. *RSC Advances, 5*(7), 4993–5003. https://doi.org/10.1039/C4RA12784F

Blanco Andujar, C. (2014). Sodium carbonate mediated synthesis of iron oxide nanoparticles to improve magnetic hyperthermia efficiency and induce apoptosis (Doctoral dissertation, University College London (University of London)).

Boroumand Moghaddam, A., Namvar, F., Moniri, M., Md Tahir, P., Azizi, S., & Mohamad, R. (2015a September). Nanoparticles biosynthesised by fungi and yeast: A review of their preparation, properties, and medical applications. *Molecules, 20*(9), 16540–16565. https://doi.org/10.3390/molecules200916540

Boroumand Moghaddam, A., Namvar, F., Moniri, M., Md Tahir, P., Azizi, S., & Mohamad, R. (2015b September 11). Nanoparticles biosynthesized by fungi and yeast: A review of their preparation, properties, and medical applications. *Molecules, 20*(9), 16540–16565. https://doi.org/10.3390/molecules200916540

Cai, L., Liu, C., Fan, G., Liu, C., & Sun, X. (2019). Preventing viral disease by ZnONPs through directly deactivating TMV and activating plant immunity in Nicotiana benthamiana. *Environmental Science: Nano, 6*(12), 3653–3669. https://doi.org/10.1039/C9EN00850K

Castillo, R. R., Lozano, D., González, B., Manzano, M., Izquierdo-Barba, I., & Vallet-Regí, M. (2019 April 3). Advances in mesoporous silica nanoparticles for targeted stimuli-responsive drug delivery: An update. *Expert Opinion on Drug Delivery, 16*(4), 415–439. https://doi.org/10.1080/17425247.2019.1598375

Chakravorty, A., Rather, G. A., Ali, A., Bhat, B. A., Sana, S. S., Abhishek, N., & Nanda, A. (2020). Nano approach: Indian spices as antimicrobial agents. In *Ethnopharmacological investigation of Indian spices* (pp. 205–241). IGI Global.

Chandra, H., Patel, D., Kumari, P., Jangwan, J. S., & Yadav, S. (2019 September 1). Phyto-mediated synthesis of zinc oxide nanoparticles of Berberis aristata: Characterisation, antioxidant activity and antibacterial activity with special reference to urinary tract pathogens. *Materials Science and Engineering, 102*, 212–220. https://doi.org/10.1016/j.msec.2019.04.035

Chaudhuri, S. K., & Malodia, L. (2017 November 1). Biosynthesis of zinc oxide nanoparticles using leaf extract of Calotropis gigantea: Characterisation and its evaluation on tree seedling growth in nursery stage. *Applied Nanoscience, 7*(8), 501–512. https://doi.org/10.1007/s13204-017-0586-7

Chauhan, R., Reddy, A., & Abraham, J. (2014). Biosynthesis of silver and zinc oxide nanoparticles using pichia fermentans JA2 and their antimicrobial property. *Applied Nanoscience*.

Chemingui, H., Missaoui, T., Mzali, J. C., Yildiz, T., Konyar, M., Smiri, M., … Yatmaz, H. C. (2019 September 11). Facile green synthesis of zinc oxide nanoparticles (ZnO NPs): Antibacterial and photocatalytic activities. *Materials Research Express, 6*(10), 1050b4. https://doi.org/10.1088/2053-1591/ab3cd6

Demissie, M. G., Sabir, F. K., Edossa, G. D., & Gonfa, B. A. (2020). Synthesis of zinc oxide nanoparticles using leaf extract of lippia adoensis (koseret) and evaluation of its antibacterial activity. *Journal of Chemistry, 2020 October 20*, 1–9. https://doi.org/10.1155/2020/7459042

Dhand, C., Dwivedi, N., Loh, X. J., Jie Ying, A. N., Verma, N. K., Beuerman, R. W., … Ramakrishna, S. (2015). Methods and strategies for the synthesis of diverse nanoparticles and their applications: A comprehensive overview. *RSC Advances, 5*(127), 105003–105037. https://doi.org/10.1039/C5RA19388E. PubMed: 105003.

Dhandapani, K. V., Anbumani, D., Gandhi, A. D., Annamalai, P., Muthuvenkatachalam, B. S., Kavitha, P., & Ranganathan, B. (2020 March 1). Green route for the synthesis of zinc oxide nanoparticles from Melia azedarach leaf extract and evaluation of their antioxidant and antibacterial activities. *Biocatalysis and Agricultural Biotechnology, 24*. https://doi.org/10.1016/j.bcab.2020.101517. PubMed: 101517.

Divya, M., Vaseeharan, B., Abinaya, M., Vijayakumar, S., Govindarajan, M., Alharbi, N. S., ... & Benelli, G. (2018). Biopolymer gelatin-coated zinc oxide nanoparticles showed high antibacterial, antibiofilm and anti-angiogenic activity. *Journal of Photochemistry and Photobiology B: Biology, 178*, 211–218.

Dizaj, S. M., Lotfipour, F., Barzegar-Jalali, M., Zarrintan, M. H., & Adibkia, K. (2014 November 1). Antimicrobial activity of the metals and metal oxide nanoparticles. *Materials Science and Engineering C, Materials for Biological Applications, 44*, 278–284. https://doi.org/10.1016/j.msec.2014.08.031

Faizan, M., Faraz, A., Yusuf, M., Khan, S. T., & Hayat, S. (2018 June 1). Zinc oxide nanoparticle-mediated changes in photosynthetic efficiency and antioxidant system of tomato plants. *Photosynthetica, 56*(2), 678–686. https://doi.org/10.1007/s11099-017-0717-0

Fricke, M., Voigt, A., Veit, P., & Sundmacher, K. (2015). Miniemulsion-based process for controlling the size and shape of zinc oxide nanoparticles. *Industrial and Engineering Chemistry Research, 54*(42), 10293–10300. https://doi.org/10.1021/acs.iecr.5b01149

Gahlawat, G., & Choudhury, A. R. (2019). A review on the biosynthesis of metal and metal salt nanoparticles by microbes. *RSC Advances, 9*(23), 12944–12967. https://doi.org/10.1039/C8RA10483B

García-López, J. I., Zavala-García, F., Olivares-Sáenz, E., Lira-Saldívar, R. H., Díaz Barriga-Castro, E., Ruiz-Torres, N. A., … Niño-Medina, G. (2018 October). Zinc oxide nanoparticles boosts phenolic compounds and antioxidant activity of Capsicum annuum L. during germination. *Agronomy, 8*(10), 215. https://doi.org/10.3390/agronomy8100215

Golmohammadi, M., Honarmand, M., & Ghanbari, S. (2020 March 15). A green approach to synthesis of ZnO nanoparticles using jujube fruit extract and their application in photocatalytic degradation of organic dyes. *Spectrochimica Acta. Part A: Molecular and Biomolecular Spectroscopy*, 229. https://doi.org/10.1016/j.saa.2019.117961. PubMed: 117961.

Gwo, S., Chen, H. Y., Lin, M. H., Sun, L., & Li, X. (2016). Nanomanipulation and controlled self-assembly of metal nanoparticles and nanocrystals for plasmonics. *Chemical Society Reviews, 45*(20), 5672–5716. https://doi.org/10.1039/c6cs00450d

Happy, A., Soumya, M., Venkat Kumar, S., Rajeshkumar, S., Sheba, R. D., Lakshmi, T., & Deepak Nallaswamy, V. (2019 March 1). Phyto-assisted synthesis of zinc oxide nanoparticles using Cassia alata and its antibacterial activity against Escherichia coli. *Biochemistry and Biophysics Reports, 17*, 208–211. https://doi.org/10.1016/j.bbrep.2019.01.002

Hasan, S. (2015). A review on nanoparticles: Their synthesis and types. *Research Journal of Recent Sciences, 2277*, 2502.

Helaly, M. N., El-Metwally, M. A., El-Hoseiny, H., Omar, S. A., & El-Sheery, N. I. (2014 April). Effect of nanoparticles on biological contamination of'in vitro'cultures and organogenic regeneration of banana. *Australian Journal of Crop Science, 8*(4), 612.

Hendel, T., Wuithschick, M., Kettemann, F., Birnbaum, A., Rademann, K., & Polte, J. (2014 November 18). In situ determination of colloidal gold concentrations with UV–vis spectroscopy: Limitations and perspectives. *Analytical Chemistry, 86*(22), 11115–11124. https://doi.org/10.1021/ac502053s

Hossain, A., Hong, X., Ibrahim, E., Li, B., Sun, G., Meng, Y., … An, Q. (2019). Green synthesis of silver nanoparticles with culture supernatant of a bacterium Pseudomonas rhodesiae and their antibacterial activity against soft rot pathogen Dickeya dadantii. *Molecules, 24*(12), 2303. https://doi.org/10.3390/molecules24122303

Iqbal, J., Abbasi, B. A., Mahmood, T., Kanwal, S., Ahmad, R., & Ashraf, M. (2019 August 5). Plant-extract mediated green approach for the synthesis of ZnONPs: Characterisation and evaluation of cytotoxic, antimicrobial and antioxidant potentials. *Journal of Molecular Structure, 1189*, 315–327. https://doi.org/10.1016/j.molstruc.2019.04.060

Ishwarya, R., Vaseeharan, B., Kalyani, S., Banumathi, B., Govindarajan, M., & Alharbi, N. (2017). Km, SN Al-Anbr M Khaled J, Benelli G. Facile green synthesis of zinc oxide nanoparticles using Ulva Lactuca seaweed extract and its evaluation of photocatalytic, antibiofilm and larvicidal activity: Impact on mosquito morphology and biofilm architecture. Journal of Photochemistry and Photobiology, Part B, 178.

Jamdagni, P., Khatri, P., & Rana, J. S. (2018 April 1). Green synthesis of zinc oxide nanoparticles using flower extract of Nyctanthes arbor-tristis and their antifungal activity. *Journal of King Saud University – Science, 30*(2), 168–175. https://doi.org/10.1016/j.jksus.2016.10.002

Jamdagni, P., Rana, J. S., Khatri, P., & Nehra, K. (2018 April 1). Comparative account of antifungal activity of green and chemically synthesised zinc oxide nanoparticles in combination with agricultural fungicides. *International Journal of Nano Dimension, 9*(2), 198–208.

Jayarambabu, N., Kumari, B. S., Rao, K. V., & Prabhu, Y. T. (2015). Beneficial role of zinc oxide nanoparticles on green crop production. *IJMART, 10*, 273–282.

Kalpana, V. N., Kataru, B. A. S., Sravani, N., Vigneshwari, T., Panneerselvam, A., & Devi Rajeswari, V. (2018). Biosynthesis of zinc oxide nanoparticles using culture filtrates of aspergillus Niger: Antimicrobial textiles and dye degradation studies. *OpenNano, 3*, 48–55. https://doi.org/10.1016/j.onano.2018.06.001

Keat, C. L., Aziz, A., Eid, A. M., & Elmarzugi, N. A. (2015 December). Biosynthesis of nanoparticles and silver nanoparticles. *Bioresources and Bioprocessing, 2*(1), 1–1.

Kolenčík, M., Ernst, D., Komár, M., Urík, M., Šebesta, M., & Dobročka, E. (2019 November). Effect of foliar spray application of zinc oxide nanoparticles on quantitative, nutritional, and physiological parameters of foxtail millet (Setaria italica l.) under field conditions. *Nanomaterials, 9*(11), 1559. https://doi.org/10.3390/nano9111559

Król, A., Pomastowski, P., Rafińska, K., Railean-Plugaru, V., & Buszewski, B. (2017). Zinc oxide nanoparticles: Synthesis, antiseptic activity and toxicity mechanism. *Advances in Colloid and Interface Science, 249*, 37–52. https://doi.org/10.1016/j.cis.2017.07.033

Krupa, A. N., & Vimala, R. (2017 July 19). AgNPs doped TEOS sol–gel coatings to prevent the adhesion of marine fouling organisms. *IET Nanobiotechnology, 12*(2), 99–105.

Laborda, F., Bolea, E., Cepriá, G., Gómez, M. T., Jiménez, M. S., Pérez-Arantegui, J., & Castillo, J. R. (2016 January 21). Detection, characterisation and quantification of inorganic engineered nanomaterials: A review of techniques and methodological approaches for the analysis of complex samples. *Analytica Chimica Acta, 904*, 10–32. https://doi.org/10.1016/j.aca.2015.11.008

Lacerda, J. S., Martinez, H. E. P., Pedrosa, A. W., Clemente, J. M., Santos, R. H. S., Oliveira, G. L., & Jifon, J. L. (2018). Importance of zinc for arabica coffee and its effects on the chemical composition of raw grain and beverage quality. *Crop Science, 58*(3), 1360–1370. https://doi.org/10.2135/cropsci2017.06.0373

Lakshmi, S. J., Bai, R. S., R., H, S., & Nidoni, U. K. (2017). A review study of zinc oxide nanoparticles synthesis from plant extracts. *Green Chemistry and Technology Letters, 3*(2), 26–37. [Several studies]. https://doi.org/10.18510/gctl.2017.321

Laware, S. L., & Raskar, S. (2014). Influence of zinc oxide nanoparticles on growth, flowering and seed productivity in onion. *International Journal of Current Microbiology Science, 3*(7), 874-881.

Lingaraju, K., Raja Naika, H. R., Manjunath, K., Basavaraj, R. B., Nagabhushana, H., Nagaraju, G., & Suresh, D. (2016 June). Biogenic synthesis of zinc oxide nanoparticles using Ruta graveolens (L.) and their antibacterial and antioxidant activities. *Applied Nanoscience, 6*(5), 703–710. https://doi.org/10.1007/s13204-015-0487-6

Lu, J., Ali, H., Hurh, J., Han, Y., Batjikh, I., Rupa, E. J., … Yang, D. C. (2019 May 1). The assessment of photocatalytic activity of zinc oxide nanoparticles from the roots of Codonopsis lanceolata synthesised by one-pot green synthesis method. *Optik, 184*, 82–89. https://doi.org/10.1016/j.ijleo.2019.03.050

Machado, S., Pacheco, J. G., Nouws, H. P., Albergaria, J. T., & Delerue-Matos, C. (2015 November 15). Characterization of green zero-valent iron nanoparticles produced with tree leaf extracts. *Science of the Total Environment, 533*, 76–81. https://doi.org/10.1016/j.scitotenv.2015.06.091

Malfatti, L., Pinna, A., Enzo, S., Falcaro, P., Marmiroli, B., & Innocenzi, P. (2015). Tuning the phase transition of ZnO thin films through lithography: An integrated bottom-up and top-down processing. *Journal of Synchrotron Radiation, 22*(1), 165–171. https://doi.org/10.1107/S1600577514024047

Mankad, M., Patil, G., Patel, S., Patel, D., & Patel, A. (2016). Green synthesis of zinc oxide nanoparticles using Azadirachta indica A. Juss. Leaves extractand its antibacterial activity against Xanthomonas orzyae pv. Oryzae. *Annals of Phytomedicine, 5*(2), 76–86. https://doi.org/10.21276/ap.2016.5.2.9

Mashrai, A., Khanam, H., & Aljawfi, R. N. (2017). Biological synthesis of ZnO nanoparticles using C. albicans and studying their catalytic performance in the synthesis of steroidal pyrazolines. *Arabian Journal of Chemistry, 10*, S1530–S1536.

Moghaddam, A. B., Moniri, M., Azizi, S., Rahim, R. A., Ariff, A. B., Saad, W. Z., & Navaderi, M. (2017 June). Biosynthesis of ZnO nanoparticles by a new pichia kudriavzevii yeast strain and evaluation of their antimicrobial and antioxidant activities. *Molecules, 22*(6), 872. https://doi.org/10.3390/molecules22060872

Moog, D., Schmitt, J., Senger, J., Zarzycki, J., Rexer, K. H., . . . Maier, U. G. (2020 December). Correction to: Using a marine microalga as a chassis for polyethylene terephthalate (PET) degradation. *Microbial Cell Factories, 19*(1), 1. https://doi.org/10.1186/s12934-019-1269-8.

Nandhini, M., Rajini, S. B., Udayashankar, A. C., Niranjana, S. R., Lund, O. S., Shetty, H. S., & Prakash, H. S. (2019 July 1). Biofabricated zinc oxide nanoparticles as an eco-friendly alternative for growth promotion and management of downy mildew of pearl millet. *Crop Protection, 121*, 103–112. https://doi.org/10.1016/j.cropro.2019.03.015

Naveed Ul Haq, A., Nadhman, A., Ullah, I., Mustafa, G., Yasinzai, M., & Khan, I. (2017). Synthesis approaches of zinc oxide nanoparticles: The dilemma of ecotoxicity. *Journal of Nanomaterials, 2017*, 1–14. https://doi.org/10.1155/2017/8510342

Nivetha, R., Gothandapani, K., Raghavan, V., Jacob, G., Sellappan, R., Bhardwaj, P., … Grace, A. N. (2020 July 21). Highly porous MIL-100 (fe) for the hydrogen evolution reaction (HER) in acidic and basic media. *ACS Omega, 5*(30), 18941–18949. https://doi.org/10.1021/acsomega.0c02171

Ogunyemi, S. O., Abdallah, Y., Zhang, M., Fouad, H., Hong, X., Ibrahim, E., … Li, B. (2019). Green synthesis of zinc oxide nanoparticles using different plant extracts and their antibacterial activity against Xanthomonas oryzae pv. oryzae. *Artificial Cells, Nanomedicine, and Biotechnology, 47*(1), 341–352. https://doi.org/10.1080/21691401.2018.1557671

Paramelle, D., Sadovoy, A., Gorelik, S., Free, P., Hobley, J., & Fernig, D. G. (2014). A rapid method to estimate the concentration of citrate capped silver nanoparticles from UV-visible light spectra. *Analyst, 139*(19), 4855–4861. https://doi.org/10.1039/c4an00978a

Pati, R., Mehta, R. K., Mohanty, S., Padhi, A., SenGupta, M., Vaseeharan, B., … Sonawane, A. (2014 August 1). Topical application of zinc oxide nanoparticles reduces bacterial skin infection in mice and exhibits antibacterial activity by inducing oxidative stress response and cell membrane disintegration in macrophages. *Nanomedicine: Nanotechnology, Biology, and Medicine, 10*(6), 1195–1208. https://doi.org/10.1016/j.nano.2014.02.012

Patil, S., & Chandrasekaran, R. (2020 December). Biogenic nanoparticles: A comprehensive perspective in synthesis, characterisation, application and its challenges. *Journal of Genetic Engineering and Biotechnology, 18*(1), 1–23.

Pulit-Prociak, J., Chwastowski, J., Kucharski, A., & Banach, M. (2016 November 1). Functionalization of textiles with silver and zinc oxide nanoparticles. *Applied Surface Science, 385*, 543–553. https://doi.org/10.1016/j.apsusc.2016.05.167

Raigond, P., Raigond, B., Kaundal, B., Singh, B., Joshi, A., & Dutt, S. (2017). Effect of zinc nanoparticles on antioxidative system of potato plants. *Journal of Environmental Biology, 38*(3), 435.

Raj, L. F. A., & E, J. (2015 January 1). Biosynthesis and characterisation of zinc oxide nanoparticles using root extract of Zingiber officinale. *Oriental Journal of Chemistry, 31*(1), 51–56. https://doi.org/10.13005/ojc/310105

Raja, A., Ashokkumar, S., Pavithra Marthandam, R., Jayachandiran, J., Khatiwada, C. P., Kaviyarasu, K., … Swaminathan, M. (2018). Eco-friendly preparation of zinc oxide nanoparticles using Tabernaemontana divaricata and its photocatalytic and antimicrobial activity. *Journal of Photochemistry and Photobiology, Part B, 181*, 53–58.

Rajabairavi, N., Raju, C. S., Karthikeyan, C., Varutharaju, K., Nethaji, S., Hameed, A. S. H., & Shajahan, A. (2017). Biosynthesis of novel zinc oxide nanoparticles (ZnO NPs) using endophytic bacteria Sphingobacterium thalpophilum. *Springer Proceedings in Physics*, 245–254. https://doi.org/10.1007/978-3-319-44890-9_23

Rajeev, P. V., Gnanasekar, S., Sellappan, R., & Grace, A. N. (2019 July 23). *Synthesis and analysis of Mo2N as efficient counter electrodes for dye sensitised solar cells*. Materials Today: Proceedings.

Ramesh, M., Palanisamy, K., Babu, K., & Sharma, N. K. (2014). Effects of bulk and nano-titanium dioxide and zinc oxide on physio-morphological changes in Triticum aestivum Linn. *Journal of Global Biosciences, 3*(2), 415–422.

Rather, G. A., Chakravorty, A., Bhat, B. A., Malik, I. M., Mir, F. H., Sana, S. S., … Choudhury, M. (2020 December 4). Routes of synthesis and characterisations of nanoparticles. In *Applications of nanomaterials in agriculture, food science and medicine* (pp. 288–309). IGI Global.

Rauf, M. A., Owais, M., Rajpoot, R., Ahmad, F., Khan, N., & Zubair, S. (2017). Biomimetically synthesised ZnO nanoparticles attain potent antibacterial activity against less susceptible S. aureus skin infection in experimental animals. *RSC Advances, 7*(58), 36361–36373. https://doi.org/10.1039/C7RA05040B

Rizwan, M., Ali, S., Zia Ur Rehman, M. Z., Adrees, M., Arshad, M., Qayyum, M. F., … Imran, M. (2019 May 1). Alleviation of cadmium accumulation in maise (Zea mays L.) by foliar spray of zinc oxide nanoparticles and biochar to contaminated soil. *Environmental Pollution, 248*, 358–367. https://doi.org/10.1016/j.envpol.2019.02.031

Rossi, L., Fedenia, L. N., Sharifan, H., Ma, X., & Lombardini, L. (2019 February 1). Effects of foliar application of zinc sulfate and zinc nanoparticles in coffee (Coffea arabica L.) plants. *Plant Physiology and Biochemistry, 135*, 160–166. https://doi.org/10.1016/j.plaphy.2018.12.005

Sabir, S., Arshad, M., & Chaudhari, S. K. (2014 October). Zinc oxide nanoparticles for revolutionising agriculture: Synthesis and applications. *Scientific World Journal, 2014*, 1–8. https://doi.org/10.1155/2014/925494

Safawo, T., Sandeep, B. V., Pola, S., & Tadesse, A. (2018 January 1). Synthesis and characterisation of zinc oxide nanoparticles using tuber extract of anchote (Coccinia abyssinica (lam.) Cong.) for antimicrobial and antioxidant activity assessment. *OpenNano, 3*, 56–63. https://doi.org/10.1016/j.onano.2018.08.001

Salem, S. S., & Fouda, A. (2021 May 6). Green synthesis of metallic nanoparticles and their prospective biotechnological applications: An overview. *Biological Trace Element Research, 199*(1), 344–370. https://doi.org/10.1007/s12011-020-02138-3

Sarkar, J., Ghosh, M., Mukherjee, A., Chattopadhyay, D., & Acharya, K. (2014 February). Biosynthesis and safety evaluation of ZnO nanoparticles. *Bioprocess and Biosystems Engineering, 37*(2), 165–171. https://doi.org/10.1007/s00449-013-0982-7

Seetharaman, P. K., Chandrasekaran, R., Gnanasekar, S., Chandrakasan, G., Gupta, M., Manikandan, D. B., & Sivaperumal, S. (2018 October 1). Antimicrobial and larvicidal activity of eco-friendly silver nanoparticles synthesised from endophytic fungi Phomopsis liquidambaris. *Biocatalysis and Agricultural Biotechnology, 16*, 22–30. https://doi.org/10.1016/j.bcab.2018.07.006

Selim, Y. A., Azb, M. A., Ragab, I., & Abd El-Azim, M. H. (2020 February 26). Green synthesis of zinc oxide nanoparticles using aqueous extract of Deverra tortuosa and their cytotoxic activities. *Scientific Reports, 10*(1), 1.

Shao, F., Yang, A., Yu, D. M., Wang, J., Gong, X., & Tian, H. X. (2018 December 1). Bio-synthesis of Barleria gibsoni leaf extract mediated zinc oxide nanoparticles and their formulation gel for wound therapy in nursing care of infants and children. *Journal of Photochemistry and Photobiology. B, Biology, 189*, 267–273. https://doi.org/10.1016/j.jphotobiol.2018.10.014

Sharmila, G., Thirumarimurugan, M., & Muthukumaran, C. (2019 March 1). Green synthesis of ZnO nanoparticles using Tecoma castanifolia leaf extract: Characterisation and evaluation of its antioxidant, bactericidal and anticancer activities. *Microchemical Journal, 145*, 578–587. https://doi.org/10.1016/j.microc.2018.11.022

Singh A. P., Dixit G., Kumar A., Mishra S., Singh P. K., Dwivedi S., et al. (2016). Nitric oxide alleviated arsenic toxicity by modulation of antioxidants and thiol metabolism in rice (Oryza sativa L.). *Frontiers in Plant Science*, 6:1272. 10.3389/fpls.2015.01272

Singh, A., Neelam, & Kaushik, M. (2019). Physicochemical investigations of zinc oxide nanoparticles synthesised from Azadirachta indica (Neem) leaf extract and their interaction with CalfThymus DNA. *Results in Physics, 13*. doi:https://doi.org/10.1016/j.rinp.2019.102168, PubMed: 102168

Singh, A., Singh, N. B., Afzal, S., Singh, T., & Hussain, I. (2018 January). Zinc oxide nanoparticles: A review of their biological synthesis, antimicrobial activity, uptake, translocation and biotransformation in plants. *Journal of Materials Science, 53*(1), 185–201. https://doi.org/10.1007/s10853-017-1544-1

Singh, A., Singh, N. B., Hussain, I., Singh, H., Yadav, V., & Singh, S. C. (2016 September 10). Green synthesis of Nano zinc oxide and evaluation of its impact on germination and metabolic activity of Solanum Lycopersicum. *Journal of Biotechnology, 233*, 84–94. https://doi.org/10.1016/j.jbiotec.2016.07.010

Singh, P., Kim, Y. J., Zhang, D., & Yang, D. C. (2016 July 1). Biological synthesis of nanoparticles from plants and microorganisms. *Trends in Biotechnology, 34*(7), 588–599. https://doi.org/10.1016/j.tibtech.2016.02.006

Soliman, A. S. (2015 February 28). El-feky SA, Darwish E. alleviation of salt stress on Moringa peregrina using foliar application of nanofertilizers. *Journal of Horticulture and Forestry, 7*(2), 36–47.

Sundrarajan, M., Ambika, S., & Bharathi, K. (2015 September 1). Plant-extract mediated synthesis of ZnO nanoparticles using Pongamia pinnata and their activity against pathogenic bacteria. *Advanced Powder Technology, 26*(5), 1294–1299. https://doi.org/10.1016/j.apt.2015.07.001

Suresh, J., Pradheesh, G., Alexramani, V., Sundrarajan, M., & Hong, S. I. (2018 February 2). Green synthesis and characterisation of zinc oxide nanoparticle using insulin plant (Costus pictus D. Don) and investigation of its antimicrobial as well as anticancer activities. *Advances in Natural Sciences: Nanoscience and Nanotechnology, 9*(1). PubMed: 015008.

Tarafdar, J. C., Raliya, R., Mahawar, H., & Rathore, I. (2014). Development of zinc nanofertilizer to enhance crop production in pearl millet (Pennisetum americanum). *Agricultural Research, 3*(3), 257–262.

Taran, M., Rad, M., & Alavi, M. (2018). Biosynthesis of TiO2 and ZnO nanoparticles by Halomonas elongata IBRC-M 10214 in different conditions of medium. *BioImpacts: BI, 8*(2), 81–89. https://doi.org/10.15171/bi.2018.10.

Taran, N., Storozhenko, V., Svietlova, N., Batsmanova, L., Shvartau, V., & Kovalenko, M. (2017 December). Effect of zinc and copper nanoparticles on drought resistance of wheat seedlings. *Nanoscale Research Letters, 12*(1), 60. https://doi.org/10.1186/s11671-017-1839-9

Taranath, T. C., & Patil, B. N. (2016 June 1). Limonia acidissima L. leaf mediated synthesis of zinc oxide nanoparticles: A potent tool against Mycobacterium tuberculosis. *International Journal of Mycobacteriology, 5*(2), 197–204. https://doi.org/10.1016/j.ijmyco.2016.03.004

Taziwa, R., Meyer, E., Katwire, D., & Ntozakhe, L. (2017 January 1). Influence of carbon modification on the morphological, structural, and optical properties of zinc oxide nanoparticles synthesised by pneumatic spray pyrolysis technique. *Journal of Nanomaterials, 2017*, 1–11. https://doi.org/10.1155/2017/9095301

Torabian, S., Zahedi, M., & Khoshgoftarmanesh, A. (2018). Effect of foliar spray of zinc oxide on some antiox idant enzymes activity of sunflower under salt stress.

Tripathi, D. K., Shweta, Singh, S., Singh, S., Pandey, R., . . . Chauhan, D. K. (2017 January 1). An overview on manufactured nanoparticles in plants: Uptake, translocation, accumulation and phytotoxicity. *Plant Physiology and Biochemistry*, 110, 2–12. doi:https://doi.org/10.1016/j.plaphy.2016.07.030.

Venkatachalam, P., Jayaraj, M., Manikandan, R., Geetha, N., Rene, E. R., Sharma, N. C., & Sahi, S. V. (2017 January 1). Zinc oxide nanoparticles (ZnONPs) alleviate heavy metal-induced toxicity in Leucaena leucocephala seedlings: A physiochemical analysis. *Plant Physiology and Biochemistry, 110*, 59–69. https://doi.org/10.1016/j.plaphy.2016.08.022

Wang, X.P., Li, Q.Q., Pei, Z.M. et al. (2018). Effects of zinc oxide nanoparticles on the growth, photosynthetic traits, and antioxidative enzymes in tomato plants. Biologia plantarum 62, 801–808. https://doi.org/10.1007/s10535-018-0813-4

Yue, J., Feliciano, T. J., Li, W., Lee, A., & Odom, T. W. (2017). Gold nanoparticle size and shape effects on cellular uptake and intracellular distribution of siRNA nanoconstructs. *Bioconjugate Chemistry, 28*(6), 1791–1800. https://doi.org/10.1021/acs.bioconjchem.7b00252

Yusefi-Tanha, E., Fallah, S., Rostamnejadi, A., & Pokhrel, L. R. (2020 January 1). Zinc oxide nanoparticles (ZnONPs) as nanofertilizer: Improvement on seed yield and antioxidant defense system in soil grown soybean (Glycine max cv. Kowsar). *bioRxiv*.

Yusof, H. M., Mohamad, R., & Zaidan, U. H. (2019 December). Microbial synthesis of zinc oxide nanoparticles and their potential application as an antimicrobial agent and a feed supplement in animal industry: A review. *Journal of Animal Science and Biotechnology, 10*(1), 1–22.

Yuvakkumar, R., Suresh, J., Nathanael, A. J., Sundrarajan, M., & Hong, S. I. (2014 August 1). Novel green synthetic strategy to prepare ZnO nanocrystals using rambutan (Nephelium lappaceum L.) peel extract and its antibacterial applications. *Materials Science and Engineering C, Materials for Biological Applications, 41*, 17–27. https://doi.org/10.1016/j.msec.2014.04.025

Chapter 8
Biochar: A Game Changer for Sustainable Agriculture

Muhammad Mahroz Hussain, Waqas Mohy-Ud-Din, Fazila Younas, Nabeel Khan Niazi, Irshad Bibi, Xing Yang, Fahad Rasheed, and Zia Ur Rahman Farooqi

Abstract The non-sustainability of soil management and agricultural practices that can damage the environment and enhance the risk of food insecurity are the emerging concerns globally. Sustainable agriculture endeavours to use a few conservation practices that can reduce the hostile effects of land-use intensification. Among the different conservation practices, micronutrient-enriched biochar (BC) is an emerging soil amendment that has proven as an accessible and critical input for sustainable agriculture. It could efficiently sequester a vast amount of carbon in the soil, thus increasing soil productivity, repairing soil erosion and minimising agriculture-related greenhouse gas emissions. Besides, it can also act as a reservoir of micro- and macronutrients. Research has also shown that using biochar can increase the crop productivity capacity of soil under various stresses and advance world food security. Over the last decade, many experiments have been carried out on biochar to see if it can be used to boost sustainable agriculture. This chapter discusses biochar's important properties, its relationship with soil microflora and its ability to boost crop productivity.

M. M. Hussain (✉) · W. Mohy-Ud-Din · F. Younas · I. Bibi · Z. U. R. Farooqi
Institute of Soil and Environmental Sciences, University of Agriculture Faisalabad, Faisalabad, Pakistan

N. K. Niazi
Institute of Soil and Environmental Sciences, University of Agriculture Faisalabad, Faisalabad, Pakistan

Southern Cross GeoScience, Southern Cross University, Lismore, NSW, Australia

X. Yang
University of Wuppertal, School of Architecture and Civil Engineering, Institute of Foundation Engineering, Water- and Waste-Management, Laboratory of Soil- and Groundwater-Management, Wuppertal, Germany

F. Rasheed
Department of Forestry & Range Management, Faculty of Agriculture, University of Agriculture Faisalabad, Faisalabad, Pakistan

S. A. Bandh (ed.), *Sustainable Agriculture*,
https://doi.org/10.1007/978-3-030-83066-3_8

Keywords Biochar · Pyrolysis · Direct and indirect · Nutrient · Impacts · Microbiota · Enrichment

8.1 Introduction

With the growing danger to environmental protection and global food security posed by non-sustainable soil conservation and agriculture activities, it has become important to develop a sustainable solution to viable agricultural practices (Amen et al., 2020). Sustainable agriculture is a growing field of interest because it entails long-term sustainable strategies for crop production that are environmentally beneficial, economically viable and socially adaptable. Sustainable agriculture (SA) is a method in which agro-farming can nourish itself by maintaining and preserving all its natural resources (Zhang et al., 2016) by generating renewable energy sources and involving solutions to safe farming practices (Yadav et al., 2020). Nowadays, biochar has become an emerging approach that has proven as a potentially vital input for sustainable agriculture to improve the fertility of the soil, increase sustainable production of agriculture and reduce the severe effects of abiotic and biotic stresses. Low nitrogen levels and increased mineralisation of soil organic matter (Renner, 2007) are the two critical constraints to sustainable agriculture. Applying organic matter such as compost and manure to the soil may enhance nutrient retention (Hussain, Bibi, et al., 2021), and biochar (BC) is more efficient than other forms of organic matter for retaining and replenishing the supply of readily available nutrients to the plants. Bacteria and fungi living inside the pores of biochar further provide additional help to plants to absorb more nutrients from the soil (Sohi et al., 2010).

Therefore, biochar is a more efficient and consistent source of nutrients than synthetic fertiliser, biowastes and farmyard manure (Jeffery et al., 2011). Application of BC amendment can increase agricultural soil sustainability in various ways, as it can serve as macro- and micronutrient retention and serve as a short-term source of more available plant nutrients and enhance long-term nutrient cycling processes (DeLuca et al., 2015). Biochar has established itself as a cost-effective, environmentally friendly and long-term sustainable alternative to other commercially available slow-release plant nutrients, such as nano- and coated fertilisers (Peiris et al., 2019).

Biochar has been studied both in the greenhouse and in the field to see how it affects crop yields (Khalid et al., 2020). Biochar made from various renewable wastes enhances soil fertility and increases crop production (Renner, 2007; Sohi et al., 2010). Crop yields are confirmed to improve dramatically following the application of biochar (Lehmann & Joseph, 2015). The various effects on crop yields seem to depend on different variables such as biochar quality, quantity, soil type and the crop studied. This chapter covers the essential aspects to use biochar for sustainable agriculture and meet UN's important sustainable development goals (SDGs: 2

(zero hunger), 12 (reasonable consumption and production) and 13 (climate action)) (Pedersen, 2018). Additionally, it discusses how biochar application can support two major facets of the soil environment: soil enzymes and nutrient dynamics.

8.2 Formulation, Properties and Biochemistry of Biochar

Produced by pyrolysis or thermal decomposition of biomass in a low-oxygen setting (Amen et al., 2020; Hu & Gholizadeh, 2019), biochar is a black, highly soluble, fine-grained, light-coloured substance that has a wide surface area and a high pH value. Various elements including carbon, hydrogen, sulphur, oxygen, phosphorus and other components, and minerals in the ash, make up the distinctive biochar. All these biochar properties and biochemistry provide benefit to soil (Khalid et al., 2020). Biochar is added to the ground to increase its condition to resolve significant agricultural soil degradation issues. It is stable biomass applied to the soil for making selective improvements to the soil's properties to enhance plant production and mitigate environmental contamination. The properties of BC are determined by the raw materials used (biomass) and the manufacturing parameters (Peiris et al., 2019).

8.2.1 Feedstock for the Production of Biochar

To make BC, a variety of organic materials, including hay, cow dung, wood chips, wheat straw, rice husk, cassava roots and other agricultural residues (Kambo & Dutta, 2015; Ronsse et al., 2013), are used as raw material. Both raw materials used and pyrolysis conditions affect the development of high-nutrient biochar (Chan et al., 2008; Younas et al., 2021). Agricultural wastes, urban wastes and municipal wastes are currently used on an industrial scale for biochar production (Xie et al., 2015). Cellulosic, hemicellulosic and lignin polymers make up most of the biomass used to produce biochar.

8.2.2 Pyrolysis Methods for Biochar Production

Biochar can be developed on a small scale with low-cost ovens or on a large scale with high-cost pyrolysis machines. It is generated through pyrolysis from a variety of biomass feedstocks, with the process producing oil and natural gas as by-products (Zhu et al., 2018). The method includes splitting the dry waste into tiny parts (no larger than 3 cm) followed by heating it to 350–700 °C in the absence of oxygen or less oxygen. At temperatures above 500 °C, rapid pyrolysis happens in seconds and maximises the bio-oil output. On the other hand, weak pyrolysis requires longer time periods from 30 minutes to many hours, to fully pyrolyse the raw materials (Peiris et al., 2019; Ronsse et al., 2013).

Low-temperature biochar (550 °C) has an amorphous carbon composition and is less aromatic than high-temperature biochar (Khalid et al., 2020). In both pyrolysis reactions, the elevated temperature reduces the yield of BC (Joseph et al., 2010). The pyrolysis method significantly impacts biochar efficiency and its future agricultural use in terms of agronomic production and carbon sequestration. BC yields from sluggish pyrolysis of biomass are estimated to be between 24% and 77% (Gangil & Wakudkar, 2013).

8.2.3 Biochar Properties

Biochar is a long-lasting source of carbon that can persist in the soil for thousands of years (Alhashimi & Aktas, 2017). It aims to apply something to the earth that sequesters carbon and increases soil quality. Biochar's physical properties aid in its use as an environmental protection instrument. It can increase soil pH and water retention, attract more beneficial fungi and other microorganisms and increase cation exchange capability when used as a soil improver (Ajema, 2018; Haider et al., 2020). Biochar also aids in the recycling of eroded soil. It adsorbs cations per carbon unit better than other organic soil substances since it has a greater negative surface charge and charge density (Liang et al., 2006), thus providing the potential of enhancing yield (Lehmann & Joseph, 2015).

Soil has its physical properties, which are determined by minerals and organic matter present, their relative content and the mineral-organic matter relationship (Shaaban et al., 2018). As BC is added to the soil mix, it directly impacts the physical properties, influencing depth, shape, composition, porosity and consistency by altering surface region, porosity and particle size distribution, mass and aggregation (Yu et al., 2019). Since the extent of penetration of the root zone and the accessibility of air and water are primarily dictated by the physical structure of the soil substrate, the influence of BC on the physical properties of the soil has a direct effect on plant development (de Jesus Duarte et al., 2019). This influences the soil's ability to maintain cations and react to changes in ambient temperature, as well as its ability to accumulate and function during soil preparation, kinetics and permeability during swelling and ability to retain cations and adapt to changes in ambient temperature (Sohi et al., 2010; Tomczyk et al., 2020).

Due to BC's ability to retain water, its addition to a drought-affected area will help mitigate the drought's impact on crop productivity. It has been demonstrated that it reduces soil restrictions that hinder plant growth and neutralises acidic soils (Atkinson et al., 2010). Carbon dioxide and oxygen may be biologically adsorbed on the surface or space in the pores of the BC (Hammes & Schmidt, 2009). It can stay in the soil for a longer time than other fertilisers and avoid being absorbed into water bodies, including rivers and reservoirs. Biochar decreases soil acidity and aids in storing nutrients and fertilisers in chemical properties (Lehmann et al., 2011;

Lehmann & Joseph, 2015). Biochar has been shown to improve soil microbe respiration by providing habitat for soil microorganisms (Terekhova et al., 2021; Tomczyk et al., 2020), resulting in increased soil biodiversity and density. Outside of the roots, biochar acts as a shelter for fungal hyphae. Biochar is also thought to have a long residence period on earth, ranging from 1000 to 10,000 years (Semenov et al., 2018; Sohi et al., 2010), thus helping the soil significantly longer.

8.3 The Role of Biochar in Sustainable Agriculture

8.3.1 Biochar and Nutrients Dynamics

Nutritional dynamics is usually influenced by the biological (such as the composition of plants, microorganisms and soil animals in the community) and non-biological (such as temperature, soil type and organic matter (OM)) factors. Biochar can change nutrient patterns by increasing nutrient bioavailability, altering the physical and chemical properties of the soil and influencing the soil environment (Khalid et al., 2020; Parker & Schimel, 2011). Nutrients may be continuously applied to the soil in synthetic fertilisers or other fertility enhancement technologies (Biederman & Harpole, 2013). However, since sterile soils have a low nutrient potential, retention of inorganic fertilisers is decreased. While these techniques are expensive, they have been used to solve this problem: utilising slow-release nutrient forms, multiple fertilising and covering plants with intact roots during fallow times (Haider et al., 2020; Laird et al., 2010; Parker & Schimel, 2011). Because of its solid nutritional content, which can be used explicitly or indirectly for plants, biochar is already recognised as a comparatively inexpensive and efficient substitute (Fig. 8.1). The direct contribution is providing nutrients to plants, whereas the indirect contribution is in the form of enhancing soil quality, thereby increasing fertiliser production (Xiao et al., 2018).

8.3.1.1 Direct and Indirect Nutrient Values of Biochar

Biochar has been used to enrich the soils' nutrient pool by including some essential macronutrients and micronutrients. It is also used as a slow-release fertiliser (Ding et al., 2016). Since a certain proportion of the elements in the raw material are unnecessarily lost by volatilisation due to hight temperature at which BC is prepared. N (VTB 200 °C) and S (VTB 375 °C) have the lowest volatilisation temperatures (VTs), while P and K have medium VTs (B 700–800 °C). At average biochar processing temperatures (VT 1000 °C), nutrients including calcium, magnesium and manganese are thermally stable (DeLuca et al., 2015). The volume of carbon lost increases as the pyrolysis temperature rises and the level of thermally stable nutrients rises. Sludge has more quantity of phosphorus that are in thermally stable

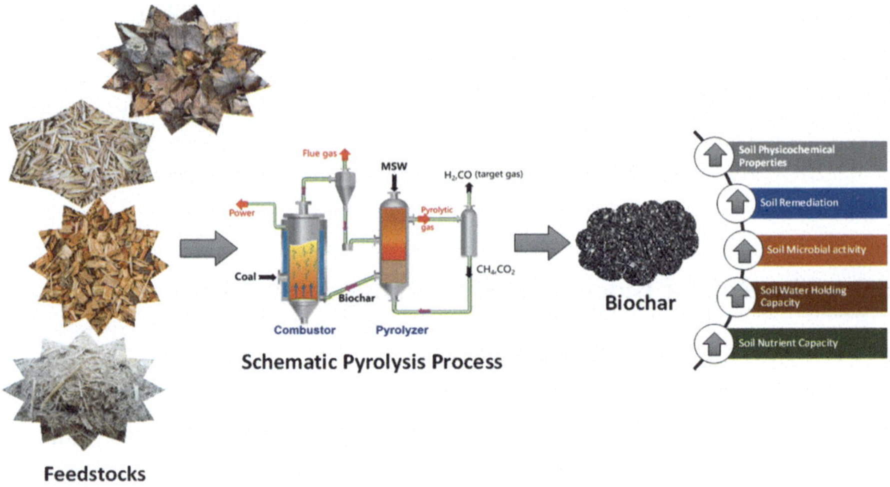

Fig. 8.1 A schematic diagram of biochar production, pyrolysis and impact on sustainable agriculture

form and develops a strong coorelation between available phosphorus temperatures (up to 800 °C) during pyrolysis (Peiris et al., 2019).

It has been documented that fresh BC made from nutrient-rich raw materials has a higher direct nutrient supply (de Jesus Duarte et al., 2019; DeLuca et al., 2015). According to various studies, freshly made BC has good bioavailability and a high potential to release N and P (Cayuela et al., 2014; Mukherjee & Zimmerman, 2013). Biochar manufactured from animal manure (such as waste sludge, liquid manure and broiler litter) has been shown to produce more P and N than vegetable BC (Irfan et al., 2017). Furthermore, the nitrogen content of pig manure biochar is slightly greater than that of sugarcane biochar at the same pyrolysis temperature (400 °C) (Ding et al., 2016). While using the same pyrolysis conditions, the same form of raw material can generate biochar with different nutrient contents. The volume and nature of stabilisers used during sludge treatment, for example, affect the bioavailability of phosphorus in sludge-based BC (Hossain et al., 2011). It has been stated that under the same circumstances, the total nitrogen content of two forms of biochar made from separate poultry litter differs significantly. However, according to various reports, the nutritional benefit of BC declines after a year of usage, rendering long-term crops challenging to produce, and the loss of nutrients is the primary downside of their direct use (Backer et al., 2018). Determining the bioavailability sequence of nutrients has been a difficult challenge in the long run.

Biochar can improve the soil's intrinsic consistency and wellbeing by indirectly influencing the physical and chemical properties such as total organic carbon, soil bulk density, pH and CEC (Liang et al., 2006). The negative surface charge of BC, generated both by OM and oxidised surface functional groups (O-SFG), is primarily responsible for CEC (Zhu et al., 2018; Zhu et al., 2017). The increased cation-nutrient retention ability of metal ions such as potassium, calcium, sodium and magnesium in the soil has also been documented due to biochar application.

Based on the built-in pyrolysis conditions, the functional groups present on the surface of BC change. There are several lactone and carboxyl groups on the surface of low-temperature BC, which are deprotonated in the soil, resulting in negatively charged anions that mix electrostatically with the cations (Ok et al., 2015). Novak et al. (2009) documented that in soils with enhanced BC, multivalent cations (such as Ca, Mg, Zn and Mn) retain more monovalent cations (such as Na and K). The retention of ammonium is primarily determined by the soil CEC, while the retention of nitrate is determined by the anion exchange capability (AEC) and the pore filling process. When utilising BC, higher retention of ammonium is always found compared to nitrite since the CEC of BC is higher than the AEC (Novak et al., 2009; Ok et al., 2015).

On the other hand, biochar is said to be more durable than fertilisers because of its carbon structure. Fidel et al. (2017) found that improving BC quality and overall organic carbon in the soil was greatly enriched instead of fertiliser application. It is worth noting that high-temperature biochar (800 °C) is less soluble in the soil than low-temperature BC, which is usually disordered and difficult to decompose.

Biochar enriches the unstable nutrient reservoir by altering the pH of the soil, improving nutrient bioavailability and encouraging microbial development to keep water and nutrients available in the soil. Via the lime effect, biochar can increase nutrient bioavailability and trap trivalent species after being applied to the soil. Weathered soils high in iron and aluminium appear to be acidic when hydroxide ions are emitted through the production of hydroxides. In certain soils, insoluble iron and aluminium phosphate limit phosphorus bioavailability (Yadav et al., 2020; Yuan et al., 2015). A lime impact is produced when the calcium oxide in the biochar combines with the phosphorus in the soil to form calcium phosphate. Because of its solubility, inorganic phosphorus is bioavailable to plants. Biochar's exchangeable sites have a strong preference for trivalent aluminium and iron, limiting their capacity to shape insoluble P complexes. Furthermore, the usage of biochar can reduce the presence of harmful elements such as Al, Cu and Mn in the soil, which can be detrimental to plants when the pH is too acidic (Hameed et al., 2021; Rawat et al., 2019).

8.3.1.2 Biochar as a Soil Amendment

The adoption of new technology or agricultural systems is driven by food stability, diminishing soil productivity, climate change and viability. Soil remediation aims to reduce the possibility of contaminants spreading to surrounding water bodies or recipient species. Organic materials (such as biochar) are a widespread alternative for this reason since their source is biological and they can be added directly to the soil without being pre-treated (Atkinson et al., 2010; Beesley et al., 2011). Biochar modifiers are preferable to other organic materials due to their strong decay tolerance, which allows them to remain in the soil for a longer time and offer long-term advantages to the soil. Biochar increases soil pH, water preservation, cation exchange and microbial flora, both of which enhance the soil quality (Mensah & Frimpong, 2018).

The usage of alkaline cations and elevated phosphorus and total nitrogen concentrations have been observed when biochar is added to the soil (Tan et al., 2015). Biochar's alkaline pH and mineral content (ash, which contains nitrogen, phosphorus, potassium and trace elements) can have important agronomic benefits for a wide variety of soils, at least in the short to medium term. As biochar is applied at a higher pH, the soil usually becomes less acidic (Yuan et al., 2015; Yuan et al., 2011). As applied in soils with a poor pH, acidic biochar can also raise the pH of the soil. Similar to other properties, the pH of biochar is influenced by the raw material, the processing temperature and time. Another advantageous feature of biochar is that it reduces greenhouse gas levels in the soil.

Spokas et al. (2009) found that adding different BC in concentrations fluctuating from 2% to 60% (w/w) decreased CO_2 and N_2O emission to a degree of more than 20% (w/w) and that all levels of environmental methane became oxidised on unenhanced land. Biochar is protective against both airborne pathogens and soil-borne pathogens (Bonanomi et al., 2015). Despite the lack of publicly available data about the effect of BC on soil-borne pathogens, Elmer et al. (2010) showed that such pathogens are controlled.

Biochar application to soil has also been seen to increase the soil's disease-fighting ability, as the presence of calcium compounds and the enhancement of the soil's physical, chemical and biological properties is believed to be the cause for disease inhibition. Biochar use in the soil may also have an effect on a variety of soil-limiting factors, including the high use of aluminium (Van Zwieten et al., 2010), soil composition and nutrient consumption, bioavailability of organic compounds and inorganic contaminants, cation exchange capacity (CEC) and nutrient retention (Haider et al., 2020; Luo et al., 2020). Biochar may also uptake toxins, nutrients and minerals in the soil, blocking these contaminants from migrating to the surface or groundwater and preventing these waters from rotting due to farming practices. BC use in the soil has been found to have a beneficial, neutral or even detrimental effect on crop yield. Therefore, it is very critical to consider the mechanism of action in the soil before its application (Peiris et al., 2019), e.g. an experiment carried out in the USA found that the application rates of peanut shell and pine BC at 11 and 22 tonnes per hectare, respectively, lowered the maize yield than the yield on the areas cultivated with regular fertilisers (Fidel et al., 2017; Gaskin et al., 2010).

8.3.2 Biochar's Impact on Soil Microbiota and Plant Growth

Numerous experiments have shown that BC can activate soil microorganisms, resulting in enhanced carbon build-up in the soil. Biochar may serve as a home for bacteria, actinomycetes and fungi in addition to absorbing organic compounds, nutrients and gases (Jeffery et al., 2011; Joseph & Lehmann, 2009). Biochar application also improves the water retention ability of soil (Busscher et al., 2010), thus altering the soil's microbial population. Apart from the fact that the interaction between BC and soil microorganisms is complex, biochar provides an ideal habitat for a diverse range of soil microorganisms. Numerous experiments have shown that combining BC with phosphate-dissolving fungi will significantly increase the growth and yield of beans and soybeans (Saxena et al., 2017).

It promotes mycorrhizal development in clover bioassay plants by favouring plant root colonisation (Solaiman et al., 2010). Carrots and legumes grown on steep slopes or in soils with a pH less than 5.2 grew significantly healthier when treated with biochar. Biochar has also improved the biological nitrogen fixation of kidney beans, owing to the increased usage of micronutrients after biochar application (Renner, 2007; Rondon et al., 2007; Rondon et al., 2004). Biochar prevents leaching by promoting NH_4^+ in the topsoil that plants can absorb (Lehmann & Joseph, 2015).

It is obvious that when biochar and mycorrhizal fungi are used in accordance with management standards, synergistic effects can improve soil quality. Predators such as collembola, mites, larger protozoa and nematodes (diameter > 16 m) can assist fungal hyphae and bacteria settling on biochar particles (or other porous substances) (Ezawa et al., 2002). Biochar has the ability to increase the value of unharvested agricultural products and to promote plant growth (Major et al., 2005; Gao, Viry, Maugey, Poulin, & Mano, 2010).

The direct benefit of adding biochar for nutrient supply is attributed to higher potassium, phosphorus, zinc consumption and lower calcium and copper content. Few experiments have investigated the possibility of transformed soil biochar influencing pathogen tolerance in plants. Yao et al. (2017) demonstrated that the correction of charcoal has an inhibitory impact on pathogens when it comes to soil pathogens. This primarily concerns the effects of AM fungal inoculation on the susceptibility of asparagus to *Fusarium* (a soil-borne root rot pathogen). Another research that backed up these findings found that applying ground hardwood biochar to the soil of asparagus fields decreased the amount of *Fusarium oxysporum*, *Fusarium asparagus* and spores relative to uncorrected controls that cause root injury (Elmer et al., 2010). Biochar eliminates the need for fertilisers and, as a result, fertiliser-related pollution. By converting agricultural waste to biochar, the amount of CH_4 (methane) generated during natural decomposition can also be decreased.

8.3.3 The Effect of Biochar on Soil Enzymes

To reliably quantify soil fertility, fluctuations in biological properties must be taken into account (Sherene, 2017). Enzyme activity has an effect on the bioavailability of nutrients in the rhizosphere, which has an effect on the health and growth of plants (Abubakar & Attanda, 2013; Ajema, 2018). Soil enzymes have a strong correlation with the changes in the soil environment, and they respond quickly, thus making them a valuable tool for determining soil content. These enzymes are extracted from microorganisms (MO) found in soil, animal and plant roots and are typically stabilised in the soil matrix through complex formation with organic matter (OM), humus colloids or clay particles (Sherene, 2017). Soil enzymes catalyse a variety of biochemical processes, including OM breakdown, mineralisation and nutrient conversion (Burns et al., 2013). Biochar can alter the physical and chemical properties of the soil's structure and the microbial environment, affecting soil fertility. The growing interest in BC as a practical tool for handling soil biota is shown by the ongoing expansion of the spectrum of soil improvements dependent on it (Ding et al., 2016).

8.3.4 The Effects of Biochar on Microorganism Extracted Soil Enzymes

There have been records of MO populations changing as a result of the implementation of BC (Anderson et al., 2011; Atkinson et al., 2010). Biochar has the ability to alter the action of a variety of enzymes (Beheshti et al., 2017). Due to the added complexity of extracellular enzyme operations, biochar application changes the physical and chemical properties of the soil, resulting in major enhancements in the composition and function of MO. Biochar contains micropores, mesopores and macropores of varying sizes that function as microorganism habitats. Saprophytic fungi may colonise these pores and cause biochar to degrade. This allows the plants to grow by making nutrients accessible to them (Thies & Rillig, 2009).

The adverse effects of biochar modification on the function of microbial enzymes have also been documented in the literature, e.g. *Thiobacillus acidophilus*' capacity to oxidise sulphur is one function that has been reduced because of this activity. Furthermore, the adsorption of organic and inorganic substrates on biochar may prevent enzymes from working correctly. Influence on MO activity and its status as a shelter for MO colonisation: MO colonisation has a subsidiary impact on soil enzyme kinetics, and the carbon and nutrient-rich properties of char aid MO colonisation (Lehmann & Joseph, 2015; Terekhova et al., 2021).

8.4 Conclusions and Future Outlook

Due to loss in productivity of agricultural land because of the population pressures, sustainable crop production practices have been increasingly important. Biochar is suggested for enhancing soil productivity by decreasing acidity and rising nutrient use. As a consequence, adding biochar to the soil can be one of the most efficient methods for mitigating biological stress and enhancing crop production. The beneficial influence of biochar on soil, plants and water contributes to improved photosynthesis and increased nitrogen and water usage quality. Biochar has been shown to improve soil properties, microbial abundance, biological nitrogen fixation and plant growth. Biochar improves plant productivity by supplying nutrients directly or altering the physical and chemical properties of the soil.

Biochar becomes an essential source of slow-release nutrients when added to the soil. In the long run, though, it does not add to soil productivity. Biochar may be used as a safe carbon supply in the soil, raising the CEC and saving several traces and macronutrients in the process. Biochar increases the overall function of soil enzymes originating from MO, plants and livestock, resulting in improved OM decomposition and nutrient cycle. Biochar has ignited scientific attention worldwide due to its versatility, cost-effectiveness and physical and chemical properties. To promote global acceptance of this valuable technology, we also advocate reinforcing partnerships between researchers and biochar production facilities.

References

Abubakar, M. S., & Attanda, M. L. (2013). The concept of sustainable agriculture: Challenges and prospects. *IOP Conference Series: Materials Science and Engineering*. IOP Publishing, 53, 012001. https://doi.org/10.1088/1757-899X/53/1/012001

Ajema, L. (2018). Effects of biochar application on beneficial soil organism. International journal of research studies in science. *Engineering Technology, 5*, 9–18.

Alhashimi, H. A., & Aktas, C. B. (2017). Life cycle environmental and economic performance of biochar compared with activated carbon: A meta-analysis. *Resources, Conservation and Recycling, 118*, 13–26. https://doi.org/10.1016/j.resconrec.2016.11.016

Amen, R., Bashir, H., Bibi, I., Shaheen, S. M., Niazi, N. K., … Rinklebe, J. (2020). A critical review on arsenic removal from water using biochar-based sorbents: The significance of modification and redox reactions. *Chemical Engineering Journal*, 396. https://doi.org/10.1016/j.cej.2020.125195, PubMed: 125195.

Anderson, C. R., Condron, L. M., Clough, T. J., Fiers, M., Stewart, A., Hill, R. A., & Sherlock, R. R. (2011). Biochar induced soil microbial community change: Implications for biogeochemical cycling of carbon, nitrogen and phosphorus. *Pedobiologia, 54*(5–6), 309–320. https://doi.org/10.1016/j.pedobi.2011.07.005

Atkinson, C. J., Fitzgerald, J. D., & Hipps, N. A. (2010). Potential mechanisms for achieving agricultural benefits from biochar application to temperate soils: A review. *Plant and Soil, 337*(1–2), 1–18. https://doi.org/10.1007/s11104-010-0464-5

Backer, R., Rokem, J. S., Ilangumaran, G., Lamont, J., Praslickova, D., Ricci, E., … Smith, D. L. (2018). Plant growth-promoting rhizobacteria: Context, mechanisms of action, and road map to commercialisation of biostimulants for sustainable agriculture. *Frontiers in Plant Science, 9*, 1473. https://doi.org/10.3389/fpls.2018.01473

Beesley, L., Moreno-Jiménez, E., Gomez-Eyles, J. L., Harris, E., Robinson, B., & Sizmur, T. (2011). A review of biochars' potential role in the remediation, revegetation and restoration of contaminated soils. *Environmental Pollution, 159*(12), 3269–3282. https://doi.org/10.1016/j.envpol.2011.07.023

Beheshti, M., Etesami, H., & Alikhani, H. A. (2017). Interaction study of biochar with phosphate-solubilising bacterium on phosphorus availability in calcareous soil. *Archives of Agronomy and Soil Science, 63*(11), 1572–1581. https://doi.org/10.1080/03650340.2017.1295138

Biederman, L. A., & Harpole, W. S. (2013). Biochar and its effects on plant productivity and nutrient cycling: A meta-analysis. *GCB Bioenergy, 5*(2), 202–214. https://doi.org/10.1111/gcbb.12037

Bonanomi, G., Ippolito, F., & Scala, F. (2015). A" black" future for plant pathology? Biochar as a new soil amendment for controlling plant diseases. *Journal of Plant Pathology, 97*, 223–234.

Burns, R. G., DeForest, J. L., Marxsen, J., Sinsabaugh, R. L., Stromberger, M. E., Wallenstein, M. D., … Zoppini, A. (2013). Soil enzymes in a changing environment: Current knowledge and future directions. *Soil Biology and Biochemistry, 58*, 216–234. https://doi.org/10.1016/j.soilbio.2012.11.009

Busscher, W. J., Novak, J. M., Evans, D. E., Watts, D. W., Niandou, M. A. S., & Ahmedna, M. (2010). Influence of pecan biochar on physical properties of a Norfolk loamy sand. *Soil Science, 175*(1), 10–14. https://doi.org/10.1097/SS.0b013e3181cb7f46

Cayuela, M. L., Van Zwieten, L., Singh, B. P., Jeffery, S., Roig, A., & Sánchez-Monedero, M. A. (2014). Biochar's role in mitigating soil nitrous oxide emissions: A review and meta-analysis. *Agriculture, Ecosystems and Environment, 191*, 5–16. https://doi.org/10.1016/j.agee.2013.10.009

de Jesus Duarte, S., Glaser, B., & Pellegrino Cerri, C. E. (2019). Effect of biochar particle size on physical, hydrological and chemical properties of loamy and sandy tropical soils. *Agronomy, 9*(4), 165. https://doi.org/10.3390/agronomy9040165

DeLuca, T. H., Gundale, M. J., MacKenzie, M. D., & Jones, D. L. (2015). Biochar effects on soil nutrient transformations. *Biochar for Environmental Management: Science, Technology and Implementation, 2*, 421–454.

Ding, Y., Liu, Y., Liu, S., Li, Z., Tan, X., Huang, X., … Zheng, B. (2016). Biochar to improve soil fertility. A review. *Agronomy for Sustainable Development, 36*, 1–18.

Elmer, W., White, J., & Pignatello, J. (2010). *Impact of biochar addition to soil on the bioavailability of chemicals important in agriculture [Report]*. University of Connecticut.

Ezawa, T., Yamamoto, K., & Yoshida, S. (2002). Enhancement of the effectiveness of indigenous arbuscular mycorrhizal fungi by inorganic soil amendments. *Soil Science and Plant Nutrition, 48*(6), 897–900. https://doi.org/10.1080/00380768.2002.10408718

Fidel, R. B., Laird, D. A., Thompson, M. L., & Lawrinenko, M. (2017). Characterisation and quantification of biochar alkalinity. *Chemosphere, 167*, 367–373. https://doi.org/10.1016/j.chemosphere.2016.09.151

Gangil, S., & Wakudkar, H. M. (2013). Generation of bio-char from crop residues. *International Journal of Emerging Technology and Advanced Engineering, 3*, 566–570.

Gaskin, J. W., Speir, R. A., Harris, K., Das, K. C., Lee, R. D., Morris, L. A., & Fisher, D. S. (2010). Effect of peanut hull and pine chip biochar on soil nutrients, corn nutrient status, and yield. *Agronomy Journal, 102*(2), 623–633. https://doi.org/10.2134/agronj2009.0083

Haider, F. U., Coulter, J. A., Liqun, C., Hussain, S., Cheema, S. A., Wu, J., & Zhang, R. (2020). *An overview on biochar production, its implications, and mechanisms of biocharinduced amelioration of soil and plant characteristics*. Pedosphere.

Hameed, M. A., Farooqi, Z. U. R., Hussain, M. M., & Ayub, M. A. (2021). PGPR-assisted bioremediation and plant growth: A sustainable approach for crop production using polluted soils. *Plant growth regulators: Signalling under stress conditions*, 403.

Hammes, K., & Schmidt, M. W. (2009). Changes of biochar in soil. Biochar for environmental management. *Science and Technology, 1*, 169–181.

Hossain, M. K., Strezov, V., Chan, K. Y., Ziolkowski, A., & Nelson, P. F. (2011). Influence of pyrolysis temperature on production and nutrient properties of wastewater sludge biochar. *Journal of Environmental Management, 92*(1), 223–228. https://doi.org/10.1016/j.jenvman.2010.09.008

Hu, X., & Gholizadeh, M. (2019). Biomass pyrolysis: A review of the process development and challenges from initial researches up to the commercialisation stage. *Journal of Energy Chemistry, 39*, 109–143. https://doi.org/10.1016/j.jechem.2019.01.024

Hussain, M. M., Bibi, I., Niazi, N. K., Shahid, M., Iqbal, J., Shakoor, M. B., … Zhang, H. (2021). Arsenic biogeochemical cycling in paddy soil-rice system: Interaction with various factors, amendments and mineral nutrients. *Science of the Total Environment, 773*, 145040. https://doi.org/10.1016/j.scitotenv.2021.145040

Irfan, M., Rafiullah, K., Naz, F., Rizwan, M., & Mehmood, I. (2017). Potential value of biochar as a soil amendment: A review. *Pure and Applied Biology, 6*(4), 1494–1502. https://doi.org/10.19045/bspab.2017.600161

Jeffery, S., Verheijen, F. G. A., van der Velde, M., & Bastos, A. C. (2011). A quantitative review of the effects of biochar application to soils on crop productivity using meta-analysis. *Agriculture, Ecosystems and Environment, 144*(1), 175–187. https://doi.org/10.1016/j.agee.2011.08.015

Joseph, S., & Lehmann, J. (2009). *Biochar for environmental management: Science and technology*. EarthScan.

Joseph, S. D., Camps-Arbestain, M., Lin, Y., Munroe, P., Chia, C. H., & Hook, J. (2010). An investigation into the reactions of biochar in soil. *Soil Research, 48*(7), 501–515. https://doi.org/10.1071/SR10009

Kambo, H. S., & Dutta, A. (2015). A comparative review of biochar and hydrochar in terms of production, physico-chemical properties and applications. *Renewable and Sustainable Energy Reviews, 45*, 359–378. https://doi.org/10.1016/j.rser.2015.01.050

Khalid, S., Shahid, M., Murtaza, B., Bibi, I., Natasha, A. N., & M., & Niazi, N. K. (2020). A critical review of different factors governing the fate of pesticides in soil under biochar application. *Science of the Total Environment, 711*. https://doi.org/10.1016/j.scitotenv.2019.134645, PubMed: 134645.

Laird, D., Fleming, P., Wang, B., Horton, R., & Karlen, D. (2010). Biochar impact on nutrient leaching from a Midwestern agricultural soil. *Geoderma, 158*(3–4), 436–442. https://doi.org/10.1016/j.geoderma.2010.05.012

Lehmann, J., & Joseph, S. (2015). *Biochar for environmental management: Science, technology and implementation*. Routledge.

Lehmann, J., Rillig, M. C., Thies, J., Masiello, C. A., Hockaday, W. C., & Crowley, D. (2011). Biochar effects on soil biota–a review. *Soil Biology and Biochemistry, 43*(9), 1812–1836. https://doi.org/10.1016/j.soilbio.2011.04.022

Liang, B., Lehmann, J., Solomon, D., Kinyangi, J., Grossman, J., & O'Neill, B. (2006). Black carbon increases cation exchange capacity in soils. *Soil Science Society of America Journal, 70*(5), 1719–1730. https://doi.org/10.2136/sssaj2005.0383

Luo, Z., Luo, Y., Wang, G., Xia, J., & Peng, C. (2020). Warming-induced global soil carbon loss attenuated by downward carbon movement. *Global Change Biology, 26*(12), 7242–7254. https://doi.org/10.1111/gcb.15370

Major, J., Steiner, C., Ditommaso, A., Falcão, N. P. S., & Lehmann, J. (2005). Weed composition and cover after three years of soil fertility management in the central Brazilian Amazon: Compost, fertiliser, manure and charcoal applications. *Weed Biology and Management, 5*(2), 69–76. https://doi.org/10.1111/j.1445-6664.2005.00159.x

Mensah, A. K., & Frimpong, K. A. (2018). Biochar and/or compost applications improve soil properties, growth, and yield of maise grown in acidic rainforest and coastal savannah soils in Ghana. *International Journal of Agronomy, 2018*, 1–8. https://doi.org/10.1155/2018/6837404

Mukherjee, A., & Zimmerman, A. R. (2013). Organic carbon and nutrient release from a range of laboratory-produced biochars and biochar–soil mixtures. *Geoderma, 193–194*, 122–130. https://doi.org/10.1016/j.geoderma.2012.10.002

Novak, J. M., Lima, I., Xing, B., Gaskin, J. W., Steiner, C., Das, K., . . . Busscher, W. J. (2009). Characterisation of designer biochar produced at different temperatures and their effects on a loamy sand. *Annals of Environmental Science*.

Ok, Y. S., Chang, S. X., Gao, B., & Chung, H.-J. (2015). SMART biochar technology—A shifting paradigm towards advanced materials and healthcare research. *Environmental Technology and Innovation, 4*, 206–209. https://doi.org/10.1016/j.eti.2015.08.003

Parker, S. S., & Schimel, J. P. (2011). Soil nitrogen availability and transformations differ between the summer and the growing season in a California grassland. *Applied Soil Ecology, 48*(2), 185–192. https://doi.org/10.1016/j.apsoil.2011.03.007

Pedersen, C. S. (2018). The UN sustainable development goals (SDGs) are a great gift to business! *Procedia CIRP, 69*, 21–24. https://doi.org/10.1016/j.procir.2018.01.003

Peiris, C., Gunatilake, S. R., Wewalwela, J. J., & Vithanage, M. (2019). *Biochar for sustainable agriculture: Nutrient dynamics, soil enzymes, and crop growth. Biochar from biomass and waste* (pp. 211–224). Elsevier.

Rawat, J., Saxena, J., & Sanwal, P. (2019). Biochar: A sustainable approach for improving plant growth and soil properties. Biochar-an imperative amendment for soil and the environment. IntechOpen.

Renner, R. (2007). *Rethinking biochar*. ACS Publications.

Rondon, M., Ramirez, A., & Hurtado, M. (2004). *Charcoal additions to high fertility ditches enhance yields and quality of cash crops in Andean hillsides of Colombia*. CIAT Annual Report Cali.

Rondon, M. A., Lehmann, J., Ramírez, J., & Hurtado, M. (2007). Biological nitrogen fixation by common beans (Phaseolus vulgaris L.) increases with bio-char additions. *Biology and Fertility of Soils, 43*(6), 699–708. https://doi.org/10.1007/s00374-006-0152-z

Ronsse, F., Van Hecke, S., Dickinson, D., & Prins, W. (2013). Production and characterisation of slow pyrolysis biochar: Influence of feedstock type and pyrolysis conditions. *GCB Bioenergy, 5*(2), 104–115. https://doi.org/10.1111/gcbb.12018

Saxena, J., Rawat, J., & Kumar, R. (2017). Conversion of biomass waste into biochar and the effect on mung bean crop production. *CLEAN - Soil, Air, Water, 45*(7). https://doi.org/10.1002/clen.201501020, PubMed: 1501020.

Semenov, V. M., Kogut, B. M., Zinyakova, N. B., Masyutenko, N. P., Malyukova, L. S., Lebedeva, T. N., & Tulina, A. S. (2018). Biologically active organic matter in soils of European Russia. *Eurasian Soil Science, 51*(4), 434–447. https://doi.org/10.1134/S1064229318040117

Shaaban, M., Van Zwieten, L., Bashir, S., Younas, A., Núñez-Delgado, A., & Chhajro, M. A. (2018). A concise review of biochar application to agricultural soils to improve soil conditions and fight pollution. *Journal of Environmental Management, 228*, 429–440. https://doi.org/10.1016/j.jenvman.2018.09.006

Sherene, T. (2017). Role of soil enzymes in nutrient transformation: A review. *Biology Bulletin, 3*, 109–131.

Sohi, S. P., Krull, E., Lopez-Capel, E., & Bol, R. (2010). A review of biochar and its use and function in soil. *Advances in Agronomy, 105*, 47–82. https://doi.org/10.1016/S0065-2113(10)05002-9

Solaiman, Z. M., Blackwell, P., Abbott, L. K., & Storer, P. (2010). Direct and residual effect of biochar application on mycorrhizal root colonisation, growth and nutrition of wheat. *Soil Research, 48*(7), 546–554. https://doi.org/10.1071/SR10002

Spokas, K. A., Koskinen, W. C., Baker, J. M., & Reicosky, D. C. (2009). Impacts of woodchip biochar additions on greenhouse gas production and sorption/degradation of two herbicides in a Minnesota soil. *Chemosphere, 77*(4), 574–581. https://doi.org/10.1016/j.chemosphere.2009.06.053

Tan, X., Liu, Y., Zeng, G., Wang, X., Hu, X., Gu, Y., & Yang, Z. (2015). Application of biochar for the removal of pollutants from aqueous solutions. *Chemosphere, 125*, 70–85. https://doi.org/10.1016/j.chemosphere.2014.12.058

Terekhova, V. A., Prudnikova, E. V., Kulachkova, S. A., Gorlenko, M. V., Uchanov, P. V., Sushko, S. V., & Ananyeva, N. D. (2021). Microbiological indicators of heavy metals and carbon-containing preparations applied to Agrosoddy-podzolic soils differing in humus content. *Eurasian Soil Science, 54*(3), 448–458. https://doi.org/10.1134/S1064229321030157

Thies, J. E., & Rillig, M. C. (2009). Characteristics of biochar: Biological properties. Biochar for environmental management. *Science and Technology, 1*, 85–105.

Tomczyk, A., Sokołowska, Z., & Boguta, P. (2020). Biochar physicochemical properties: Pyrolysis temperature and feedstock kind effects. *Reviews in Environmental Science and Bio/Technology, 19*(1), 191–215. https://doi.org/10.1007/s11157-020-09523-3

Van Zwieten, L., Kimber, S., Downie, A., Morris, S., Petty, S., Rust, J., & Chan, K. Y. (2010). A glasshouse study on the interaction of low mineral ash biochar with nitrogen in a sandy soil. *Soil Research, 48*(7), 569–576. https://doi.org/10.1071/SR10003

Xiao, X., Chen, B., Chen, Z., Zhu, L., & Schnoor, J. L. (2018). Insight into multiple and multi-level structures of biochars and their potential environmental applications: A critical review. *Environmental Science and Technology, 52*(9), 5027–5047. https://doi.org/10.1021/acs.est.7b06487

Xie, T., Reddy, K. R., Wang, C., Yargicoglu, E., & Spokas, K. (2015). Characteristics and applications of biochar for environmental remediation: A review. *Critical Reviews in Environmental Science and Technology, 45*(9), 939–969. https://doi.org/10.1080/10643389.2014.924180

Yadav, A. N., Rastegari, A. A., Yadav, N., & Kour, D. (2020). *Advances in plant microbiome and sustainable agriculture*. Springer.

Yao, Q., Liu, J., Yu, Z., Li, Y., Jin, J., Liu, X., & Wang, G. (2017). Three years of biochar amendment alters soil physiochemical properties and fungal community composition in a black soil of Northeast China. *Soil Biology and Biochemistry, 110*, 56–67. https://doi.org/10.1016/j.soilbio.2017.03.005

Younas, F., Mustafa, A., Farooqi, Z. U. R., Wang, X., Younas, S., Mohy-Ud-Din, W., … Hussain, M. M. (2021). Current and emerging adsorbent technologies for wastewater treatment: Trends, limitations, and environmental implications. *Water, 13*(2), 215. https://doi.org/10.3390/w13020215

Yu, H., Zou, W., Chen, J., Chen, H., Yu, Z., Huang, J., … Gao, B. (2019). Biochar amendment improves crop production in problem soils: A review. *Journal of Environmental Management, 232*, 8–21. https://doi.org/10.1016/j.jenvman.2018.10.117

Yuan, H., Lu, T., Huang, H., Zhao, D., Kobayashi, N., & Chen, Y. (2015). Influence of pyrolysis temperature on physical and chemical properties of biochar made from sewage sludge. *Journal of Analytical and Applied Pyrolysis, 112*, 284–289. https://doi.org/10.1016/j.jaap.2015.01.010

Yuan, J. H., Xu, R. K., & Zhang, H. (2011). The forms of alkalis in the biochar produced from crop residues at different temperatures. *Bioresource Technology, 102*(3), 3488–3497. https://doi.org/10.1016/j.biortech.2010.11.018

Zhang, D., Yan, M., Niu, Y., Liu, X., van Zwieten, L., & Chen, D. (2016). Is current biochar research addressing global soil constraints for sustainable agriculture? *Agriculture, Ecosystems and Environment, 226*, 25–32. https://doi.org/10.1016/j.agee.2016.04.010

Zhu, L. X., Xiao, Q., Shen, Y. F., & Li, S. Q. (2017). Effects of biochar and maise straw on the short-term carbon and nitrogen dynamics in a cultivated silty loam in China. *Environmental Science and Pollution Research International, 24*(1), 1019–1029. https://doi.org/10.1007/s11356-016-7829-0

Zhu, L., Lei, H., Zhang, Y., Zhang, X., Bu, Q., & Wei, Y. (2018). A review of biochar derived from pyrolysis and its application in biofuel production. *SF Journal of Materials Chemistry Enineering, 1*(1), 1007.

Chapter 9
Production of Biochar Using Top-Lit Updraft and Its Application in Horticulture

Chandan Singh, Priya Pathak, Neelam Chaudhary, and Deepak Vyas

Abstract Biochar is a charcoal, rich in carbon, produced by typically burning organic residues of plants and animals to more than 250 °C in a low-oxygen environment. It could be efficiently produced by the various methods, but the top-lit updraft (TLUD) method is the most affordable at each farm level in agriculture. Several controlling factors determine the distinctive quality of biochar; however, the agricultural application of biochar is precisely beneficial if applied appropriately. It increases the water retention capability of the soil and cation exchange rates and holds the nutrient-holding capacity and reclamation of acidic soils. Moreover, biochar could also endure an efficient way to sequestrate carbon and a valuable agent for sustainable agriculture.

Keywords Biochar · Horticulture · Sustainable agriculture · Carbon · Charcoal

9.1 Introduction

Biochar is a carbon-charcoal product obtained by combusting biomasses like wood, manure, leaves, or animal debris in a closed container with little or no available air, or biochar is the substance obtained by thermoconversion of organic substrates in the limited presence of oxygen and at a temperature range of 250 °C–700 °C (Nartey & Zhao, 2014; Pathak et al., 2020). The production of biochar and its application in agriculture is thought to be a suitable method of mitigating climate change while fertilizing soil (Nartey & Zhao, 2014). Conversion of biomass to biochar is reported to be producing sustainable renewable energy and is found to reduce carbon dioxide content in the atmosphere (Carpenter & Nair, 2012; Kavitha et al., 2018; Lehmann, 2007). Historically, biochar has been traced in the Amazonian River basin (with highly dark fertile soil) that indicates the uses of biochar in ancient agricultural

C. Singh (✉) · P. Pathak · N. Chaudhary · D. Vyas
Lab of Microbial Technology and Plant Pathology, Department of Botany, Dr Harisingh Gour Central University, Sagar, Madhya Pradesh, India

S. A. Bandh (ed.), *Sustainable Agriculture*,
https://doi.org/10.1007/978-3-030-83066-3_9

practices (Leverett, 2008). These dark, fertile soils were called terra preta (USBI, n.d.). The older people used to set a pile of organic material on fire directly and cover the pile with clays before burning to delimit oxygen but hold the heat to bake the piled-up organic materials. Charcoal is also formed naturally due to forest fire and intentionally by humans in burn pits or handmade structures. When this charcoal is added to the soil as a soil amendment, it is termed "biochar." Though biochar production using the traditional methods is beneficial, it has demerit too, because it is not an environmentally friendly practice as it releases a huge smoke and dust particulates in the atmosphere (USBI, n.d.). The modern technology of biochar production is based on organic materials and the nature of the application (Nartey & Zhao, 2014). Usage of biochar in the soil as a fertilizing agent is beneficial for plant health. It improves the physical and chemical properties of the soil by preventing the leaching of nutrient from the soil (Jien, 2018; Jyoti Rawat & Sanwal, 2019; Sánchez-Monedero et al., 2019). Production methods, chemical properties, physical properties, and combined application of biochar in horticulture crops have been discussed here, with an emphasis on the production of biochar using the TLUD method at the farm level.

9.2 Methods of Biochar Production

It has already been mentioned that biochar is typically obtained from various biomasses by thermal degradation under different operating conditions. The process like pyrolysis and carbonization converts biomasses into bioenergy. Biochar is produced economically by three pyrolysis modes, i.e., fast, intermediate, and slow (Panwar et al., 2019). It is a fact that biochar yield is higher in slow methods than in other pyrolysis modes (Kung et al., 2015). The biochar production system can be classified as shown in diagram below (Fig. 9.1).

Depending upon the need for biochar and the costs of biochar production plant and efficacy, the specialized production process is carefully selected. The efficient TLUD method of biochar production is cost-effective, portable, and locally available for any farmer who ideally wants to produce biochar (Fig. 9.2). Though biochar production yields 10–22% by employing the TLUD method, it is a simplified and widely used method. Further, TLUD biochar is inevitably varied in its properties for a considerable variety of reasons (Masís-Meléndez et al., 2020; Panwar et al., 2019).

9.2.1 Properties and Characteristics of Biochar

Properties and distinguishing characteristics of biochar depend on many factors that affect the nature of biochar. The physiochemical properties of the biochar significantly depend on the types of feedstock used for its production and the method adopted for pyrolysis. The temperature plays a substantial role in biochar physical

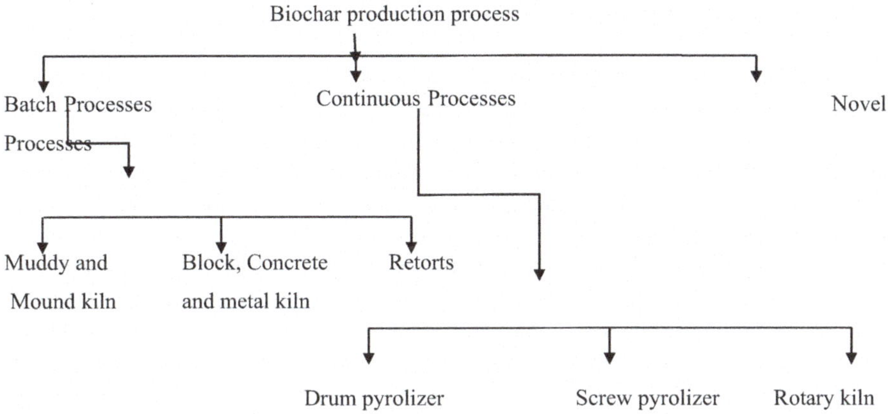

Fig. 9.1 Classification of the biochar production process (Source: Panwar et al., 2019)

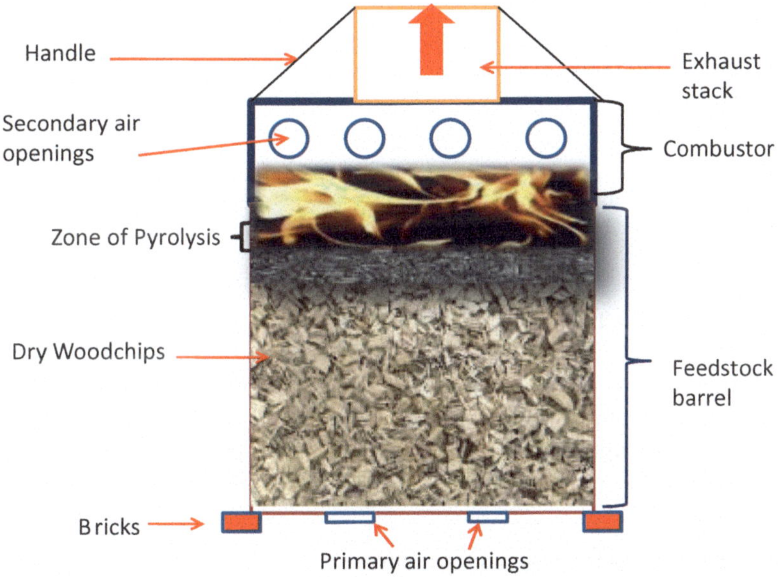

Fig. 9.2 Showing schematic diagram of TLUD workings and construction method using a barrel

and chemical nature, e.g., biochar produced at low temperature has a small pore size and low hydrophobicity compared to the biochar produced at high temperature (Jien, 2018; Masís-Meléndez et al., 2020; Suman & Gautam, 2017). The biochar production parameters depend on what is desired. Higher the processing temperature, lesser biochar will be produced but will have higher carbon stability (Retrieved from. https://biochar.international/the-biochar-opportunity/biochar-production-and-by-products, n.d.). A detailed account of the physical and chemical properties has been discussed in the subsections below.

9.2.1.1 Physical Characters

Every matter has its own physical and chemical properties, so biochar too possesses physical and chemical properties. The properties like the surface area, charge, density, structure of pores, and distributions are the essential physical features of biochar (Jien, 2018). Lehmann and Joseph (2009) mentioned that the physical properties of biochar are influenced by operating parameters, like processing heating rate, highest treatment temperature, pressure, reaction residence time, and the flow rate of ancillary inputs, irrespective of the type of feedstock. Morphological analysis of biochar by SEM microscopy reveals that more pores are present in the biochar produced under high pyrolization temperature than low pyrolization temperature (Jien, 2018). Research shows that increasing pyrolysis temperature increases the BET (Brunauer-Emmett-Teller) surface area and enhances pore development (Billa et al., 2019; Jien, 2018; Major, 2010; Suman & Gautam, 2017). The most common physical properties of fresh and aged biochar are:

I. Bulk density
II. Particle density
III. Porosity (micro and macro)
IV. Water-holding capacity
V. Grindability
VI. Surface area
VII. Hydraulic conductivity
VIII. Playability

9.2.1.2 Chemical Characters

If a user wants to use biochar as a soil amendment, he must be aware of the high variability of its chemical properties (Evans et al., 2017). The chemical properties of individual and mixed-feedstock derived biochar possess significant spatio-temporal variabilities (Nartey & Zhao, 2014). Some of the chemical characteristics of biochar are:

I. It contains macro- and micronutrients.
II. Soluble in organic solvents.
III. Shows proton activity.
IV. Has variability in EC.
V. Contribute in liming.
VI. Has cation and anion exchange capacity.
VII. Has high absorptivity.

Chemical properties of the biochar must be recognized to check whether the biochar to be used as a root substrate is under a suitable range of applications; otherwise, biochar may negatively impact plants. A study by Evans et al. (2017) compared the chemical properties of biochar manufactured from poultry litter produced at 400 °C for 2 hours in a muffle furnace which has higher macro- and microelements than the biochar made from mixed hardwood species. The chemical properties of biochar also vary with the type of biochar produced from their respective feedstocks (Evans et al., 2017; Nartey & Zhao, 2014; Panwar et al., 2019). Therefore, it is imperative to know the chemical properties of biochar before its application in the soil for better results.

9.3 Biochar as a Soil Amendment

The soil physicochemical property determines the growth of the plants and their nutrient availability. A balanced ratio of the macro- and micronutrient is essential for nutrient mobility, and soil microflora plays a substantial role in this regard. Therefore, before the amendment of soil, the user must grasp the underlying principles of soil requirement and the physicochemical properties of the amendment agents. The application of the amendment agent also executes a leading and crucial driver in the effectiveness of the amendment agent. When it comes to applying biochar to the soil to improve its fertility, the ideal application of biochar is nearer to the soil surface of the root zone, where the recycling of the nutrients and uptake is high and actively used by the plants. Besides this, it is equally important to select the specific cropping systems. The purpose of biochar application determines the application method; for purposes like carbon dioxide sequestration and moisture management, biochar must be applied in layers below the root zone. If it is to be used solely for carbon sequestration, it must be placed deeper in the soil to obtain good results (Major, 2010). The oxygen-to-carbon ratio of biochar and feedstocks, along with the condition of biochar production, determines the stability of biochar (Panwar et al., 2019). Biochar may have more than 100 years half-life time if the oxygen-to-carbon molar ratio (O:C) is more significant than 0.6, and when the ratio lies between 0.2 and 0.6, then half-life is between 100 and 1000 years; if it is less than 0.2, then the half-life is more significant than 1000 years (Spokas, 2010).

9.3.1 Biochar Impact on Soil Physicochemical Properties

Depending upon the nutrient content of the soil, the physiochemical soil properties change when biochar is added to it. The increasing population has influenced the agricultural systems, elevating the disintegration of humus and ultimately destroying soil physical properties (Aslam et al., 2014). Moreover, the non-judicious use of inorganic fertilizer has polluted the soil and has altered the physiochemical balance of the soil (Massah & Azadegan, 2016; Bista et al., 2019). Therefore, it is a matter of concern to improve the soil physicochemical properties and the fertility of the arable soil by adding substances like carbon-rich biochar (60–80%). It could enhance soil properties (Fig. 9.3) and affect soil components (Mensah & Frimpong,

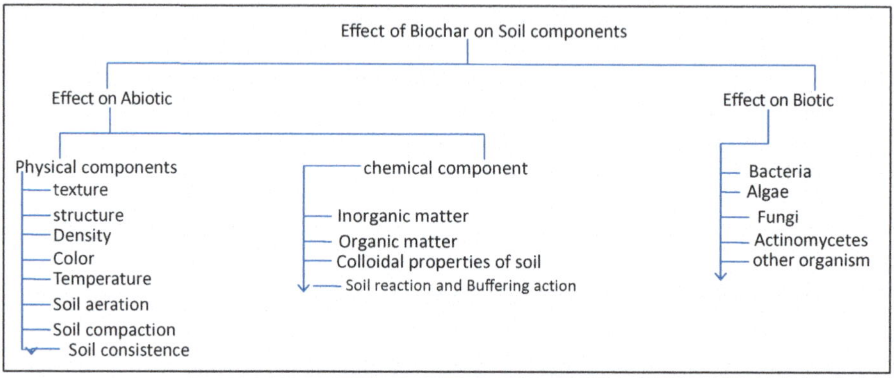

Fig. 9.3 Different components of soil where biochar could interact with each other to execute soil functions

2018). Recent experimental research has established that biochar could be an excellent soil conditioner if applied to agricultural soils (Adekiya et al., 2020; Egamberdieva et al., 2019; Egamberdieva et al., 2017). When it comes to the soil physical properties, the integration of biochar has been reported to elevate its aggregation ability, water-holding capacity, saturated hydraulic capacity, water retention and porosity (Kavitha et al., 2018; Bista et al., 2019). Generally, coarse-textured soil is more benefited by the addition of biochar than fine-textured soil. Sandy soils also show more response than clay-rich soils (Blanco-Canqui, 2017). The particle size of biochar and the depth of its application highly affect the overall water-holding capacity (Kavitha et al., 2018). According to Ibrahim et al. (2017), the particle size of biochar ranging from 0.5 to 1.0 mm increases the water-holding capacity when added to sandy soils. It also affects the soil water retention capacity (Blanco-Canqui, 2017). In a study carried out by Kameyama et al. (2016), it was observed that the greater than 3% concentration of biochar applications could increase the water-holding capacity of clay soil by 60%, therefore playing a pivotal role in the water-holding capacity and as a soil moderator. When >5% concentration is added to sandy loam soils, it decreases the size of the pores and affects the hydraulic conductivity (Kavitha et al., 2018). Amendment of biochar in the soil is found to improve soil fertility by facilitating the biochemical cycling of nitrogen and phosphorus (Gul & Whalen, 2016). Due to the high residence time and stability of biochar, it has a slow rate of decomposition in the soil and resides in the soil for a longer time. In addition, to the benefits described above, biochar also affects various other physical properties of a soil, such as swelling/shrinkage, tensile strength, surface area, and cracking density (Aslam et al., 2014; Blanco-Canqui, 2017; Kavitha et al., 2018). It imparts a positive response to the activity of soil enzymes; however, the repercussions of biochar on soil enzyme levels are yet to be assessed (Kavitha et al., 2018). Having absorptive properties, biochar suck up the heavier metal from the contaminated and toxic soils (Kameyama et al., 2016). The surface area of biochar acts as an interaction site for many of the organic as well as inorganic ions of soil and prevents leaching of the biologically available nutrient while making them available for plant growth and development (Mensah & Frimpong, 2018). Moreover, biochar provides a good niche for microbial flora of soil and nutrients to thrive in the soil which maintains the soil complex system. The impact of biochar on the soil microorganisms has been discussed below.

9.3.2 Impact of Biochar on Soil Microorganisms

Since biochar has several pores, it provides a good niche for the microbes of the soil. However, till today the mechanism of biochar, soil organic matter, and soil biota interaction has not been thoroughly analyzed. Much literature explains the mechanism of interaction of soil microorganisms and biochar (Gorovtsov et al., 2020). Still, the possible biochar-microorganism interaction mechanism includes the toxicity and volatile organic compounds that act on the soil microorganisms and the other mechanism. It influences the soil microorganism indirectly by affecting the soil properties, managing the nutrient availability and modifying the enzymatic activities (Ameloot, 2013). These interactions do not work separately but influence each other to some extent. The hydrophobicity and surface chemistry of biochar play a significant role in the attachment of the microorganisms. It has also been reported that microorganisms attach quicker to hydrophobic nonpolar surfaces than hydrophilic ones (Gorovtsov et al., 2020). The composition of biochar determines the colonization of microbes over the biochar, and the structure and composition of biochar are greatly influenced by the feedstocks, residence time, pyrolysis reactor temperature, etc. Other factors such as soil physiochemical properties, the abundance and composition of the consortia of the pre-existing microbes in the soil, and the biochar-soil contact time are the primary factors that affect the microorganism in the soil when biochar is added (Agegnehu et al., 2016; Hussain et al., 2018; Nartey & Zhao, 2014). It has been seen that aged biochar favors the abundance of microorganisms; this is because biochar provides shelter and carbon sources and maintains the favorable conditions for microbes growth (Fig. 9.4). Egamberdieva et al. (2016) observed that when biochar was incorporated in the soil to check its effects on the development of soya beans, the microbiome shifted in root-associated beneficial bacteria and resulted in improvement of plant growth. Similarly, in a study where biochar was added consecutively for 4 years in the soil, increased microbial biomass carbon and nitrogen was observed. Despite the positive effects of biochar on microbes, it has some negative impact due to the toxic chemical components of biochar (Gorovtsov et al., 2020; Spokas, 2010).

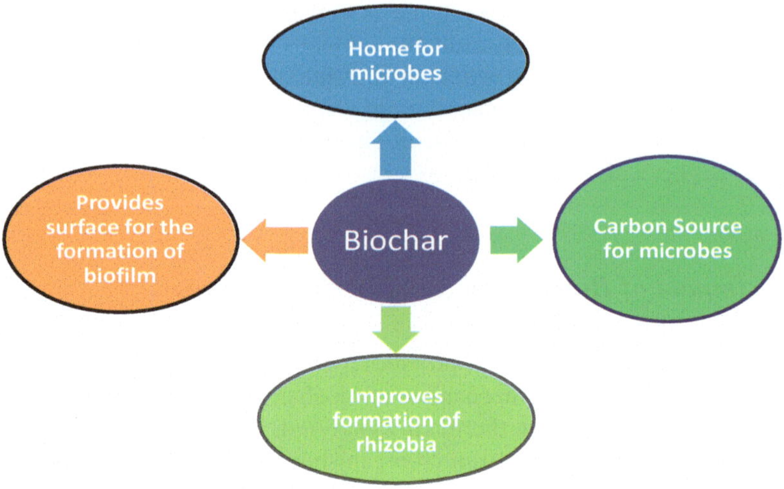

Fig. 9.4 Role of biochar for the soil microbes

9.3.3 Application of Biochar in Horticulture

"Horticulture is the science and art of development, sustainable production, marketing, and use of high-value, intensively cultivated food and ornamental plants" (Michigan State University, n.d.). Horticultural crops are varied, including annual and perennial species, fruits and vegetables, and decorative indoor and landscape plants. We are aware of the physical and chemical parameter of biochar, production methods, and how it facilitates soil physicochemical properties and its role on soil microbiome. Knowing all these facts, biochar could be an effective soil conditioner in the horticultural sector. Its use in horticulture got huge attention in recent times due to its positive effects on pH and its ability to enhance the cation exchange capacity (Blok et al., 2016). Mixtures of biochar and other different substrates such as peat, compost, and other bioagents have been successfully used to cultivate different horticulture crops (Agegnehu et al., 2016; Blok et al., 2017; Oustriere et al., 2017; Trupiano et al., 2017). Biochar shows a good response in a potting soil mixture agent as it retains water, supplies nutrients, provides a niche to microbial life, and suppresses diseases (Blok et al., 2017). When biochar was applied with *Bradyrhizobium* inoculums to Lupin (*L. angustifolius* L.), it improved its growth under drought stress conditions (Egamberdieva et al., 2017). Soya bean nodulation increases when biochar is used along with *Rhizobium* bacteria (Ma et al., 2019). When biochar is used as a soil conditioner, it shows better growth and production in broad bean (Egamberdieva et al., 2020), improves maize yield and biomass production (Zhu et al., 2014), increases endophytic bacteria that suppress diseases (Egamberdieva et al., 2020), increases avocado yields (Joseph et al., 2020), increases tomato yields (Priya et al., 2020), and enhances vegetable production (Jia et al., 2012). Despite the positive effect of biochar on plant health, it also has undesirable effects on plant growth. It may be due to high salt content and high pH value and contain phytotoxic compounds that affect the soil enzymes and microbes and, in return, adversely affect plant health. Therefore, the use of biochar as a soil amendment depends on the properties of the biochar and the ratio of the biochar with other composts or substrates. It has been established that biochar must have low salt content and pH if it is to be used in horticulture. Other important factors like water-holding capacity, stability, and nutrient content and the absence of phytotoxic compound must be considered before applying to the soil. Thus, to use biochar as a soil amendment, its production process must be optimized to make it favorable for horticulture crops. Biochar has a low oxygen uptake rate; in consequence of this, a free carbon source is hardly available for microbes; therefore, if anyone wishes to stimulate microbial activity in the soil, an additional source of carbon must be added (Blok et al., 2016). There are many ways through which biochar works on soil, and among them the most probable ones are listed below:

- Improves soil quality by improvising pH
- Increases soil water-holding capacity
- Stimulates activity of beneficial fungi and microbes
- Improves EC and cation exchange capacity
- Retains nutrients
- Sequestrates carbon from the atmosphere-biosphere pool and transfer it to the soil

9.4 Sustainable Agriculture and Biochar

There has been a paradigm shift in the agriculture of developed countries from traditional practices to modern practices, with the rising demands of food for the over-expanding population. It has transfigured the face of agricultural practices, with farmers relying more on high-level inorganic fertilizers and pesticides (Edwards, 2019). The higher inputs of inorganic fertilizers and pesticides along with specialized breed crops have responded well. They have increased yields dramatically, but in due course of time, the soil's inherent fertility has been degraded. The heavy use of chemical fertilizer has created many changes to soil physical and chemical properties.

Moreover, the applied pesticides absorbed by the crops enter into the food chain and get accumulated in the higher consumers to get biologically magnified, resulting in interference in the ecological cycle, causing harmful effects on the environment and arable land and consumer health. Therefore, a sustainable approach could be a practical step to reduce the vulnerability of land to degradation. The concept of agroecology could define the sustainability of agriculture, and the elements of agroecology could establish a relationship between the (Fig. 9.5) application of biochar and agricultural sustainability (FAO, 2018).

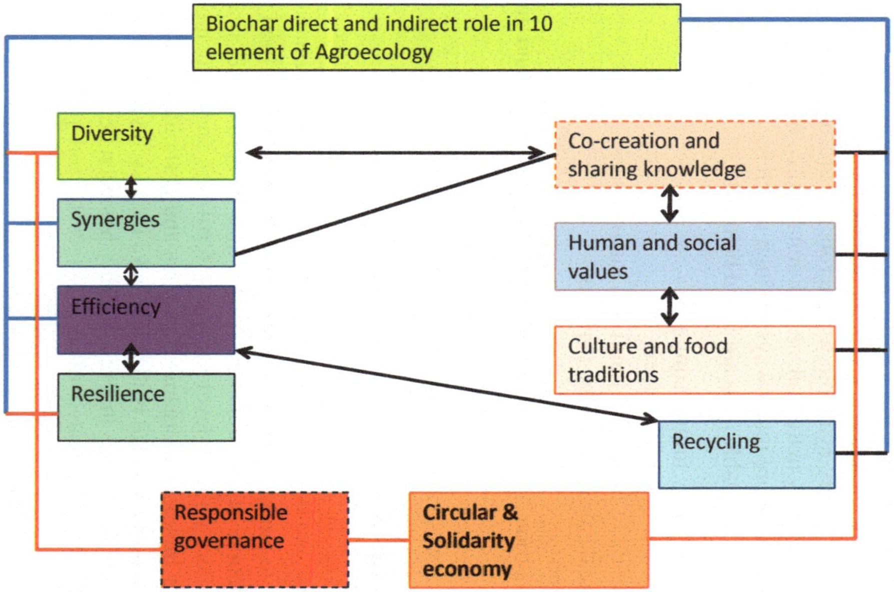

Fig. 9.5 Showing integration of biochar with ten elements: The ten elements of agroecology have been defined by the FAO

Biochar delivers various practical impacts on the environment, and many studies have mentioned the role of biochar in sustainable agriculture (Lehmann & Joseph, 2009; Jyoti Rawat & Sanwal, 2019).

9.5 Conclusions

Biochar is obtained from various biomasses through the thermochemical process by numerous techniques and methods, but all the production techniques or methods are not farmer-friendly. Due to the outrageous cost of setting up a unit for biochar production, the most efficient method conceivably is the TLUD method because this method is user-friendly and portable and has a meagre production cost. Successful application of biochar is an ancient practice; however, its application in horticultural is not a very old practice. Long-term application of biochar would help in the reclamation of contaminated soil, reduce soil toxicity and sequestration of carbon, reduce nutrient leaching, provide a niche for microbes, and assist in the effective management of agri-waste. However, proper dosages based on soil type and specific requirement should be optimized, as an accurate characterization of biochar, and its probable fate in the soil needs extensive research.

Acknowledgments The authors would dearly like to thank the Head of Botany Department, Dr. Harisingh Gour Central University, for his continuous support and encouragement in writing this chapter, and the UGC Non-Net fellowship Grant (Delhi, India) for providing financial support.

References

Adekiya, A. O., Agbede, T. M., Olayanju, A., Ejue, W. S., Adekanye, T. A., Adenusi, T. T., & Ayeni, J. F. (2020). Effect of Biochar on Soil Properties, Soil Loss, and Cocoyam Yield on a Tropical Sandy Loam Alfisol. Scientific World Journal, 2020. https://doi.org/10.1155/2020/9391630

Agegnehu, G., Bass, A. M., Nelson, P. N., & Bird, M. I. (2016). Benefits of biochar, compost and biochar-compost for soil quality, maize yield and greenhouse gas emissions in a tropical agricultural soil. *Science of the Total Environment, 543*, 295–306. https://doi.org/10.1016/j.scitotenv.2015.11.054

Ameloot, N. (2013). Biochar additions to soils : effects on soil microorganisms and carbon stability.

Aslam, Z., Khalid, M., & Aon, M. (2014). Impact of Biochar on Soil Physical Properties. Scholarly *Journal of Agricultural Science, 4*(5), 280–284. https://doi.org/10.1111/j.1365-2486.2009.02044.x.Novak

Retrieved from. https://biochar.international/the-biochar-opportunity/biochar-production-and-by-products. (n.d.). Retrieved from https://biochar.international/the-biochar-opportunity/biochar-production-and-by-products

Billa, S. F., Angwafo, T. E., & Ngome, A. F. (2019). Agro-environmental characterization of biochar issued from crop wastes in the humid forest zone of Cameroon. *International Journal of Recycling of Organic Waste in Agriculture, 8*(1), 1–13. https://doi.org/10.1007/s40093-018-0223-9

Bista, P., Ghimire, R., Machado, S., & Pritchett, L. (2019). Biochar effects on soil properties and wheat biomass vary with fertility management. *Agronomy, 9*(10). https://doi.org/10.3390/agronomy9100623

Blanco-Canqui, H. (2017). Biochar and soil physical properties. *Soil Science Society of America Journal, 81*(4), 687–711. https://doi.org/10.2136/sssaj2017.01.0017

Blok, C., Regelink, I. C., Hofl, J., & Streminska, M. (2016). Perspectives for the use of biochar in horticulture. In *Wageningen ur Greenhouse Horticulture*.

Blok, C., Van Der Salm, C., Hofland-Zijlstra, J., Streminska, M., Eveleens, B., Regelink, I., … Visser, R. (2017). Biochar for horticultural rooting media improvement: Evaluation of biochar from gasification and slow pyrolysis. *Agronomy, 7*(1), 6. https://doi.org/10.3390/agronomy7010006

Carpenter, B. H., & Nair, A. (2012). Biochar as a soil amendment for vegetable production. *Iowa State Research Farm Progress Reports*, 34–36, paper 1917.

Edwards, C. A. (2019). The concept of integrated systems in lower input. *Journal of Sustainable Agriculture*. https://doi.org/10.1017/S0889189300009255

Egamberdieva, D., Li, L., Ma, H., Wirth, S., & Bellingrath-Kimura, S. D. (2019). Soil amendment with different maize biochars improves chickpea growth under different moisture levels by improving symbiotic performance with Mesorhizobium ciceri and soil biochemical properties to varying degrees. *Frontiers in Microbiology, 10*(OCT), 1–14. https://doi.org/10.3389/fmicb.2019.02423

Egamberdieva, D., Reckling, M., & Wirth, S. (2017). Biochar-based Bradyrhizobium inoculum improves growth of lupin (Lupinus angustifolius L.) under drought stress. *European Journal of Soil Biology, 78*, 38–42. https://doi.org/10.1016/j.ejsobi.2016.11.007

Egamberdieva, D., Shurigin, V., Alaylar, B., Ma, H., Müller, M. E. H., Wirth, S., … Bellingrath-Kimura, S. D. (2020). The effect of biochars and endophytic bacteria on growth and root rot disease incidence of fusarium infested narrow-leafed lupin (Lupinus angustifolius L.). *Microorganisms, 8*(4), 496. https://doi.org/10.3390/microorganisms8040496

Egamberdieva, D., Wirth, S., Behrendt, U., Abd Allah, E. F., & Berg, G. (2016). Biochar treatment resulted in a combined effect on soybean growth promotion and a shift in plant growth promoting rhizobacteria. *Frontiers in Microbiology, 7*, 209. https://doi.org/10.3389/fmicb.2016.00209

Egamberdieva, D., Zoghi, Z., Nazarov, K., Wirth, S., & Bellingrath-Kimura, S. D. (2020). Plant growth response of broad bean (Vicia faba L.) to biochar amendment of loamy sand soil under irrigated and drought conditions. *Environmental Sustainability, 3*(3), 319–324. https://doi.org/10.1007/s42398-020-00116-y

Evans, M. R., Jackson, B. E., Popp, M., & Sadaka, S. (2017). Chemical properties of biochar materials manufactured from agricultural products common to the Southeast United States. *HortTechnology, 27*(1), 16–23. https://doi.org/10.21273/HORTTECH03481-16

Food and Agriculture Organization. (2018). *Guiding the transition to sustainable food and agricultural systems the 10 elements of agroecology*. Food and Agriculture Organization of the United Nations.

Gorovtsov, A. V., Minkina, T. M., Mandzhieva, S. S., Perelomov, L. V., Soja, G., Zamulina, I. V., … Yao, J. (2020). The mechanisms of biochar interactions with microorganisms in soil. *Environmental Geochemistry and Health, 42*(8), 2495–2518. https://doi.org/10.1007/s10653-019-00412-5

Gul, S., & Whalen, J. K. (2016). Biochemical cycling of nitrogen and phosphorus in biochar-amended soils. *Soil Biology and Biochemistry, 103*(August), 1–15. https://doi.org/10.1016/j.soilbio.2016.08.001

Hussain, F., Hussain, I., Khan, A. H. A., Muhammad, Y. S., Iqbal, M., Soja, G., … Yousaf, S. (2018). Combined application of biochar, compost, and bacterial consortia with Italian ryegrass enhanced phytoremediation of petroleum hydrocarbon contaminated soil. *Environmental and Experimental Botany, 153*(May), 80–88. https://doi.org/10.1016/j.envexpbot.2018.05.012

Ibrahim, A., Usman, A. R. A., Al-Wabel, M. I., Nadeem, M., Ok, Y. S., & Al-Omran, A. (2017). Effects of conocarpus biochar on hydraulic properties of calcareous sandy soil: Influence of particle size and application depth. *Archives of Agronomy and Soil Science, 63*(2), 185–197. https://doi.org/10.1080/03650340.2016.1193785

Jia, J., Li, B., Chen, Z., Xie, Z., & Xiong, Z. (2012). Effects of biochar application on vegetable production and emissions of n2o and ch4. *Soil Science and Plant Nutrition, 58*(4), 503–509. https://doi.org/10.1080/00380768.2012.686436

Jien, S. H. (2018). Physical characteristics of biochars and their effects on soil physical properties. In Y. S. Ok, D. C. W. Tsang, N. Bolan & J. M. Novak (Eds.), *Biochar from biomass and waste: Fundamentals and applications* (pp. 21–35). https://doi.org/10.1016/B978-0-12-811729-3.00002-9.

Joseph, S., Pow, D., Dawson, K., Rust, J., Munroe, P., Taherymoosavi, S., … Solaiman, Z. M. (2020). Biochar increases soil organic carbon, avocado yields and economic return over 4 years of cultivation. *Science of the Total Environment, 724*, 138153. https://doi.org/10.1016/j.scitotenv.2020.138153

Jyoti Rawat, J. S., & Sanwal, P. (2019). Biochar: A sustainable approach for improving plant growth and soil properties. In *Biochar—An imperative amendment for soil and the environment* (pp. 1–9) Retrieved from https://www.intechopen.com/online-first/biochar-a-sustainable-approach-for-improving-plant-growth-and-soil-properties

Kameyama, K., Miyamoto, T., Iwata, Y., & Shiono, T. (2016). Influences of feedstock and pyrolysis temperature on the nitrate adsorption of biochar. *Soil Science and Plant Nutrition, 62*(2), 180–184. https://doi.org/10.1080/00380768.2015.1136553

Kavitha, B., Reddy, P. V. L., Kim, B., Lee, S. S., Pandey, S. K., & Kim, K. H. (2018). Benefits and limitations of biochar amendment in agricultural soils: A review. *Journal of Environmental Management, 227*(August), 146–154. https://doi.org/10.1016/j.jenvman.2018.08.082

Kung, C. C., Kong, F., & Choi, Y. (2015). Pyrolysis and biochar potential using crop residues and agricultural wastes in China. *Ecological Indicators, 51*, 139–145. https://doi.org/10.1016/j.ecolind.2014.06.043

Lehmann, J. (2007). A handful of carbon. *Nature, 447*(7141), 143–144. https://doi.org/10.1038/447143a

Lehmann, J., & Joseph, S. (2009). Biochar for environmental management. In *Biochar for environmental management*. https://doi.org/10.4324/9780203762264

Leverett, F. (2008). Black is the new green. *National Interest, 442*(93), 624–626. https://doi.org/10.1038/442624a

Ma, H., Egamberdieva, D., Wirth, S., & Bellingrath-Kimura, S. D. (2019). Effect of biochar and irrigation on soybean- Rhizobium symbiotic performance and soil. *Agronomy, 9*, 626.

Major, J. (2010). Guidelines on practical aspects of biochar application to field soil in various soil management systems.

Masís-Meléndez, F., Segura-Chavarría, D., García-González, C. A., Quesada-Kimsey, J., & Villagra-Mendoza, K. (2020). Variability of physical and chemical properties of TLUD stove derived biochars. *Applied Sciences, 10*(2), 1–20. https://doi.org/10.3390/app10020507

Massah, J., & Azadegan, B. (2016). Effect of chemical fertilizers on soil compaction and degradation. *AMA, Agricultural Mechanization in Asia, Africa and Latin America, 47*(1), 44–50.

Mensah, A. K., & Frimpong, K. A. (2018). Biochar and/or compost applications improve soil properties, growth, and yield of maize grown in acidic rainforest and coastal Savannah Soils in Ghana. *International Journal of Agronomy, 2018*, 1–8. https://doi.org/10.1155/2018/6837404

Michigan State University. (n.d.). What is horticulture? A modern applied Plant Science! Retrieved from https://www.canr.msu.edu/hrt/about-us/horticulture_is#:~:text=Horticulture is the science and, decorative indoor and landscape plants.

Nartey, O. D., & Zhao, B. (2014). Biochar preparation, characterization, and adsorptive capacity and its effect on bioavailability of contaminants: An overview. *Advances in Materials Science and Engineering, 2014*, 1–12. https://doi.org/10.1155/2014/715398

Oustriere, N., Marchand, L., Rosette, G., Friesl-Hanl, W., & Mench, M. (2017). Wood-derived-biochar combined with compost or iron grit for in situ stabilization of cd, Pb, and Zn in a contaminated soil. *Environmental Science and Pollution Research International, 24*(8), 7468–7481. https://doi.org/10.1007/s11356-017-8361-6

Panwar, N. L., Pawar, A., & Salvi, B. L. (2019). Comprehensive review on production and utilization of biochar. *SN Applied Sciences, 1*(2), 1–19. https://doi.org/10.1007/s42452-019-0172-6

Priya, P., Singh, C., Chaudhary, N., & Vyas, D. (2020). A comparative study of biochar, leaf compost and spent mushroom compost for tomato growth. *Research Journal of Agricultural Sciences, 11*(6), 1362–1366.

Sánchez-Monedero, M. A., Cayuela, M. L., Sánchez-García, M., Vandecasteele, B., D'Hose, T., López, G., … Mondini, C. (2019). Agronomic evaluation of biochar, compost and biochar-blended compost across different cropping systems: Perspective from the European project FERTIPLUS. *Agronomy, 9*(5), 225. https://doi.org/10.3390/agronomy9050225

Spokas, K. A. (2010). Review of the stability of biochar in soils: Predictability of O:C molar ratios. *Carbon Management, 1*(2), 289–303. https://doi.org/10.4155/cmt.10.32

Suman, S., & Gautam, S. (2017). Effect of pyrolysis time and temperature on the characterization of biochars derived from biomass. *Energy Sources, Part A: Recovery, Utilization, and Environmental Effects, 39*(9), 933–940. https://doi.org/10.1080/15567036.2016.1276650

Trupiano, D., Cocozza, C., Baronti, S., Amendola, C., Vaccari, F. P., Lustrato, G., … Scippa, G. S. (2017). The effects of biochar and its combination with compost on lettuce (Lactuca sativa L.) growth, soil properties, and soil microbial activity and abundance. *International Journal of Agronomy, 2017*, 1–12. https://doi.org/10.1155/2017/3158207

USBI. (n.d.). How is biochar made? Retrieved from Montana The Magazine of Western History website. Retrieved from https://biochar-us.org/biochar-production

Zhu, Q. H., Peng, X. H., Huang, T. Q., Xie, Z., & Holden, N. M. (2014). Effect of biochar addition on maize growth and nitrogen use efficiency in acidic red soils. *Pedosphere, 24*(6), 699–708. https://doi.org/10.1016/S1002-0160(14)60057-6

Chapter 10
The Use of Genomics and Precise Breeding to Genetically Improve the Traits of Agriculturally Important Organisms

Aehsan Ul Haq, Mohammad Lateef Lone, Sumira Farooq, Shazia Parveen, Foziya Altaf, and Inayatullah Tahir

Abstract Achieving food sufficiency for the increasing population is a global concern in contemporary times. According to the latest world summit on food security, it is important to increase food production by more than 70% by 2050 to meet the demands of the growing population. Besides population increase, extreme weather events like floods, droughts, untimely rains and pest outbreaks due to climate change negatively affect agricultural productivity. Moreover, expanding human settlements have led to the shrinkage of available farmland. Under this scenario, newly emerging technologies in crop breeding like gene editing provide a tremendous potential for sustainable agriculture and food security. Different gene-editing techniques, including zinc finger, TALEN and the widely used CRISPR/Cas system, are worthy to note. These techniques are used both in plant and animal systems to breed for desirable agronomic traits, leading to increased crop yields, reduced use of chemical fertilisers and pesticides and increased resistance of crops to climatic stress, with decreased post-harvest losses.

Furthermore, understanding genetic diversity with the help of genome sequencing technology has led to identifying agronomically important traits for breeding purposes. The key catalysts for the current genomic revolution are developing next-generation DNA sequencing technology that recently crossed a $1000 human genome barrier. This technology revolutionises crop production as quickly as it revolutionised medicine. This enables the sequencing of several crop genomes and facilitates the association of genomic variation and agronomic characteristics, laying the groundwork for genomic-assisted breeding. Thus, genomics and precision breeding could act as a game-changer in achieving food security by improving traits of agriculturally important organisms.

A. U. Haq (✉) · M. L. Lone · S. Farooq · S. Parveen · F. Altaf · I. Tahir
Plant Physiology and Biochemistry Research Laboratory, Department of Botany, University of Kashmir, Srinagar, Jammu and Kashmir, India

S. A. Bandh (ed.), *Sustainable Agriculture*,
https://doi.org/10.1007/978-3-030-83066-3_10

Keywords CRISPR/Cas · Gene editing · Food security · Climate change · Genome sequencing

10.1 Introduction

The increased food requirements for swiftly expanding human populations have led to severe global food quantity and quality issues. Moreover, extreme weather events such as drought and heat stress due to climate change have caused substantial yield losses and reduced crop output. Besides, demographic growth and urbanisation rates have significantly increased over the recent years, whereas the ratio of food production to consumption has drastically declined. By 2050, the global population is projected to hit 10 billion, but there are no good plans in place to feed this large-scale population (Kc & Lutz, 2017). About 800 million people are chronically affected by hunger, and 2 billion are micronutrient deficient (FAO, 2019). Under this climate change scenario and rapid population expansion, developing improved varieties of plants through precise breeding and genomics can prove instrumental in ensuring food security for such massive populations. These innovative technologies could improve crop output, encourage less pesticide and chemical fertilisers, improve crop resilience to climatic stress, alleviate post-harvest losses and improve food quality (Bailey-Serres et al., 2019; Eshed & Lippman, 2019; Zaidi et al., 2019). Precision breeding enables breeders to target particular parts of the genome much more rapidly and helps achieve breeding goals much promptly. It involves gene-editing techniques such as DNA insertion, modification, replacement or deletion at specific loci in an organism's genome. Usually, targeted genetic scissors are employed to induce loci-specific double-stranded breaks, which are later repaired by cell's repair systems (Metzker, 2010). The most commonly used gene-editing techniques include zinc finger, TALEN and the recently developed CRISPR/Cas system in 2012 (Schindele et al., 2020). Gene editing is a highly dynamic field of research with constantly emerging improved methods (Hickey et al., 2019). The scope of these editing techniques is confined to plants and used in humans to diagnose and repair genetic disorders. Gene-editing techniques may induce genetic alterations that include complex or straightforward mutations or alien- and species-specific genes. However, most of the genetically edited crops produced so far don't involve incorporating foreign genes but simple point mutations with the desired traits (Zaidi et al., 2019). Gene editing has opened a new era of precision plant breeding in agriculture and is expected to drive the second green revolution. This technology is regarded as the best innovative breeding technique and offers alternate means to escape the route of stringent norms of "genetically modified organisms" or GMOs. Besides precise breeding techniques, recent advances in genomics offer new methods and tools to plant breeders that may lead to a great breakthrough in plant breeding like genetic dissection, breeding for complex traits and "super domestication" of crops (Pérez-de-Castro et al., 2012). Next-generation sequencing, including Roche 454 sequencing, Illumina (Solexa) sequencing, solid-state sequencing and Ion Torrent (NGS), is widely regarded as the cornerstone of genomic reproduction, as it enables the full-genome sequencing of crop plants. Among these, Roche 454 and Illumina are the most frequently used for seed genome sequencing. The whole-genome sequencing of target plants reveals an unparalleled abundance of knowledge that enables breeders to identify and exploit crop enhancement variants (Bevan & Uauy, 2013). In addition, it also facilitates the study of the interrelationship between genotypes and phenotypes (Tester & Langridge, 2010). According to FAO, genomics is crucial to produce efficient cultivars of plants, the key to the new green revolution required to feed the expanding global populations while conserving natural resources. Additionally, genomic tools allow the identification of QTL and the mining of already existing advantageous alleles with a limited impact, which are often ignored and hence do not contribute to the gene pool used for breeding purposes (Morgante & Salamini, 2003). Furthermore, functional genomics enables the discovery of gene networks implicated in the regulation of beneficial agronomic trait variation in elite breeding populations. Similarly, combining novel genomic knowledge with traditional breeding methods is critical for improving response to selection and thus crop improvement (Tuberosa et al., 2011). This chapter provides a summary of various genomic and precision breeding methods, their application to crop enhancement and the regulatory environment in which they operate.

10.2 Genomic and Precise Breeding Techniques

The latest developments in genetics and genomics have significantly increased our knowledge of the structural and functional features of plant genomes. Nonetheless, they have presented us with many convincing lines of inquiry. The full-genome sequences of rice, *Arabidopsis*, poplar and sorghum have been published, as well as an enormous amount of plant expressed gene tags (ESTs). Over the next few years, the most big crops' whole genomes, or at least their gene space, will likely be sequenced. However, it is improved varieties, not sequence, that add to the farmer's economic return. Functional genomics and system biology studies are promoting the discovery of gene networks involved in regulating genetic variance for economically important traits in elite breeding populations.

Additionally, integrating modern genomic expertise with traditional breeding methods is critical for improving selection response and thus crop improvement. Superior varieties can be generated as a result of the discovery of novel genetic variation, the development of improved selection techniques and the identification of genotypes with improved characteristics as a result of superior combinations of alleles at multiple loci assembled through marker-assisted selection. While it is clear that genomics research has the potential to revolutionise the discipline of plant breeding, the high costs associated with genomics research currently prevent the widespread use of genomics-assisted crop improvement, especially for inbreeding and minor crops.

The inception of next-generation sequencing technologies has changed the field of genomics (Metzker, 2010). These technologies have facilitated de novo and resequencing of several crop species and revolutionised the field of plant breeding. These technologies are cost-effective and facilitate rapid sequencing of DNA and RNA fragments than Sanger sequencing. The following is a short overview of these sequencing technologies:

10.2.1 454 Pyrosequencing

Pyrosequencing was invented in Sweden and was later bought by Qiagen, which licensed it to 454 Life Sciences. The first and second" generation sequencing techniques focused on the identification of pyrophosphate, a by-product of nucleotide incorporation that indicates whether a certain base was incorporated into an elongating chain of DNA (Ronaghi et al., 1996). Typically, 400–700 bp DNA fragments are ligated to adapters and then amplified by PCR in an individual emulsion "bead" (emPCR) reaction. The beads contain complementary DNA sequences to the adaptors, facilitating direct binding of DNA fragments to the beads (usually one fragment binds to each bead). Following DNA synthesis, chemical detection of the reactions occurs in a picoliter-sized chamber where the amount of pyrophosphate released is quantified. By constantly filling the chambers with sequencing reagents containing one of the four nucleotides, pyrophosphate release is quantified using a light-generating reaction when the right nucleotide is injected into the DNA chain. Additionally, the light intensity contains details about homopolymer "runs" of nucleotides in the chain, while issues arise as longer tracts of the same nucleotide are used.

10.2.2 Ion Torrent

Ion Torrent technology involves direct conversion of nucleotide sequence into digital information on a semiconductor chip (Rothberg et al., 2011). In this technology, insertion of correct nucleotide in an elongating DNA chain during DNA synthesis reaction causes hydrogen ion release, which triggers a pH change of the solution, recorded as voltage change by an ion sensor, much like a pH metre. However, no voltage spike happens when no nucleotide is added. Additionally, when two neighbouring nucleotides introduce the same nucleotide, two hydrogen atoms are released, doubling the voltage. Thus, it is possible to calculate "runs" of a single nucleotide.

10.2.3 Illumina Sequencing

Illumina sequencing is the most advanced technology for next-generation sequencing focused on bridge amplification. DNA fragments of approximately 500 bp with suitable ligated adapters on both ends are used as substrates in this technology for repeated amplification synthesis reactions on a solid support (glass slide) containing oligonucleotide sequences complementary to a ligated adapter. The oligonucleotides are arranged on the slide in such a way that the DNA undergoes continuous rounds of amplification, resulting in the formation of clonal "clusters" of approximately 1000 copies of each oligonucleotide fragment. During DNA synthesis reactions, modified nucleotides corresponding to each of the four bases are fluorescently labelled differently. These labels enable their detection during incorporation into the growing DNA chain.

10.3 Applications of Genomics

The advent of next-generation sequencing technology has changed the pace of the current genome sequencing projects, and scientists are rapidly adopting this technology to gain detailed insights into the desired crop genomes. The inception of this technology has facilitated the sequencing of new genomes and resequencing of already sequenced genomes at a higher pace, hence unravelling enormous information useful for crop improvement. The 430 Mbp genome of *Theobroma cacao* has been sequenced by Roche 454 technology (Scheffler et al., 2009). However, Roche 454 and Sanger sequencing have been used in combination to sequence the apple genome (Velasco, 2009; Velasco et al., 2009). Likewise, the cotton genome has been characterised by a combination of Roche 454 and Illumina Solexa sequencing (Wilkins et al., 2009). *Miscanthus* genome has been surveyed by Roche 454 sequencing technology (Swaminathan et al., 2009). The combination of Illumina Solexa, Sanger and Roche 454 sequencing was used to investigate the banana genome (Hribova et al., 2009). Similarly, sequencing of complex BAC from barley has been performed by Roche 454 technology (Stein, 2009; Wicker et al., 2006). Furthermore, Ion Torrent technology has been used to detect induced mutations in *Linum usitatissimum* (Galindo-González et al., 2015).

10.4 Precision Breeding Techniques

The recent breakthroughs in genome-editing technology using CRISPR/Cas, zinc finger nucleases and TALENs have opened a new era of plant breeding. Thus, to produce generations of the crop with the desired traits, these novel techniques are being used by plant breeders and researchers throughout the globe (Ahmar et al., 2020). A brief overview of these techniques is mentioned below:

10.4.1 Zinc Finger Nucleases

Zinc finger nucleases are synthetic type II restriction enzymes that are used to digest any sequence in a double-stranded DNA stretch (Carroll, 2011; Osakabe et al., 2010; Zhang et al., 2010). ZFN monomer is a synthetic nuclease synthesised by combining a Cys2-His2 zinc finger domain and a non-specific DNA cleavage domain from the DNA restriction enzyme *Flavobacterium okeanokoites* I (FokI) (Curtin et al., 2011). To split DNA, the FokI cleavage domain must dimerise (Bitinaite et al., 1998; Smith et al., 2000). Thus, after the binding of two ZFN monomers to their respective target sequences, target DNA is cleaved. The two ZFN monomers encircle a 5- to 6-bp-long spacer chain within the target site DNA, facilitating FokI dimer digestion within the spacer sequence. The FokI dimer generates double-stranded breaks in the spacer sequence, which are flanked by an array of two zinc finger binding sites (Curtin et al., 2012; Puchta & Fauser, 2013). These breaks are then repaired by the cell's endogenous DNA repair machinery via error-prone homologous recombination or non-homologous end joining. When no homologous sequences are present, the cell moves to a non-homologous end-joining mechanism. The separated ends are processed and directly joined, resulting in nucleotide deletion or insertion, leading to frameshift mutations and lack of function in the gene (Qi et al., 2013b). Despite the complexity of their modular structure, ZFNs have been commonly used to modify desirable genes in a variety of plants, including *Arabidopsis* (Qi et al., 2013a), *Nicotiana tobacum* (Townsend et al., 2009), maize (*Zea mays*), soybean (*Glycine max*) and canola (*Brassica napus*) (Curtin et al., 2011; Gupta et al., 2012; Shukla et al., 2009). ZFNs have been used to cause resistance to bialaphos and herbicides in maize and tobacco, respectively (Shukla et al., 2009; Townsend et al., 2009), as well as an ABA-insensitive phenotype in *Arabidopsis* (Osakabe et al., 2010). ZFNs have been shown to significantly enhance antiviral tolerance in plants by preventing viral replication proteins from interacting with DNA-binding sites (Sera, 2005; Takenaka et al., 2007). Additionally, by attacking a particular site in the viral DNA, synthetic zinc finger proteins have been used to confer various resistances against many begomoviruses, including *Tobacco curly shoot virus* (TbCSV) and *Tomato yellow leaf curl China virus* (TYLCCNV) (Chen et al., 2014).

10.4.2 TALENs

TALENs are formed by the fusion of transcriptional activator-like effector (TALE) repeats and the FokI restriction enzyme (Boch et al., 2009). These type II effector TALE proteins are secreted by *Xanthomonas* spp. into plant cells to modulate host gene expression. These proteins comprise a transcriptional activation domain, nuclear localisation signal and a central DNA-binding domain (Boch & Bonas, 2010). The most common methods of delivering TALENs into plants include PEG-mediated transformation, *Agrobacterium*-mediated transformation and biolistic methods. The nuclear localisation signal mediates the entry of TALEs into the nucleus of a plant cell, and the activation domain stimulates gene expression by activating the transcriptional machinery (Hsu et al., 2013). The DNA-binding central domain of each TALE comprises variable repeat units, recognising a single nucleotide. The terminal repeat unit is called a "half-repeat" as it is shorter, comprising of only 20 amino acids, except two variable amino acids at the 12th and 13th position. These repeat units are highly conserved (Osakabe et al., 2010; Zhang et al., 2010).

The variable residues, called repeat variable di-residues, determine the repeat region's DNA-binding specificity (RVDs). TALENs are exceptionally precise in their binding to DNA sequences due to their unusual RVD variations. TALENs act as molecular scissors by binding to specific DNA sequences and causing double-strand breaks (DSBs) in the DNA at a specific location. These breaks are then repaired by the endogenous repair systems of the cells through homologous recombination (HR) or non-homologous end joining (NHEJ). This method of chromosomal end joining can result in deletions, replacements, insertions or larger chromosomal rearrangements. TALENs are the most exciting and successful genome-editing methods due to their ease of design and low cost (Zhang et al., 2016). Thus, TALENS can be conveniently changed to modify preferred DNA sequences, making them effective next-generation gene-editing methods.

10.4.2.1 Application of TALENs in Crop Plants

TALEN is an extremely advanced and flexible method for modifying the genome at various locations, providing an enormous potential for crop enhancement. TALENs have been extensively used to increase crop yields in a variety of plants and were initially used to confer resistance on rice against the blight pathogen *Xanthomonas oryzae* (Li et al., 2013). Similarly, TALENs were used to cause INDELS in the promoter region of the HvPAPhya phytase gene in barley (Wendt et al., 2013). TALENs were used to cause site-directed mutagenesis in soybean fatty acid desaturase genes (FAD2-1A and FAD2-1B), resulting in the conversion of oleic acid to linoleic acid. After mutation, mutated plants produced fourfold the amount of fatty acid (oleic acid) produced by parents (Haun et al., 2014).

Moreover, the three homoalleles (TaMLO-A1, TaMLO-B1 and TaMLO-D1) in hexaploid bread wheat have been altered using TALEN resistance against powdery mildew causing fungus. TALENs have additionally been effectively utilised for targeting various genes in different plant species, including maize, *Arabidopsis*, tobacco and *Brachypodium*, for numerous applications (Christian et al., 2013; Liang et al., 2014; Shan et al., 2013a, b). The altered gene via TALEN technology in different plant species is mentioned in the table below (Table 10.1).

10.4.3 CRISPR/Cas

In comparison to ZFNs and TALENs, CRISPR/Cas is a cost-effective genome-editing technique with many gene targets (Cong et al., 2013; Mali et al., 2013).

Table 10.1 List of the reported targeted gene(s) via TALEN technology in different plant species

Plant species	Genes	TALEN assembly method	Delivery methods	References
Arabidopsis thaliana	ADH1, TT4, MAPKKK1, DSK2Ba, DSK2Bb, NATA2a, NATA2b	Golden Gate	Protoplast transformation *Agrobacterium*-mediated floral dip transformation	Christian et al. (2013)
Brachypodium	BdABA1, BdCKX2, BdMC6, BdSPL, BdHRT, BdSPP, BdHTA1, BdCO11	Golden Gate	Protoplast and *Agrobacterium*-mediated transformation	Shan et al. (2013b)
Tobacco	SurA, SurB	Golden Gate	Protoplast transformation	Zhang et al. (2013)
Rice	OsBADH2, OsDEP1,OsSD1, OsCKX2, Os11N3	Golden Gate	Protoplast and *Agrobacterium*-mediated transformation	Shan et al. (2013b)
Barley	HvPAPhy	Golden Gate	*Agrobacterium*-mediated transformation	Wendt et al. (2013)
Soybean	FAD2-1A, FAD2-1B	Golden Gate	*Agrobacterium*-mediated transformation	Haun et al. (2014)
Wheat	TaMLO-A1, TaMLO-B1, TaMLO-D1	Golden Gate	Protoplast transformation and biolistic transformation	Wang et al. (2014)
Tomato	PROCERA	Golden Gate	*Agrobacterium*-mediated transformation	Lor et al. (2014)
Zea mays	ZmPDS, ZmIPK1A, ZmIPK, ZmMRP4	Golden Gate	Protoplast and *Agrobacterium*-mediated transformation	Liang et al. (2014)

However, some bacterial cells have identified a variety of CRISPR-based defence mechanisms (Gilles & Averof, 2014; Haft et al., 2005). The *Streptococcus pyogenes* type II CRISPR/SpCas9 method is a highly flexible genome-editing platform with a wide range of applications (Hsu et al., 2014). The *Streptococcus pyogenes* CRISPR/SpCas9 system is composed of three genes that encode CRISPR RNA (crRNA), trans-activating crRNA (tracrRNA) and Cas9 protein. A simplified version of the CRISPR/Cas system comprises cas9 protein complex and single-guide RNA comprising CRISPR tracerRNA and short, mature crRNA. The guide RNA recognises and binds to target DNA sequences based on complementarity. After binding, the Cas9 digests the target DNA sequences at the desired locus (Graham & Root, 2015). After cleavage, the breaks created by the nuclease are repaired by the endogenous repair system of the cell either by non-homologous end-joining process or by homologous recombination (Shukla et al., 2009). The CRISPR/Cas system is widely used to produce null alleles, or gene knockouts, via insertion or deletion of nucleotides or by insertion of premature stop codons. Owing to its effectiveness and low cost, it is a widely used editing tool in plant systems (Li et al., 2013; Nekrasov et al., 2013; Shan et al. 2013a, b). Moreover, it has also proven as an effective solution to numerous problems about crop breeding (Gao, 2018). CRISPR/Cas has been used to boost crop yields, induce biotic resistance and enhance the nutritional value of key crops (Zhang et al., 2018). CRISPR/Cas9-mediated mutation of the Gn1a, DEP1 and GS3 genes in the Zhonghua11 rice cultivar boosts grain size and number. Similarly, CRISPR/Cas9-mediated deletion of the LAZY1 gene in rice may increase crop yield (Miao et al., 2013). CRISPR/Cas9 technology was used to increase the amount of oleic acid in *Camelina sativa* while decreasing the amount of polyunsaturated fatty acids (Jiang et al., 2017). Additionally, the waxy gene Wx1 was deleted from maize using CRISPR/Cas9 to mask the expression of the granule-bound starch synthase (GBSS) gene, resulting in (waxy) maize with increased digestibility. Researchers at the Swedish Agricultural University have targeted the same gene to produce waxy potatoes (Andersson et al., 2017). Similarly, non-browning mushrooms were produced by knocking out the polyphenol oxidase (PPO) gene, responsible for browning (Waltz, 2016a). Besides improving the yield and nutritional value of important food crops, this technology has also been used to produce disease-resistant crops. Zhang et al. (2017) utilised CRISPR/Cas9 technology to produce powdery mildew-resistant wheat plants by altering three *EDR1* homologs. Similarly, resistance to blast and bacterial blight was induced in rice by mutagenesis of OsERF922 and OsSWEET13 (Wang et al., 2016; Zhou et al., 2015). Additionally, tomatoes resistant to powdery mildew and bacterial speck were developed in plants by editing the SlMLO1 (Nekrasov et al., 2017) and SlJAZ2 (Ortigosa et al., 2018) genes, respectively. Similarly, to avoid economic losses caused by citrus canker in grapes, the CsLOB1 gene's coding area was disrupted in Duncan grapefruits to generate canker-resistant plants (Jia et al., 2017). Additionally, through using CRISPR/Cas9 technology, multiple antiviral resistances was induced in cucumber against cucumber vein yellowing virus, yellow mosaic virus and potyviruses zucchini and papaya ringspot mosaic virus-W (Chandrasekaran et al., 2016).

10.5 Regulation of Genome-Edited Crops

There is no clear consensus regarding the regulation, production and consumption of genome-edited plants. Some countries willingly grow and consume them, while others reject their production and consumption (Garcia Ruiz et al., 2018). In the USA, genome editing has been declared equivalent to conventional breeding by US Department of Agriculture (USDA), which doesn't require any regulatory framework (Waltz, 2016b). The decision was based on the fact that gene editing doesn't involve the insertion of foreign DNA (transgene) and the altered genome doesn't possess resistance to pesticides or herbicides. However, Canada declared that any novel product produced due to gene-editing technology should be subjected to stringent regulatory norms to check its toxicity, allergenicity and effects on other organisms (Smyth, 2017). In Argentina, genome-edited plants are approved under a regulatory framework based on Cartagena Protocol (Whelan & Lema, 2015). A similar regulatory protocol was established by Chile and Brazil, while as in European Union (EU) countries, genetically modified crops are politically opposed (Waltz, 2016a). In New Zealand, the Hazardous Substances and New Organisms Act 1996 (HSNO) was amended in 2016, stating that genome-edited plants are subjected to the same GMOs (Shimatani et al., 2017). As a result of this, no GMO crop is grown in the country. India has already established a regulatory framework in 1989 for research and development, including GMOs, their products and novel gene techniques. Thus all the gene-editing techniques are regulated under this regulatory framework (Friedrichs et al., 2019). The foregoing discussion suggests that regulating and deregulating genetically edited plants are determined by the already existing regulatory framework in the country.

10.6 Technological Risks

There may be risks associated with new crop varieties released into the environment and consumed by humans and animals. These risks include risks due to the breeding process and risks due to the development of specific traits. As the well-documented safety record is not available for gene-editing technologies and the point mutations produced due to editing are no longer genetically distinguishable from induced or natural mutations (Grohmann et al., 2019). Therefore, it is not possible to expect new types of risks. However, gene editing may lead to off-target effects with a frequency much less than GMOs and induced mutagenesis (Holme et al., 2019). The second type of risk associated with the editing technology may be due to the development of new trait; however, such types of risk can't be assessed for gene-edited crops as every unique trait developed may have different effects.

10.7 Conclusions and Future Perspectives

In conclusion, genome-editing techniques, especially CRISPR-Cas9, hold a great potential for agricultural transformation by conferring biotic and abiotic stress resistance to plants and improving their yield and nutritional value. Besides editing techniques, next-generation sequencing technologies have a vital role in mining the desired genes from different crop plants for breeding purpose. Together, these features are essential to meet the food demands of the increasing global population. Moreover, C_3 plants such as rice and barley can be engineered by gene-editing technology to improve yield losses due to inefficient photorespiration. However, to channelise this technology effectively for crop improvement, the various societal concerns and biosafety issues need to be addressed by the scientific society. In addition, the regulatory framework concerning genome-edited crops need to be revised and general awareness to be generated about their properties.

References

Ahmar, S., Gill, R. A., Jung, K. H., Faheem, A., Qasim, M. U., Mubeen, M., & Zhou, W. (2020). Conventional and molecular techniques from simple breeding to speed breeding in crop plants: Recent advances and future outlook. *International Journal of Molecular Sciences, 21*(7), 2590. https://doi.org/10.3390/ijms21072590.

Andersson, M., Turesson, H., Nicolia, A., Fält, A. S., Samuelsson, M., & Hofvander, P. (2017). Efficient targeted multiallelic mutagenesis in tetraploid potato (Solanum tuberosum) by transient CRISPR-Cas9 expression in protoplasts. *Plant Cell Reports, 36*(1), 117–128. https://doi.org/10.1007/s00299-016-2062-3.

Bailey-Serres, J., Parker, J. E., Ainsworth, E. A., Oldroyd, G. E. D., & Schroeder, J. I. (2019). Genetic strategies for improving crop yields. *Nature, 575*(7781), 109–118. https://doi.org/10.1038/s41586-019-1679-0.

Bevan, M. W., & Uauy, C. (2013). Genomics reveals new landscapes for crop improvement. *Genome Biology, 14*(6), 206. https://doi.org/10.1186/gb-2013-14-6-206.

Bitinaite, J., Wah, D. A., Aggarwal, A. K., & Schildkraut, I. (1998). FokI dimerisation is required for DNA cleavage. *Proceedings of the National Academy of Sciences, 95*(18), 10570–10575. https://doi.org/10.1073/pnas.95.18.10570.

Boch, J., & Bonas, U. (2010). Xanthomonas AvrBs3 family-type III effectors: Discovery and function. *Annual Review of Phytopathology, 48*, 419–436. https://doi.org/10.1146/annurev-phyto-080508-081936.

Boch, J., Scholze, H., Schornack, S., Landgraf, A., Hahn, S., Kay, S., … Bonas, U. (2009). Breaking the code of DNA binding specificity of TAL-type III effectors. *Science, 326*(5959), 1509–1512. https://doi.org/10.1126/science.1178811.

Carroll, D. (2011). Genome engineering with zinc-finger nucleases. *Genetics, 188*(4), 773–782. https://doi.org/10.1534/genetics.111.131433.

Chandrasekaran, J., Brumin, M., Wolf, D., Leibman, D., Klap, C., Pearlsman, M., … Gal-On, A. (2016). Development of broad virus resistance in non-transgenic cucumber using CRISPR/Cas9 technology. *Molecular Plant Pathology, 17*(7), 1140–1153. https://doi.org/10.1111/mpp.12375.

Chen, W., Qian, Y., Wu, X., Sun, Y., Wu, X., & Cheng, X. (2014). Inhibiting replication of begomoviruses using artificial zinc finger nucleases that target viral-conserved nucleotide motif. *Virus Genes, 48*(3), 494–501. https://doi.org/10.1007/s11262-014-1041-4.

Christian, M., Qi, Y., Zhang, Y., & Voytas, D. F. (2013). Targeted mutagenesis of Arabidopsis thaliana using engineered TAL effector nucleases. *G3, 3*(10), 1697–1705. https://doi.org/10.1534/g3.113.007104.

Cong, L., Ran, F. A., Cox, D., Lin, S., Barretto, R., Habib, N., … Zhang, F. (2013). Multiplex genome engineering using CRISPR/Cas systems. *Science, 339*(6121), 819–823. https://doi.org/10.1126/science.1231143.

Curtin, S. J., Zhang, F., Sander, J. D., Haun, W. J., Starker, C., Baltes, N. J., … Stupar, R. M. (2011). Targeted mutagenesis of duplicated genes in soybean with zinc-finger nucleases. *Plant Physiology, 156*(2), 466–473. https://doi.org/10.1104/pp.111.172981.

Curtin, S. J., Voytas, D. F., & Stupar, R. M. (2012). Genome engineering of crops with designer nucleases. *Plant Genome, 5*(2), 42–50. https://doi.org/10.3835/plantgenome2012.06.0008.

Eshed, Y., & Lippman, Z. B. (2019). Revolutions in agriculture chart a course for targeted breeding of old and new crops. *Science, 366*(6466). https://doi.org/10.1126/science.aax0025

Food and Agriculture Organization. (2019). *The state of food security and nutrition in the world.* Rome: Food and Agriculture Organization of the United Nations.

Friedrichs, S., Takasu, Y., Kearns, P., Dagallier, B., Oshima, R., Schofield, J., & Moreddu, C. (2019). Meeting report of the OECD conference on "Genome Editing: Applications in Agriculture-Implications for Health, Environment and Regulation". *Transgenic Research.* Meeting report of the OECD conference on "genome editing: applications in agriculture—implications for health, environment and regulation",, *28*(3–4), 419–463. https://doi.org/10.1007/s11248-019-00154-1.

Galindo-González, L., Pinzón-Latorre, D., Bergen, E. A., Jensen, D. C., & Deyholos, M. K. (2015). Ion Torrent sequencing as a tool for mutation discovery in the flax (Linum usitatissimum L.) genome. *Plant Methods, 11*(1), 19. https://doi.org/10.1186/s13007-015-0062-x.

Gao, C. (2018). The future of CRISPR technologies in agriculture. *Nature Reviews Molecular Cell Biology, 19*(5), 275–276. https://doi.org/10.1038/nrm.2018.2.

Garcia Ruiz, M. T., Knapp, A. N., & Garcia-Ruiz, H. (2018). Profile of genetically modified plants authorised in Mexico. *GM Crops and Food, 9*(3), 152–168. https://doi.org/10.1080/21645698.2018.1507601.

Gilles, A. F., & Averof, M. (2014). Functional genetics for all: Engineered nucleases, CRISPR and the gene editing revolution. *EvoDevo, 5*(1), 43. https://doi.org/10.1186/2041-9139-5-43.

Graham, D. B., & Root, D. E. (2015). Resources for the design of CRISPR gene editing experiments. *Genome Biology, 16*(1), 260. https://doi.org/10.1186/s13059-015-0823-x.

Grohmann, L., Keilwagen, J., Duensing, N., Dagand, E., Hartung, F., Wilhelm, R., … Sprink, T. (2019). Detection and identification of genome editing in plants: Challenges and opportunities. *Frontiers in Plant Science, 10*, 236. https://doi.org/10.3389/fpls.2019.00236.

Gupta, M., DeKelver, R. C., Palta, A., Clifford, C., Gopalan, S., Miller, J. C., … Petolino, J. F. (2012). Transcriptional activation of *Brassica napus* beta-ketoacyl-ACP synthase II with an engineered zinc finger protein transcription factor. *Plant Biotechnology Journal, 10*(7), 783–791. https://doi.org/10.1111/j.1467-7652.2012.00695.x.

Haft, D. H., Selengut, J., Mongodin, E. F., & Nelson, K. E. (2005). A guild of 45 CRISPR-associated (Cas) protein families and multiple CRISPR/Cas subtypes exist in prokaryotic genomes. *PLoS Computational Biology, 1*(6), e60. https://doi.org/10.1371/journal.pcbi.0010060.

Haun, W., Coffman, A., Clasen, B. M., Demorest, Z. L., Lowy, A., & Ray, E. (2014). Improved soybean oil quality by targeted mutagenesis of the fatty acid desaturase 2 gene family. *Plant Biotechnology Journal, 12*(7), 934–940. https://doi.org/10.1111/pbi.12201.

Hickey, L. T., Hafeez, A. N., Robinson, H., Jackson, S. A., Leal-Bertioli, S. C. M., Tester, M., … Wulff, B. B. H. (2019). Breeding crops to feed 10 billion. *Nature Biotechnology, 37*(7), 744–754. https://doi.org/10.1038/s41587-019-0152-9.

Holme, I. B., Gregersen, P. L., & Brinch-Pedersen, H. (2019). Induced genetic variation in crop plants by random or targeted mutagenesis: Convergence and differences. *Frontiers in Plant Science, 10*, 1468. https://doi.org/10.3389/fpls.2019.01468.

Hribova, E., Neumann, P., Macas, J., & Dolezel, J. (2009). Analysis of genome structure and organisation in banana (Musa acuminata) using. In *Plant and animal genomes XVII*. San Diego, CA, 454 sequencing.

Hsu, P. D., Scott, D. A., Weinstein, J. A., Ran, F. A., Konermann, S., & Agarwala, V. (2013). DNA targeting specificity of RNA-guided Cas9 nucleases. *Nature Biotechnology, 31*(9), 827–832. https://doi.org/10.1038/nbt.2647.

Hsu, P. D., Lander, E. S., & Zhang, F. (2014). Development and applications of CRISPR-Cas9 for genome engineering. *Cell, 157*(6), 1262–1278. https://doi.org/10.1016/j.cell.2014.05.010.

Jia, H., Zhang, Y., Orbović, V., Xu, J., White, F. F., Jones, J. B., & Wang, N. (2017). Genome editing of the disease susceptibility gene *CsLOB1* in citrus confers resistance to citrus canker. *Plant Biotechnology Journal, 15*(7), 817–823. https://doi.org/10.1111/pbi.12677.

Jiang, W. Z., Henry, I. M., Lynagh, P. G., Comai, L., Cahoon, E. B., & Weeks, D. P. (2017). Significant enhancement of fatty acid composition in seeds of the allohexaploid, Camelina sativa, using CRISPR/Cas9 gene editing. *Plant Biotechnology Journal, 15*(5), 648–657. https://doi.org/10.1111/pbi.12663.

Kc, S., & Lutz, W. (2017). The human core of the shared socioeconomic pathways: Population scenarios by age, sex and level of education for all countries to 2100. *Global Environmental Change: Human and Policy Dimensions, 42*, 181–192. https://doi.org/10.1016/j.gloenvcha.2014.06.004.

Li, J. F., Norville, J. E., Aach, J., McCormack, M., Zhang, D., Bush, J., ... Sheen, J. (2013). Multiplex and homologous recombination–mediated genome editing in Arabidopsis and Nicotiana benthamiana using guide RNA and Cas9. *Nature Biotechnology, 31*(8), 688–691. https://doi.org/10.1038/nbt.2654.

Liang, Z., Zhang, K., Chen, K., & Gao, C. (2014). Targeted mutagenesis in Zea mays using TALENs and the CRISPR/Cas system. *Journal of Genetics and Genomics, 41*(2), 63–68. https://doi.org/10.1016/j.jgg.2013.12.001.

Lor, V. S., Starker, C. G., Voytas, D. F., Weiss, D., & Olszewski, N. E. (2014). Targeted mutagenesis of the tomato PROCERA gene using transcription activator-like effector nucleases. *Plant Physiology, 166*(3), 1288–1291. https://doi.org/10.1104/pp.114.247593.

Mali, P., Yang, L., Esvelt, K. M., Aach, J., Guell, M., DiCarlo, J. E., ... Church, G. M. (2013). RNA-guided human genome engineering via Cas9. *Science, 339*(6121), 823–826. https://doi.org/10.1126/science.1232033.

Metzker, M. L. (2010). Sequencing technologies—The next generation. *Nature Reviews. Genetics, 11*(1), 31–46. https://doi.org/10.1038/nrg2626.

Miao, J., Guo, D., Zhang, J., Huang, Q., Qin, G., Zhang, X., ... Qu, L. J. (2013). Targeted mutagenesis in rice using CRISPR-Cas system. *Cell Research, 23*(10), 1233–1236. https://doi.org/10.1038/cr.2013.123.

Morgante, M., & Salamini, F. (2003). From plant genomics to breeding practice. *Current Opinion in Biotechnology, 14*(2), 214–219. https://doi.org/10.1016/s0958-1669(03)00028-4.

Nekrasov, V., Staskawicz, B., Weigel, D., Jones, J. D., & Kamoun, S. (2013). Targeted mutagenesis in the model plant Nicotiana benthamiana using Cas9 RNA-guided endonuclease. *Nature Biotechnology, 31*(8), 691–693. https://doi.org/10.1038/nbt.2655.

Nekrasov, V., Wang, C., Win, J., Lanz, C., Weigel, D., & Kamoun, S. (2017). Rapid generation of a transgene-free powdery mildew resistant tomato by genome deletion. *Scientific Reports, 7*(1), 482. https://doi.org/10.1038/s41598-017-00578-x.

Ortigosa, A., Gimenez-Ibanez, S., Leonhardt, N., & Solano, R. (2018). Design of a bacterial speck resistant tomato by CRISPR/Cas9-mediated editing of s.l. JAZ 2. *Plant Biotechnology Journal, 17*(3), 665–673. https://doi.org/10.1111/pbi.13006.

Osakabe, K., Osakabe, Y., & Toki, S. (2010). Site-directed mutagenesis in *Arabidopsis* using custom-designed zinc finger nucleases. *Proceedings of the National Academy of Sciences of the United States of America, 107*(26), 12034–12039. https://doi.org/10.1073/pnas.1000234107.

Pérez-de-Castro, A. M., Vilanova, S., Cañizares, J., Pascual, L., Blanca, J. M., ... Picó, B. (2012). Application of genomic tools in plant breeding. *Current Genomics, 13*(3), 179–195. https://doi.org/10.2174/138920212800543084.

Puchta, H., & Fauser, F. (2013). Gene targeting in plants: 25 years later. *International Journal of Developmental Biology, 57*(6–8), 629–637. https://doi.org/10.1387/ijdb.130194hp.

Qi, Y., Li, X., Zhang, Y., Starker, C. G., Baltes, N. J., Zhang, F., … Voytas, D. F. (2013a). Targeted deletion and inversion of tandemly arrayed genes in Arabidopsis thaliana using zinc finger nucleases. *G3, 3*(10), 1707–1715. https://doi.org/10.1534/g3.113.006270.

Qi, Y., Zhang, Y., Zhang, F., Baller, J. A., Cleland, S. C., Ryu, Y., … Voytas, D. F. (2013b). Increasing frequencies of site-specific mutagenesis and gene targeting in *Arabidopsis* by manipulating DNA repair pathways. *Genome Research, 23*(3), 547–554. https://doi.org/10.1101/gr.145557.112.

Ronaghi, M., Karamohamed, S., Pettersson, B., Uhlén, M., & Nyrén, P. (1996). Real-time DNA sequencing using detection of pyrophosphate release. *Analytical Biochemistry, 242*(1), 84–89. https://doi.org/10.1006/abio.1996.0432.

Rothberg, J. M., Hinz, W., Rearick, T. M., Schultz, J., Mileski, W., Davey, M., … Bustillo, J. (2011). An integrated semiconductor device enabling non-optical genome sequencing. *Nature, 475*(7356), 348–352. https://doi.org/10.1038/nature10242.

Scheffler, B. E., Kuhn, D. N., Motamayor, J. C., & Schnell, R. J. (2009). Efforts towards sequencing the Cacao genome (Theobroma cacao). In *Plant and animal genomes* XVII. San Diego, CA.

Schindele, A., Dorn, A., & Puchta, H. (2020). CRISPR/Cas brings plant biology and breeding into the fast lane. *Current Opinion in Biotechnology, 61*, 7–14. https://doi.org/10.1016/j.copbio.2019.08.006.

Sera, T. (2005). Inhibition of virus DNA replication by artificial zinc finger proteins. *Journal of Virology, 79*(4), 2614–2619. https://doi.org/10.1128/JVI.79.4.2614-2619.2005.

Shan, Q., Wang, Y., Li, J., Zhang, Y., Chen, K., Liang, Z., … Gao, C. (2013a). Targeted genome modification of crop plants using a CRISPR-Cas system. *Nature Biotechnology, 31*(8), 686–688. https://doi.org/10.1038/nbt.2650.

Shan, Q., Wang, Y., Chen, K., Liang, Z., Li, J., Zhang, Y., … Gao, C. (2013b). Rapid and efficient gene modification in rice and Brachypodium using TALENs. *Molecular Plant, 6*(4), 1365–1368. https://doi.org/10.1093/mp/sss162.

Shimatani, Z., Kashojiya, S., Takayama, M., Terada, R., Arazoe, T., Ishii, H., … Kondo, A. (2017). Targeted base editing in rice and tomato using a CRISPR-Cas9 cytidine deaminase fusion. *Nature Biotechnology, 35*(5), 441–443. https://doi.org/10.1038/nbt.3833.

Shukla, V. K., Doyon, Y., Miller, J. C., DeKelver, R. C., Moehle, E. A., Worden, S. E., … Urnov, F. D. (2009). Precise genome modification in the crop species *Zea mays* using zinc-finger nucleases. *Nature, 459*(7245), 437–441. https://doi.org/10.1038/nature07992.

Smith, J., Bibikova, M., Whitby, F. G., Reddy, A. R., Chandrasegaran, S., & Carroll, D. (2000). Requirements for double-strand cleavage by chimeric restriction enzymes with zinc finger DNA-recognition domains. *Nucleic Acids Research, 28*(17), 3361–3369. https://doi.org/10.1093/nar/28.17.3361.

Smyth, S. J. (2017). Canadian regulatory perspectives on genome engineered crops. *GM Crops and Food, 8*(1), 35–43. https://doi.org/10.1080/21645698.2016.1257468.

Stein, N. (2009). Barley genome sequencing: First steps. In *Plant and animal genomes XVII*. San Diego, CA.

Swaminathan, K., Varala, K., Moose, S. P., Rokhsar, D., Ming, R., & Hudson, M. E. (2009). A genome survey of Miscanthus Giganteus. In *Plant and animal genomes XVII*. San Diego, CA.

Takenaka, K., Koshino-Kimura, Y., Aoyama, Y., & Sera, T. (2007). Inhibition of tomato yellow leaf curl virus replication by artificial zinc-finger proteins. *Nucleic Acids Symposium Series, 51*(51), 429–430. https://doi.org/10.1093/nass/nrm215.

Tester, M., & Langridge, P. (2010). Breeding technologies to increase crop production in a changing world. *Science, 327*(5967), 818–822. https://doi.org/10.1126/science.1183700.

Townsend, J. A., Wright, D. A., Winfrey, R. J., Fu, F., Maeder, M. L., Joung, J. K., & Voytas, D. F. (2009). High-frequency modification of plant genes using engineered zinc-finger nucleases. *Nature, 459*(7245), 442–445. https://doi.org/10.1038/nature07845.

Tuberosa, R., Graner, A., & Varshney, R. K. (2011). Genomics of plant genetic resources: An introduction. *Plant Genetic Resources, 9*(2), 151–154. https://doi.org/10.1017/S1479262111000700.

Velasco, R. (2009). The golden delicious apple genome: An international whole genome sequencing initiative. In *Plant and animal genomes XVII*. San Diego, CA.

Velasco, R., Zharkikh, A., Troggio, M., Salvi, S., & Pindo, M. (2009). Apple genome sequencing and post-genomic program at IASMA research center. In *Plant and animal genomes, XVII*. San Diego, CA.

Waltz, E. (2016a). Gene-edited CRISPR mushroom escapes US regulation. *Nature, 532*(7599), 293. https://doi.org/10.1038/nature.2016.19754.

Waltz, E. (2016b). CRISPR-edited crops free to enter market, skip regulation. *Nature Biotechnology, 34*(6), 582. https://doi.org/10.1038/nbt0616-582.

Wang, Y., Cheng, X., Shan, Q., Zhang, Y., Liu, J., Gao, C., & Qiu, J. L. (2014). Simultaneous editing of three homoeoalleles in hexaploid bread wheat confers heritable resistance to powdery mildew. *Nature Biotechnology, 32*(9), 947–951. https://doi.org/10.1038/nbt.2969.

Wang, F., Wang, C., Liu, P., Lei, C., Hao, W., Gao, Y., … Zhao, K. (2016). Enhanced rice blast resistance by CRISPR/Cas9-targeted mutagenesis of the ERF transcription factor gene OsERF922. *PLoS One, 11*(4), e0154027. https://doi.org/10.1371/journal.pone.0154027.

Wendt, T., Holm, P. B., Starker, C. G., Christian, M., Voytas, D. F., Brinch-Pedersen, H., & Holme, I. B. (2013). TAL effector nucleases induce mutations at a pre-selected location in the genome of primary barley transformants. *Plant Molecular Biology, 83*(3), 279–285. https://doi.org/10.1007/s11103-013-0078-4.

Whelan, A. I., & Lema, M. A. (2015). Regulatory framework for gene editing and other new breeding techniques (NBTs) in Argentina. *GM Crops and Food, 6*(4), 253–265. https://doi.org/10.1080/21645698.2015.1114698.

Wicker, T., Schlagenhauf, E., Graner, A., Close, T. J., Keller, B., & Stein, N. (2006). 454 Sequencing put to the test using the complex genome of barley. *BMC Genomics, 7*, 275. https://doi.org/10.1186/1471-2164-7-275.

Wilkins, T. A., Mudge, J., Abidi, N., Allen, R., & Auld, D. (2009). The sequencing and resequencing of cotton. In *Plant and animal genomes XVII*. San Diego, CA.

Zaidi, S. S. E. A., Vanderschuren, H., Qaim, M., Mahfouz, M. M., Kohli, A., Mansoor, S., & Tester, M. (2019). New plant breeding technologies for food security. *Science, 363*(6434), 1390–1391. https://doi.org/10.1126/science.aav6316.

Zhang, F., Maeder, M. L., Unger-Wallace, E., Hoshaw, J. P., Reyon, D., Christian, M., … Voytas, D. F. (2010). High frequency targeted mutagenesis in Arabidopsis thaliana using zinc finger nucleases. *Proceedings of the National Academy of Sciences of the United States of America, 107*(26), 12028–12033. https://doi.org/10.1073/pnas.0914991107.

Zhang, Y., Zhang, F., Li, X., Baller, J. A., Qi, Y., Starker, C. G., … Voytas, D. F. (2013). Transcription activator-like effector nucleases enable efficient plant genome engineering. *Plant Physiology, 161*(1), 20–27. https://doi.org/10.1104/pp.112.205179.

Zhang, H., Gou, F., Zhang, J., Liu, W., Li, Q., Mao, Y., … Zhu, J. K. (2016). TALEN-mediated targeted mutagenesis produces a large variety of heritable mutations in rice. *Plant Biotechnology Journal, 14*(1), 186–194. https://doi.org/10.1111/pbi.12372.

Zhang, Y., Bai, Y., Wu, G., Zou, S., Chen, Y., Gao, C., & Tang, D. (2017). Simultaneous modification of three homoeologs of Ta EDR 1 by genome editing enhances powdery mildew resistance in wheat. *Plant Journal: For Cell and Molecular Biology, 91*(4), 714–724. https://doi.org/10.1111/tpj.13599.

Zhang, Y., Li, D., Zhang, D., Zhao, X., Cao, X., Dong, L., … Wang, D. (2018). Analysis of the functions of Ta GW 2 homoeologs in wheat grain weight and protein content traits. *Plant Journal: For Cell and Molecular Biology, 94*(5), 857–866. https://doi.org/10.1111/tpj.13903.

Zhou, J., Peng, Z., Long, J., Sosso, D., Liu, B., Eom, J. S., … Yang, B. (2015). Gene targeting by the TAL effector PthXo2 reveals cryptic resistance gene for bacterial blight of rice. *Plant Journal: For Cell and Molecular Biology, 82*(4), 632–643. https://doi.org/10.1111/tpj.12838.

Chapter 11
Plant Growth-Promoting Rhizobacteria (PGPR): Strategies to Improve Heavy Metal Stress Under Sustainable Agriculture

Ananya Roy Chowdhury

Abstract Among several soil pollutants, the heavy metal effluents discharged from different industries directly or indirectly influence the global environmental balance and eventually decrease agricultural productivity. As a result of these harmful activities, soil pollution due to heavy metal toxicity is a potentially crucial environmental issue globally. The conventional methods of removing the huge metals from the environment are not eco-friendly, and these processes produce huge toxic residues. So, in this situation, bioremediation is the most preferred way to minimise the effects of heavy metals on the environment. Under such circumstances, the impact of plant growth-promoting rhizobacteria (PGPR) in remediation of metal toxicated areas has gained importance in sustainable agriculture systems. PGPRs increase plant growth by solubilising phosphate, synthesising IAA, producing enzymes, fixing the nitrogen, etc. So, the inoculation of suitable and specific heavy metal-tolerant PGPR strains associated with plants can maximise the phytoremediation. In this work, the impact of PGPR on remediation of the heavy metal contaminated zone is adequately described.

Keywords PGPR · Heavy metal · Phytoremediation · Sustainable · Agriculture

11.1 Introduction

The continued expansion of industrial activities and, more particularly, the dense industrial effluents are the main reasons contributing to soil pollution. Among various soil pollutants, heavy metals are highly phytotoxic, and their toxicity has a significant effect not only on plant growth but also on mass crop yield and health. It is a well-known fact that to enhance crop production, the deliberate application of

---

A. Roy Chowdhury (✉)
Department of Botany, Chakdaha College, Nadia, West Bengal, India

S. A. Bandh (ed.), *Sustainable Agriculture*,
https://doi.org/10.1007/978-3-030-83066-3_11

chemical fertilisers, especially nitrogen and phosphorus, has led to extreme deleterious effects on soil structure and total plant health. In this situation, rhizosphere researchers have been throwing up surprises regarding the rhizospheric microorganisms. Plant growth-promoting rhizobacteria (PGPR) are bacteria that live in plant roots. In recent years, substantial attention has been paid to the potential of PGPR to substitute agrochemicals (fertilisers and pesticides) for plant growth promotion through a variety of mechanisms including organic matter decomposition, soil structure formation, organic pollutant degradation, mineral solubilisation and biocontrol of seed-borne pathogens. Heavy metal stress has become a big issue in whole terrestrial ecosystems worldwide. Damage to soil texture means disturbing the pH of soil. Heavy metal accumulation is the chief factors responsible for the reduction of plant growth and development. The huge industrial discharge, particularly wastewater discharge, contains a heavy load of metal effluents. When these materials get accumulated in agricultural land through irrigation, they produce severe problems in human bodies and the entire living systems. Under such circumstances, PGPR can be the safest option to decrease the notorious impacts of heavy metals on these environments.

11.2 An Introduction to PGPR

The rhizosphere is a layer of soil that is tightly regulated by the root system of the plant. This area is nutrient-dense as a result of the accumulation of a variety of nutritious plant exudates, such as sugars and amino acids. It is home to a diverse array of bacteria that colonise this region. Rhizobacteria are the microorganisms and bacteria that inhabit this area. Numerous rhizobacteria genera have been classified as PGPR, but *Pseudomonas* and *Bacillus* are the most prevalent.

Due to the ever-increasing hunger of the excessively increasing human population, the use of PGPR for reducing the application of agrochemicals is a critical issue. PGPR, the beneficial root-inhabiting bacteria, stimulates plant growth and protects them from various seed-borne pathogens by establishing a symbiotic relationship.

11.3 Mechanisms of PGPR's Action

PGPRs promote plant development in a number of ways, both overt and indirect with phosphate solubilisation, nitrogen fixation, IAA synthesis and siderophore synthesis as the examples of direct pathways. Indirect pathways, on the other hand, involve the suppression of fungal, bacterial, fungal and nematode infections by the synthesis of various enzymes such as cellulases proteases and chitinases. Additional indirect pathways include quorum sensing, signal interference, mineral nutrient solubilisation, biofilm inhibition and systemic acquired tolerance (Fig. 11.1).

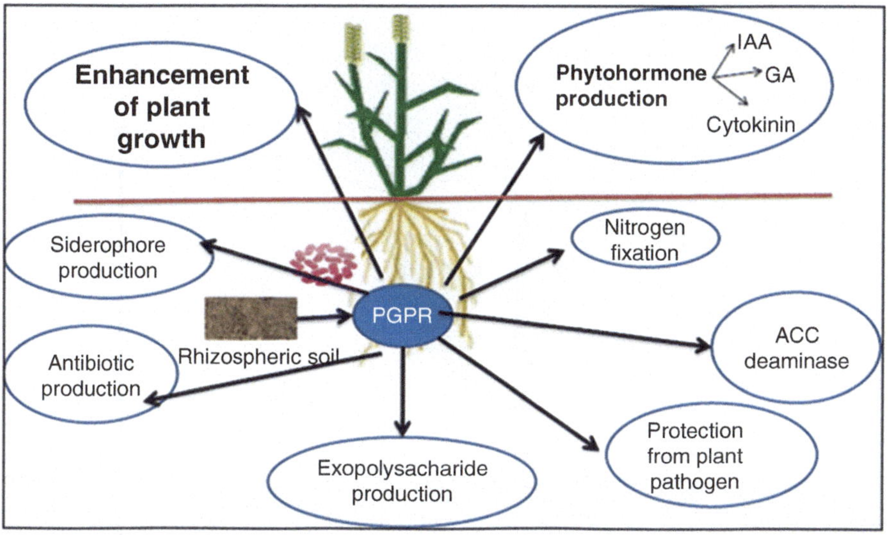

Fig. 11.1 Mechanism of PGPR's actions

The root-colonising bacteria can improve plant growth by N-fixation (Djordjevic et al., 1987; Strzelczyk et al., 1994), phosphate solubilisation (Kloepper et al., 1988), phytohormone (auxin, gibberellins, cytokinin) production and decreasing the ethylene level in plants (Glick et al., 2007; Glick et al., 1999). Promoting water absorption and nutrient translocation, promoting rhizo anatomical development (Okon & Kapulnik, 1986), improving the whole enzyme system and cooperating with other groups of beneficiary soil microbes to perform better are the other mechanisms by which they improve the plant growth.

11.3.1 Direct Mechanism

PGPRs enhance the growth of plants through the following direct (Arora et al., 2012; Bhardwaj et al., 2014).

11.3.1.1 Nitrogen Fixation

PGPRs are widely applied to fix nitrogen, the most significant nutrient for plant growth and development (Fig. 11.2). Irrespective of the presence of nitrogen in the highest concentration in air, the plants are incapable of converting it into ammonia, thus remaining unavailable to plants. PGPRs convert dinitrogen into ammonia, utilising nitrogenase enzyme (Gaby & Buckley, 2012). They fix nitrogen either by building symbiotic association or by non-symbiotic pathway. Among different

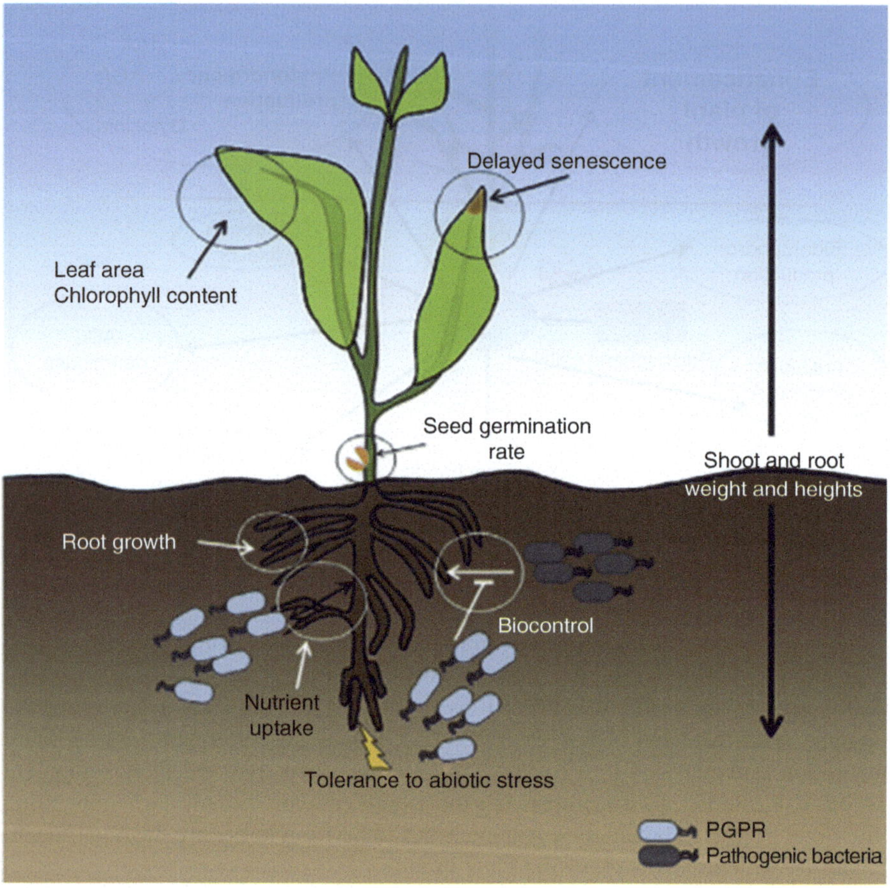

Fig. 11.2 Impact PGPR on plant growth (citation-60)

nutrients, nitrogen is one of the most essential, specifically in rice production. Every year 50–70% loss in rice yield occurs due to the failure of fulfilment of nitrogen demand of rice plants by chemical fertilisers (Ladha et al., 2005). As new varieties of rice demand a higher amount of nitrogen, it is getting impossible to provide it only by chemical fertilisers.

Among the root-colonising bacteria population that fix atmospheric nitrogen and benefit plant growth are plant growth-promoting rhizobacteria (PGPR). Alternatively, they are known as bio-enhancers or biofertilisers (Kloepper et al., 1980; Shamsuddin et al., 2014).

It is stated that PGPRs fix nitrogen in cereals, banana and grasses (Döbereiner, 1997). They also increase the nutrient absorption rate and resistance to droughts (Arzanesh et al., 2011). Among several naturally occurring host-microbe interactions, the symbiotic relationship between *Rhizobium* and leguminous plants is well established. This symbiosis is best understood and is a well-applied nitrogen-providing system to leguminous plants. Nowadays, research is going on to develop *Rhizobium*-non-legume interactions as well. This approach involves the integration of nitrogen-fixing gene into the rice-*Rhizobium* system. The rhizobial gene manipulation and modulation have lots of benefits. It develops a high level of root architecture, increases root hairs and enhances the rate of nutrient absorption (Yanni et al., 1997).

The co-inoculation of *Rhizobium* with two PGPR strains, *Pseudomonas fluorescens* P-93 and *Azospirillum lipoferum* S-21, effectively controlled the total nitrogen uptake, nutrient uptake and translocation of nutrients in *Phaseolus vulgaris* L. The PGPRs first make entry into the plants following the nodule formation where nitrogen fixation starts. The rhizo-microbial population showing symbiotic association includes several symbionts, like *Bradyrhizobium*, *Rhizobium*, *Mesorhizobium* and *Sinorhizobium* with legume-forming plants (Zahran, 2001).

Non-symbiotic nitrogen fixation is carried out by free-living diazotrophs, which concurrently promote the growth and yield of non-leguminous plants. *Azotobacter*, *Azoarcus*, *Acetobacter*, *Burkholderia*, *Cyanobacteria* and *Pseudomonas* are all examples of non-symbiotic nitrogen fixers, 2012; Vessey, 2003).

11.3.1.2 Phosphate Solubilisation

Phosphorus is another vital mineral element for a plant's nutrition. It plays a pivotal role in photosynthesis, signal transduction, respiration and energy transfer. In soil, phosphate is present in inorganic and organic form with inadequate amounts, but plants can't absorb phosphate because 98% of phosphate is present in insoluble and precipitated form (Pandey & Maheshwari, 2007). Plants can utilise phosphate only in two forms, the monobasic form (H_2PO_4) and the dibasic (HPO_4^{2-}) form (Bhattacharyya & Jha, 2012).

It remains in the soil either in mineral salt form or in organic form. Hence, irrespective of the abundance of phosphorus in the soil, the plants can't absorb it because of its insolubility which becomes a major limiting factor for the proper development of plants. That's why it gradually becomes necessary to apply phosphorus in soluble form through fertilisers in the agricultural field.

Recent research indicates that inoculating crops with phosphate-solubilising microbes (PSM) will result in a 50% reduction in phosphate fertiliser application without affecting crop output (Yazdani et al., 2009). PSB (phosphate-solubilising bacteria) can also be beneficial in the phytoremediation of soils contaminated with heavy metals or in the bioleaching of rare earth elements from mined ores.

PGPRs solubilise inorganic phosphates by releasing phosphatase enzyme during substrate degradation (Sharma et al., 2013). Phosphate-solubilising PGPRs belong to the genera *Arthrobacter*, *Beijerinckia*, *Microbacterium*, *Erwinia*, *Rhodococcus*, *Burkholderia*, *Flavobacterium*, *Enterobacter* and *Serratia* (Bhattacharyya & Jha, 2012). It is also reported that the phosphate solubilisation rate gradually increases by the application of other beneficial soil microbes along with PGPRs (Zaidi et al., 2009).

11.3.1.3 Siderophore Production

Iron is a critical micronutrient for all living species and is found in abundance in soil. Irrespective of its high presence in soil, plants can't utilise it because of its low solubility rate. Iron is the fourth in position according to its abundance on earth. It is readily assimilated neither by the plants nor by any bacteria because of its presence in the aerobic soils in ferric ion (Fe^{3+}) form that is not readily soluble in water (Ma, 2005). But some microorganisms have developed some unique mechanisms for iron assimilation, including the formation of low-molecular-weight iron-chelating products called siderophores (Arora et al., 2012; Schwyn & Neilands, 1987). Through siderophores, it enters within the plant body. The siderophores perform an important function like iron's extracellular solubilisation from minerals.

According to distinct functional groups, siderophores can be divided into three major categories: carboxylates, hydroxamates and catecholates (Cornelis, 2010), and bacteria can produce all four types of siderophores. Examples of some active siderophore-producing bacteria are *Salmonella*, *Enterobacter*, *Vibrio cholerae*, *Escherichia coli*, *Aeromonas* and *Yersinia*.

Some fungi, for examples, *A. versicolor* (Holinsworth & Martin, 2009), *Ustilago sphaerogena* (Shanmugaiah et al., 2015), *Rhizopus* (Shenker et al., 1992), etc., are also reported to produce siderophore. A large number of PGPRs, e.g. *Streptomyces* (Dimkpa et al., 2008), *Azotobacter* (Fekete et al., 1983), *Rhizobium* (Datta & Chakrabartty, 2014), *Burkholderia* (Ong et al., 2016), *Aeromonas* (Hirst et al., 1991), *Pseudomonas*, etc. (Sujatha & Ammani, 2013), also produce siderophores and improve plant growth.

11.3.1.4 Production of Phytohormone

Phytohormones are usually organic, and their impact on plant occurs in a meagre amount. They are synthesised in different tissues and are then transported to their target sites. Hormones are categorised into five groups by plant biologists: auxin, cytokinins, ethylene, gibberellins and abscisic acid (ABA). Recently, two novel hormones, brassinosteroid and strigolactones, are also reported to be produced by plants (Zwanenburg et al., 2016). Microorganisms in the rhizosphere generate growth-stimulating hormones such as indole acetic acid (IAA), cytokinins and gibberellins, among others (Arora et al., 2013), which significantly enhance plant growth.

Indole Acetic Acid (IAA)

The fungi *Rhizopus suinus* and *Absidia ramosa* were identified to produce auxin. About 80% of root-colonising microbial populations isolated from different crops and vegetables are proven to produce auxin due to secondary metabolism (Vessey, 2003). IAA is the natural auxin, and it has positive effects on the root and shoot elongation. The primary precursor of IAA is tryptophan which is found to occur in root exudates (Miransari & Smith, 2014). Different PGPRs like *Pseudomonas*, *Agrobacterium*, *Klebsiella* and *Enterobacter* produce IAA either via the formation of indole-3-pyruvic acid or via indole-3-acetic-aldehyde (Shilev, 2013).

Several free-living PGPRs like *Alckaligenes faecalis*, *Acetobacter diazotrophicous* and *Enterobacter cloacae* are also related to the low level of IAA production. IAA increases the rate of cell division, differentiation, lateral and adventitious root development and pigment production and provides resistance to stress conditions (Spaepen & Vanderleyden, 2011).

IAA is a secondary metabolite produced through either tryptophan-dependent pathway or an independent tryptophan pathway in plants and bacteria. In *Azospirillum brasilense* IAA is produced through tryptophan-independent pathway. It is reported that IAA produced in wheat plants by *Azospirillum brasilense* stimulates a high number and length of lateral roots.

Irrespective of plant growth promotion and root nodulation, IAA also helps in root proliferation and root branching. The function of IAA, produced by the PGPB *Pseudomonas putida* GR12-2, is well established by the experiment done on canola roots. The IAA-deficient mutant bacterial strain was applied on other canola plant sets and eventually showed no such root and shoot growth compared to PGPB-treated set. Inoculation in a seed with *Pseudomonas putida* GR12-2 showed root formations that were 35–50% longer than the roots grown from seeds inoculated with IAA-deficient mutant PGPB strain. In another set where the mung bean plant was used in experimentation, the same PGPB showed overproduction of IAA in plants compared to the uninoculated control set. IAA is proven in taking a role in root and shoot growth and necessary for the formation of nodules.

The production of IAA in plants is a stimulator of the cell wall loosening. It is secreted in higher amounts as a root exudate, which provides excess nutrients to root-colonising bacteria.

Gibberellins and Cytokinins

Among different rhizobacteria, *Azotobacter* sp., *Bacillus subtilis*, *Pseudomonas fluorescens*, *Paenibacillus polymyxa* and *Rhodospirillum rubrum* are known to produce either gibberellins or cytokinins or both, which play an essential role in plant growth promotion (Kang et al., 2010). PGPRs produce a lower level of cytokinins as compared to different phytopathogens. According to Barea et al. (2005), 90% of root-colonising bacteria isolated from other crops exhibit the ability of cytokinin like compound production, e.g. a free-living soil bacterium, *Pseudomonas polymyxa*, was reported to produce cytokinins (Timmusk et al., 2014).

Gibberellins are involved in seed germination, floral induction, stem and floral growth and crop and fruit development. Simultaneously, GA processing by PGPRs encourages the growth and yield of a wide variety of grain plants and vegetables.

It is also reported that *Azospirillum brasilense* and *Arthrobacter giacomelli* produce a dense concentration of cytokinins grown in mixed culture condition (Lippi et al., 1991). In 1972, Philips and Torrey experimentally proved the presence of zeatin like compounds around the *Rhizobium* nodules.

11.3.2 Indirect Mechanisms

The application of PGPR is a promising approach for sustainable agriculture, as it upsurges plant growth and soil fertility in an indirect way. Based on their indirect mechanism, researchers are trying to reduce the application of different agrochemicals. PGPR can improve soil fertility via antibiosis, lytic enzyme production, the implication of induced systemic resistance, etc. (Lugtenberg & Kamilova, 2009).

11.3.2.1 Antibiotic Production

According to researchers, PGPRs develop some antibiotics that are highly effective against a variety of soil-borne phytopathogens (Shilev, 2013). Numerous antibiotics are developed by various *Pseudomonas* bacteria, including phenazine, oomycin A, pyrrolnitrin, tensin and cyclic lipopeptides (Loper & Gross, 2007). *Streptomyces* and *Bacillus* contain antibiotics such as kanamycin, oligomycin A and zwittermicin A (Compant et al., 2005).

Pseudomonas sp. produces 2,4-DAPG in soil, which can be used to monitor *Gaeumannomyces graminis* var. *tritici* in wheat (de Souza et al., 2003). *Bacillus amyloliquefaciens* produces lipopeptides and polyketides that are effective against soil-borne pathogens (Ongena & Jacques, 2008). Certain PGPRs can also synthesise a volatile compound called HCN, which is used as a biocontrol agent against *Thielaviopsis basicola* (Sacherer et al., 1994).

11.3.2.2 Lytic Enzyme Production

PGPRs can produce different types of lytic enzymes like chitinases, lipases, proteases, dehydrogenase, phosphatase, etc. (Hayat et al., 2010; Joshi et al., 2012). They show hyperparasitic function in attacking phytopathogens by their cell wall hydrolysis. PGPRs can also tolerate different living and non-living stresses by suppressing several pathogenic fungi, including *Fusarium oxysporum*, *Pythium ultimum*, *Rhizoctonia solani*, etc. (Nadeem et al., 2013; Upadyay et al., 2012). *Azotobacter chroococcum* has been reported to give good performance on *Sesamum indicum* at field trial (Maheshwari et al., 2012). Similarly, *Trichoderma* sp. inoculation in peanut acts as a biocontrol against *Aspergillus niger* which causes collar rot disease in the plant (Rabeendran et al., 2000).

11.3.2.3 Development of Induced Systemic Resistance (ISR)

When the environmental stimuli activate, the innate defence mechanism of the plant gets power against different challenges, called ISR (Avis et al., 2008). As PGPRs provide systemic resistance against several pathogens, like bacteria, fungi, insects and nematodes, they can be applied to host plants (Naznin et al., 2013). ISR stimulates jasmonates and ethylene, which help plants against several types of pathogens (Glick, 2012).

Induced resistance is a physiological condition that occurs when plant growth-promoting rhizobacteria induce an increase in defensive capacity (PGPR). The enzymes which have so far been reported to be involved with ISR include chitinase, peroxidase, superoxide dismutase, phenylalanine lyase, catalase, polyphenol oxidase and ascorbate peroxidase (Annapurna et al., 2013).

11.4 Impact of PGPR on Plants

PGPR is a distinct group of root-colonising bacteria that facilitate rooting and promote overall plant growth (Glick et al., 1995). According to Piao et al. (1992), plant growth-improving bacteria are collectively called YIB (Yield Increasing Bacteria). Some workers referred to them as plant beneficial bacteria (PBB). Gaind and Gaur (1991) referred to them as 'direct PGPR', and Grayston and Germida (1991) supported the term 'direct PGPR', whereas described them as non-biocontrol PGPRs.

The PGPR can enhance plant growth by nitrogen fixation (Roy Chowdhury et al., 2017), phosphate solubilisation (; Yazdani et al., 2009), phytohormone production (Vejan et al., 2016), potassium solubilisation (Han & Lee, 2005; Parmar & Sindhu, 2013) and siderophore production (Beneduzi et al., 2012; Pahari & Mishra, 2017).

11.4.1 As Biofertilisers

The host plant-PGPR relationship is critical for optimising plant growth and production on a broad scale. As the PGPRs are efficient for producing different phytohormones, specifically IAA, the phytohormonal network is still understudied. Among several PGPRs, C138 is proven to supply iron in iron-starved tomato plants. Similarly, *Bacillus amyloliquefaciens* is reported as growth and yield improver in soybean in India.

Burkholderia kururiensis (Estrada-de los Santos et al., 2001) and *Burkholderia vietnamensis* (Gillis et al., 1995) are examples of nitrogen fixers. Besides nitrogen fixation, phosphate solubilisation and siderophore productions are also fulfilled by the PGPRs. For these reasons, PGPRs are used as biofertilisers.

11.4.2 As Biocontrol Agent

Few root-colonising bacteria, especially *Pseudomonads*, inhibit the growth of some soil-inhabited pathogens. These bacteria not only impart resistance but also help in the improvement of plant growth and yield. In such circumstances, the term 'biocontrol plant growth-promoting rhizobacteria' was proposed by soil microbiologists (Tilak et al., 1999).

In paddy, *P. fluorescens* strains exhibited inhibitory action on hyphal growth of *Rhizoctonia solani* (Radjacommare et al., 2004). *Stenotrophomonas marcescens* strain, another biocontrol PGPR, inhibits different soil-borne pathogens, including *Fusarium oxysporum*. Different strains of *Bacillus* spp. constitute the ability to indulge ISR in a wide variety of crops. The biocontrol power of *Bacillus* spp. is one of the critical agents that can combat rhizo- and soil-borne pathogens in the case of chickpea (Landa et al., 1997).

They offer biological control to plants either by antibiotic production or by siderophore/phytohormone production. Genetic modification has opened a new way for developing PGPRs as biocontrol agents.

11.4.3 As Environmental Stress Controller

Due to climate change, rainfall patterns become more erratic, resulting in a tremendous reduction in crop production. As a result, the plants either become exposed to severe drought condition or floods. In addition to these, the continuous rising of the pollutant level in the environment, specifically the toxic gases in the atmosphere and heavy metals in the soil, leads to a drastic reduction of crop yield. The chemical effluents from several industries are also getting mixed in the river water, which subsequently passes to the agricultural fields. All these situations are giving birth to different environmental stresses.

Under stress conditions, plants produce a high concentration of ethylene, utilising ACC as a precursor. The ethylene retards the root and shoot elongation and suppresses leaf expansion. So, it is clear that if the PGPRs can produce ACC deaminase, they can tolerate the stress to a certain level (Akhgar et al., 2014). The ACC deaminase synthesising PGPRs reduces several environmental stresses by producing different exopolysaccharides, which immediately binds with cations (Na^{++}) and eventually form a sheath on the plant roots. Few rhizobacteria also develop heavy metal tolerance mechanisms (Maxton et al., 2018) in plants.

ACC deaminase producers can relieve different stresses, such as heavy metals, drought, polyaromatic hydrocarbons and salts (Glick et al., 2007). Jacobson et al. (1994) showed that *Pseudomonas putida* contains ACC deaminase that helps reduce the adverse stress level in plants. Under water stress conditions, *Pseudomonas* sp.

can improve CAT enzyme activity in basil plants. In GPX and APX function and total chlorophyll content, combinations of three PGPRs (*Pseudomonas* sp., *Bacillus lentus*, *Azospirillum brasilense*) showed tremendously good result under water stress condition (Heidari & Golpayegani, 2012).

11.5 Reports on the Effect of PGPRs in the Role of Biofertilisers

Biofertiliser is becoming a fundamental pillar of eco-friendly organic farming and a significant component of sustainable agriculture globally. They contain products that can be inoculated to seeds, soil or epidermal plant portion. They subsequently colonise the rhizospheric area and ultimately enhance plant growth by providing nutrients to the host plant. As bio-formulation, biofertilisers contain many microorganisms responsible for enriching the plants' nutrient uptake status.

Azotobacter is a cytokinin synthesiser, which showed increased yield in cucumber (Alori et al., 2017). It fixes nitrogen in wheat, barley, rice, maize, lime, coconut, tobacco, etc. (Wani et al., 2013). *Azorhizobium* is highly efficient for nitrogen fixation in wheat, and it is applied as biofertiliser in wheat cultivation (Sabry et al., 1997). *Bacillus* bacterisation develops more lateral roots in cucumber (Sokolova et al., 2011) and synthesises gibberellins in pepper (Joo et al., 2005). It can also solubilise potassium in these crops (Han & Lee, 2005). Some other PGPRs are reported to act as biofertilisers on different crops and vegetables enlisted in Table 11.1.

From the above-mentioned examples, it is clear that PGPRs have outstandingly worked on different plant species as biofertilisers. They provide the safest and the most eco-friendly approach to sustainable agriculture. The importance of PGPRs in yield development and their capacity to elicit ISR against several abiotic stresses has been reported (Avis et al., 2008). The symbiotic association between PGPR and host plants is the most promising way for developing a new approach for sustainable agriculture.

11.6 Heavy Metal Stress in the Environment

Heavy metals are a significant cause of soil pollution. Numerous metals contribute to soil pollution, including Ni, Cd, Zn, Cr, Cu and Pb. Heavy metals exert toxic impacts on the soil microflora; hence their population size, diversity and total activities get drastically affected. Different physiological activities of plants like photosynthesis, water absorption and cell division get affected tremendously. The toxic

Table 11.1 Report of different PGPRs on plant growth enhancement

Sl. no.	Name of the PGPR	Name of the host pant	Findings	References
1.	*Burkholderia*	Paddy	Siderophore production and high nitrogen fixation	Govindarajan et al. (2008)
2.	*Rhizobium*	Legume plants	Helped in developing resistance against several stresses	El-Akhal et al. (2013)
3.	*Streptomyces*	Indian lilac	Production of IAA	Verma et al. (2011)
4.	*Pseudomonas*	Pigeon pea	Chitinase and β-glucanase production	Kumar et al. (2010)
6.	*Herbaspirillum*	Rice	Nitrogen fixation	Elbeltagy et al. (2001)
7.	*P. putida*	Maize	High percentage of seed germination	Gholami et al. (2009)
8.	*Rhizobium leguminosarum*	Wheat	Improves the yield and phosphorus uptake	Afzal and Bano (2008)
9.	*Sphingomonas*	Tomato	Gibberellin production	Khan et al. (2014)
10.	*Beijerinckia*	Sugarcane	Nitrogen fixation	
11.	*Phyllobacterium*	Strawberry	Phosphate solubilisation	Flores-Félix et al. (2015)
12.	*Mycobacterium*	Maize	Induction of resistance against environmental stresses	Egamberdiyeva (2007)
13.	*Bacillus megaterium*	Tea	Phosphate solubilisation	Chakraborty et al. (2006)
14.	*Bacillus pumilus*	Tobacco	Compete against blue mould	Zhang et al. (2003)
15.	*Bacillus subtilis CE1*	Maize	Gives resistance against *Fusarium verticillioides*	Cavaglieri et al. (2005)
16.	*Pseudomonas chlororaphis*	Soybeans	Phosphate solubilisation	
17.	*Bradyrhizobium japonicum*	Cowpeas	Nitrogen fixation	Rivas et al. (2009)
18.	*B. cereus*	Wheat	Gives resistance against *R. solani* AG 8	Ryder et al. (1999)
19.	*Bacillus circulans*	Mung bean	Phosphate solubilisation	Singh and Kapoor (1999)

symptoms include the appearance of dark green leaves, permanent wilting of plants, stunted growth, brown short leaves and roots. The plants' uptake metals from soil and these metals eventually enter the food chain and result in high health risk for living animals, including humans. The agricultural runoff containing heavy metal discharge enters the aquatic environment and leads to toxic effects on aquatic animals and plants.

Heavy metals reduce bacterial species richness in the contaminated soils. Among different heavy metals, cadmium (Cd) is considered the most toxic one to the microbial enzymes, whereas lead (Pb) decreases catalase, urease, alkaline phosphatase and acid phosphatase. The nature of sensitivity of soil enzymes to different heavy metals is quite different from each other.

11.6.1 Effects of PGPRs on Plants in Heavy Metal-Contaminated Soil

Hyperaccumulator plants can accumulate a high level of heavy metals and tolerate heavy metal stress to an extent. The plants growing in the heavy metal-polluted soils harbour a wide group of microbes that can tolerate heavy metal concentrations to a higher limit and provide several nutrients to host plants. Among the rhizospheric microbes, the plant growth-promoting rhizobacteria (PGPR) attract special attention because they can enhance the phytoremediation method by releasing chelators, synthesising different phytohormones, etc. The following table (Table 11.2) summarises the existing reports regarding the effects of PGPR on phytoremediation in metal-polluted soil.

PGPRs are known to affect the metal mobility and availability to the host plant, and it may occur through redox changing, acidification, siderophore production, mobilisation of inorganic phosphate, etc. The sensitivity and sequestration power of rhizospheric microbes towards heavy metal stress broaden the way of bioremediation. The PGPRs can also alter the plant metabolism to better withstand the heavy metal stress in the soil.

11.7 Conclusions

Phytoremediation is a new cost-effective way for sustainable agriculture. The recent trends of research on remediation of heavy metals in soil by applying PGPRs show a brilliant prospect for modern agriculture. The application of PGPRs in enhancing crop growth and development helps in heavy metal mobilisation, which is quite advantageous to applying chemical fertilisers. The microbial metabolites are less toxic, biodegradable and eco-friendly. So, to remove the harmful impact of heavy metals from the agricultural soil, it is the safest option to use the PGPR, which will open a new gateway to sustainable agriculture.

Table 11.2 Report of different PGPRs on plant growth enhancement under heavy metal stress

Sl. no.	Name of PGPR	Heavy metal stress and treated pant	Mechanism	References
1.	*Bacillus mucilaginosus HKK-1* *Bacillus megaterium HKP-1*	Zn, Pb, Cu (Indian mustard (*B. juncea*))	P, K solubilisation	Wu et al. (2006)
2.	*Brevibacillus brevis*	Cd, Ni, Pb (white clover (*Trifolium repens*))	IAA production	
3.	*Bacillus subtilis*, *Bacillus cereus*, *B. megaterium*	Cd (Chinese violet cress (*Orychophragmus violaceus*))	IAA production	Liang et al. (2014)
4.	*Pseudomonas* sp. Lk9	Cd (black nightshade (*Solanum nigrum* L.))	Siderophores, organic acids	Chen et al. (2014)
5.	*Burkholderia* sp. J62	Pb and Cd (maize (Zea mays) and tomato (*Lycopersicon esculentum*)	IAA, siderophores, ACC deaminase, P solubilisation	Jiang et al. (2008)
6.	*Achromobacter xylosoxidans* AX10	Cu (*B. juncea*)	Phytoextraction of Ni, Cr	Ma et al. (2009)
7.	*Pseudomonas chlororaphis* SZY6	Cu (*B. napus*)	Root length promotion (phytoextraction of copper)	He et al. (2010)
8.	*Microbacterium oxydans* AY509223	Ni (*Alyssum murale*)	Phytoextraction (Ni)	Aboushanab et al. (2006)
9.	*Kluyvera ascorbate* SUD 165	Ni, Pb, Zn (Indian mustard)	Overall plant growth promotion	Burd et al. (2000)

References

Aboushanab, R. A. I., Angle, J. S., & Chaney, R. L. (2006). Bacterial inoculants affecting nickel uptake by *Alyssum murale* from low, moderate and high Ni soils. *Soil Biology and Biochemistry, 38*(9), 2882–2889. https://doi.org/10.1016/j.soilbio.2006.04.045.

Afzal, A., & Bano, A. (2008). *Rhizobium* and phosphate solubilizing bacteria improve the yield and phosphorus uptake in wheat (*Triticum aestivum*). *International Journal of Agriculture and Biology, 10(8530)*, 1560.

Akhgar, A., Arzanlou, M., Bakker, A. H. M., & Hamidpour, M. (2014). Characterisation of 1-am inocyclopropane-1-carboxylate (ACC) deaminase-containing Pseudomonas spp. in the rhizosphere of salt-stressed canola. *Pedosphere, 24*(4), 461–468.

Alori, E. T., Glick, B. R., & Babalola, O. O. (2017) Microbial phosphorus solubilization and its potential for use in sustainable agriculture. *Front Microbiol 8*, 971.

Annapurna, K., Kumar, A., Kumar, L. V., Govindasamy, V., Bose, P., & Ramadoss, D. (2013). PGPR-induced systemic resistance (ISR) in plant disease management. In D. Maheshwari (Ed.), *Bacteria in agrobiology: Disease management*. Berlin, Heidelberg: Springer.

Arora, N. K., Tewari, S., & Singh, R. (2013). Multifaceted plant-associated microbes and their mechanisms diminish the concept of direct and indirect PGPRs. In N. K. Arora (Ed.), *Plant microbe symbiosis: Fundamentals and advances* (pp. 411–449). Berlin: Springer.

Arora, N. K., Tewari, S., Singh, S., Lal, N., & Maheshwari, D. K. (2012). PGPR for protection of plant health under saline conditions. In D. K. Maheshwari (Ed.), *Bacteria in agrobiology: Stress management* (pp. 239–258). Springer-Verlag.

Arzanesh, M. H., Alikhani, H. A., Khavazi, K., Rahimian, H. A., & Miransari, M. (2011). Wheat (*Triticum aestivum* L.) growth enhancement by *Azospirillum* sp. under drought stress. *World Journal of Microbiology and Biotechnology, 27*(2), 197–205. https://doi.org/10.1007/s11274-010-0444-1.

Avis, T. J., Gravel, V., Antoun, H., & Tweddell, R. J. (2008). Multifaceted beneficial effects of rhizosphere microorganisms on plant health and productivity. *Soil Biology and Biochemistry, 40*(7), 1733–1740. https://doi.org/10.1016/j.soilbio.2008.02.013.

Beneduzi, A., Ambrosini, A., & Passaglia, L. M. (2012). Plant growth-promoting rhizobacteria (PGPR): Their potential as antagonists and biocontrol agents. *Genetics and Molecular Biology, 35*(4(Suppl.)), 1044–1051. https://doi.org/10.1590/s1415-47572012000600020.

Bhardwaj, D., Ansari, M. W., Sahoo, R. K., & Tuteja, N. (2014). Biofertilisers function as key player in sustainable agriculture by improving soil fertility, plant tolerance and crop productivity. *Microbial Cell Factories, 13*, 66. https://doi.org/10.1186/1475-2859-13-66.

Bhattacharyya, P. N., & Jha, D. K. (2012). Plant growth-promoting rhizobacteria (PGPR): Emergence in agriculture. *World Journal of Microbiology and Biotechnology, 28*(4), 1327–1350. https://doi.org/10.1007/s11274-011-0979-9.

Barea, J-M., Pozo, M., Azcón, R., & Azcón-Aguilar, C., (2005). Microbial co-operation in the rhizosphere. *Journal of Experimental Botany, 56*(417):1761–1778. https://doi.org/10.1093/jxb/eri197.

Burd, G. I., Dixon, D. G., & Glick, B. R. (2000). Plant growth- promoting bacteria that decrease heavy metal toxicity in plants. *Canadian Journal of Microbiology, 46*(3), 237–245. https://doi.org/10.1139/w99-143.

Cavaglieri, L., Orlando, J., Rodríguez, M. I., Chulze, S., & Etcheverry, M. (2005). Biocontrol of *Bacillus subtilis* against *Fusarium verticillioides* in vitro and at the maize root level. *Research in Microbiology, 156*(5–6), 748–754. https://doi.org/10.1016/j.resmic.2005.03.001.

Chakraborty, U., Chakraborty, B., & Basnet, M. (2006). Plant growth promotion and induction of resistance in *Camellia sinensis* by *Bacillus megaterium*. *Journal of Basic Microbiology, 46*(3), 186–195. https://doi.org/10.1002/jobm.200510050.

Chen, L., Luo, S., Li, X., Wan, Y., Chen, J., & Liu, C. (2014). Interaction of Cd hyperaccumulator Solanum nigrum L. and functional endophyte Pseudomonas sp. *Lk9* on soil heavy metals uptake. *Soil Biology and Biochemistry, 68*, 300–308. https://doi.org/10.1016/j.soilbio.2013.10.021.

Compant, S., Reiter, B., Sessitsch, A., Nowak, J., Clément, C., & Ait Barka, E. (2005). Endophytic colonisation of *Vitis vinifera* L. by plant growth-promoting bacterium Burkholderia sp. strain 45. *Applied and Environmental Microbiology, 71*(4), 1685–1693. https://doi.org/10.1128/AEM.71.4.1685-1693.2005.

Cornelis, P. (2010). Iron uptake and metabolism in pseudomonads. *Applied Microbiology and Biotechnology, 86*(6), 1637–1645. https://doi.org/10.1007/s00253-010-2550-2.

Datta, B., & Chakrabartty, P. K. (2014). Siderophore biosynthesis genes of *Rhizobium* sp. isolated from *Cicer arietinum* L. *3. Biotech, 4*(4), 391–401. https://doi.org/10.1007/s13205-013-0164-y.

de Souza, J. T., Weller, D. M., & Raaijmakers, J. M. (2003). Frequency, diversity and activity of 2, 4-diacetylphloroglucinol producing fluorescent *Pseudomonas* spp. in Dutch take-all decline soils. *Phytopathology, 93*(1), 54–63. https://doi.org/10.1094/PHYTO.2003.93.1.54.

Djordjevic, M. A., Gabriel, D. W., & Rolfe, B. G. (1987). *Rhizobium*-the refined parasite of legumes. *Annual Review of Phytopathology, 25*(1), 145–168. https://doi.org/10.1146/annurev.py.25.090187.001045.

Dimkpa, C., Svatos, A., Merten, D., Büchel, G., & Kothe, E. (2008). Hydroxamate siderophores produced by *Streptomyces acidiscabies* E13 bind nickel and promote growth in cowpea (Vigna unguiculata L.) under nickel stress. *Canadian Journal of Microbiology, 54*(3), 163–172. https://doi.org/10.1139/w07-130.

Döbereiner, J. (1997). Biological nitrogen fixation in the tropics: Social and economic contributions. *Soil Biology and Biochemistry, 29*(5–6), 771–774. https://doi.org/10.1016/S0038-0717(96)00226-X.

Egamberdiyeva, D. (2007). The effect of plant growth promoting bacteria on growth and nutrient uptake of maise in two different soils. *Applied Soil Ecology, 36*(2–3), 184–189. https://doi.org/10.1016/j.apsoil.2007.02.005.

El-Akhal, M. R., Rincón, A., Coba de la Peña, T., Lucas, M. M., El Mourabit, N., Barrijal, S., & Pueyo, J. J. (2013). Effects of salt stress and rhizobial inoculation on growth and nitrogen fixation of three peanut cultivars. *Plant Biology, 15*(2), 415–421. https://doi.org/10.1111/j.1438-8677.2012.00634.x.

Elbeltagy, A., Nishioka, K., Sato, T., Suzuki, H., Ye, B., Hamada, T., … Minamisawa, K. (2001). Endophytic colonisation and in planta nitrogen fixation by a *Herbaspirillum* sp. isolated from wild rice species. *Applied and Environmental Microbiology, 67*(11), 5285–5293. https://doi.org/10.1128/AEM.67.11.5285-5293.2001.

Estrada-de los Santos, P., Bustillos-Cristales, R., & Caballero-Mellado, J. (2001). *Burkholderia*, a genus rich in plant-associated nitrogen fixers with wide environmental and geographic distribution. *Applied and Environmental Microbiology, 67*(6), 2790–2798. https://doi.org/10.1128/AEM.67.6.2790-2798.2001, PubMed: 27902798.

Fekete, F. A., Spence, J. T., & Emery, T. (1983). Siderophores produced by nitrogen-fixing Azotobacter vinelandii OP in iron-limited continuous culture. *Applied and Environmental Microbiology, 46*(6), 1297–1300. https://doi.org/10.1128/aem.46.6.1297-1300.1983.

Flores-Félix, J. D., Silva, L. R., Rivera, L. P., Marcos-García, M., García-Fraile, P., Martínez-Molina, E., … Rivas, R. (2015). Plants probiotics as a tool to produce highly functional fruits: The case of Phyllobacterium and vitamin C in strawberries. *PLoS One, 10*(4), e0122281. https://doi.org/10.1371/journal.pone.0122281.

Gaby, J. C., & Buckley, D. H. (2012). A comprehensive evaluation of PCR primers to amplify the nifH gene of nitrogenase. *PLoS One, 7*(7), e42149. https://doi.org/10.1371/journal.pone.0042149.

Gaind, S., & Gaur, A. C. (1991). Thermo tolerant phosphate solubilising microorganisms and their interaction with mung bean. *Plant and Soil, 133*(1), 141–149. https://doi.org/10.1007/BF00011908.

Gholami, A., Shahsavani, S., & Nezarat, S. (2009). The effect of plant growth promoting rhizobacteria (PGPR) on germination, seedling growth and yield of maize. *International Journal of Agricultural and Biosystems Engineering, 3*, 1.

Gillis, M., Kesters, K., Hoste, B., Janssens, D., Kropenstedt, R. M., Stephen, M. P., … de Ley, J. (1995). *Acetobacter diazotrophicus*sp. Nov. a nitrogen fixing acid bacterium associated with sugarcane. *International Journal of Systematic and Evolutionary Microbiology, 39*, 361–364.

Glick, B. R. (2012). Plant growth-promoting bacteria: Mechanisms and applications. *Scientifica, 2012*, 963401. https://doi.org/10.6064/2012/963401.

Glick, B. R., Cheng, Z., Czarny, J., & Duan, J. (2007). Promotion of plant growth by ACC deaminase-producing soil bacteria. *European Journal of Plant Pathology, 119*(3), 329–339. https://doi.org/10.1007/s10658-007-9162-4.

Glick, B. R., Karaturovíc, D. M., & Newell, P. C. (1995). A novel procedure for rapid isolation of plant growth promoting pseudomonads. *Canadian Journal of Microbiology, 41*(6), 533–536. https://doi.org/10.1139/m95-070.

Glick, B. R., Penrose, D. M., & Li, J. (1999). A model for the lowering of plant ethylene concentrations by plant growth promoting rhizobacteria. *Journal of Theoretical Biology, 190*, 63–68.

Govindarajan, M., Balandreau, J., Kwon, S. W., Weon, H. Y., & Lakshminarasimhan, C. (2008). Effects of the inoculation of *Burkholderia vietnamensis* and related endophytic diazotrophic bacteria on grain yield of rice. *Microbial Ecology, 55*(1), 21–37. https://doi.org/10.1007/s00248-007-9247-9.

Grayston, S. J., & Germida, J. J. (1991). Sulphur-oxidising bacteria as plant growth promoting rhizobacteria for canola. *Canadian Journal of Microbiology, 37*(7), 521–529. https://doi.org/10.1139/m91-088.

Han, H. S., & Lee, K. D. (2005). Phosphate and potassium solubilizing bacteria effect on mineral uptake, soil availability and growth of eggplant. *Research Journal of Agriculture and Biological Sciences, 1*(2), 176–180.

Hayat, R., Ali, S., Amara, U., Khalid, R., & Ahmed, I. (2010). Soil beneficial bacteria and their role in plant growth promotion: A review. *Annals of Microbiology, 60*(4), 579–598. https://doi.org/10.1007/s13213-010-0117-1.

He, L. Y., Zhang, Y. F., Ma, H. Y., Su, L. N., Chen, Z. J., & Wang, Q. Y. (2010). Characterisation of copper resistant bacteria and assessment of bacterial communities in rhizosphere soils of copper-tolerant plants. *Applied Soil Ecology, 44*(1), 49–55. https://doi.org/10.1016/j.apsoil.2009.09.004.

Heidari, M., & Golpayegani, A. (2012). Effects of water stress and inoculation with plant growth promoting rhizobacteria (PGPR) on antioxidant status and photosynthetic pigments in basil (*Ocimum basilicum* L.). *Journal of the Saudi Society of Agricultural Sciences, 11*(1), 57–61. https://doi.org/10.1016/j.jssas.2011.09.001.

Hirst, I. D., Hastings, T. S., & Ellis, A. E. (1991). Siderophore production by *Aeromonas salmonicida. Journal of General Microbiology, 137*(5), 1185–1192. https://doi.org/10.1099/00221287-137-5-1185.

Holinsworth, B., & Martin, J. D. (2009). Siderophore production by marine-derived fungi. *Biometals: an International Journal on the Role of Metal Ions in Biology, Biochemistry, and Medicine, 22*(4), 625–632. https://doi.org/10.1007/s10534-009-9239-y.

Retrieved from. https://www.google.com/search?hl=en&biw=1366&bih=577&tbm=isch&sa=1&ei=pBMZXPXgKdHbrQHRs5KoDw&q=mechanism+of+pgpr+action+in+cycle+form&oq=mechanism+of+pgpr+action+in+cycle+form&gs_l=img.3. 1.0.0.240.2902.0j11j4.. . . . 0. . . .1..gws-wiz-img.iFqJeLp4_80#imgrc=8hiQbTaie4kuXM, *42372*(48609), 49154.

Retrieved from. https://www.google.com/search?q=impact+of+pgpr+on+plant+growth&hl=en&source=lnms&tbm=isch&sa=X&ved=0ahUKEwiy4v-s26nfAhUNWX0KHTaDB3EQ_AUIDygC&biw=1366&bih=577#imgrc=kx_AXX_KjqijIM.

Jiang, C. Y., Sheng, X. F., Qian, M., & Wang, Q. Y. (2008). Isolation and characterization of a heavy metal-resistant Burkholderia sp. from heavy metal-contaminated paddy field soil and its potential in promoting plant growth and heavy metal accumulation in metal-polluted soil. *Chemosphere, 72*(2), 157–164. https://doi.org/10.1016/j.chemosphere.2008.02.006.

Joo, G. J., Kim, Y. M., Kim, J. T., Rhee, I. K., Kim, J. H., & Lee, I. J. (2005). Gibberellins-producing rhizobacteria increase endogenous gibberellins content and promote growth of red peppers. *Journal of Microbiology, 43*(6), 510–515.

Joshi, M., Shrivastava, R., Sharma, A. K., & Prakash, A. (2012). Screening of resistant verities and antagonistic *Fusarium oxysporum* for biocontrol of *Fusarium* Wilt of Chilli. *Plant Pathologia et Microbiologia, 3*, 134.

Kang, B. G., Kim, W. T., Yun, H. S., & Chang, S. C. (2010). Use of plant growth-promoting rhizobacteria to control stress responses of plant roots. *Plant Biotechnology Reports, 4*(3), 179–183. https://doi.org/10.1007/s11816-010-0136-1.

Khan, A. L., Waqas, M., Kang, S. M., Al-Harrasi, A., Hussain, J., Al-Rawahi, A., … Lee, I. J. (2014). Bacterial endophyte Sphingomonas sp. LK11 produces gibberellins and IAA and promotes tomato plant growth. *Journal of Microbiology, 52*(8), 689–695. https://doi.org/10.1007/s12275-014-4002-7.

Kloepper, J. W., Leong, J., Teintze, M., & Schroth, M. N. (1980). Enhanced plant growth by siderophores produced by plant growth-promoting rhizobacteria. *Nature, 286*(5776), 885–886. https://doi.org/10.1038/286885a0.

Kloepper, J. W., Lifshitz, R., & Schroth, M. N. (1988). *Pseudomonas* inoculants to benefit plant production. *ISI Atlas of Science – Animal and Plant Sciences, 1*, 60–64.

Kumar, H., Bajpai, V. K., Dubey, R. C., Maheshwari, D. K., & Kang, S. C. (2010). Wilt disease management and enhancement of growth and yield of *Cajanus cajan* (L.) var. Manak by bacterial combinations amended with chemical fertiliser. *Crop Protection, 29*(6), 591–598. https://doi.org/10.1016/j.cropro.2010.01.002.

Ladha, J. K., Pathak, H., Krupnik, J., Six, T., & van Kessel, C. (2005). Efficiency of fertilizer nitrogen in cereal production: Retrospects and prospects. *Advances in Agronomy, 87*, 85–156. https://doi.org/10.1016/S0065-2113(05)87003-8.

Landa, B. B., Hervás, A., Bettiol, W., & Jiménez-Díaz, R. M. (1997). Antagonistic activity of bacteria from the chickpea rhizosphere against *Fusarium oxysporum* f. sp. *ciceris*. *Phytoparasitica, 25*(4), 305–318. https://doi.org/10.1007/BF02981094.

Liang, X., He, C. Q., Ni, G., Tang, G. I., Chen, X. P., & Lei, Y. R. (2014). Growth and Cd accumulation of Orychophragmus violaceus as affected by inoculation of Cd-tolerant bacterial strains. *Pedosphere, 24*(3), 322–329. https://doi.org/10.1016/S1002-0160(14)60018-7.

Lippi, D., Cacciari, I., Paola, Q., & Pietrosanti, T. (1991). *Interactions between Azospirillum and sorghum rhizosphere isolates under different cultural conditions.*

Loper, J. E., & Gross, H. (2007). Genomic analysis of antifungal metabolite production by Pseudomonas fluorescens Pf-5. *European Journal of Plant Pathology, 119*(3), 265–278. https://doi.org/10.1007/s10658-007-9179-8.

Lugtenberg, B., & Kamilova, F. (2009). Plant-growth-promoting rhizobacteria. *Annual Review of Microbiology, 63*, 541–556. https://doi.org/10.1146/annurev.micro.62.081307.162918.

Ma, Y., Rajkumar, M., & Freitas, H. (2009). Isolation and characterisation of Ni mobilising PGPB from serpentine soils and their potential in promoting plant growth and Ni accumulation by Brassica spp. *Chemosphere, 75*(6), 719–725. https://doi.org/10.1016/j.chemosphere.2009.01.056, PubMed: 19232424.

Ma, J. F. (2005). Plant root responses to three abundant soil minerals: Silicon, aluminum and iron. *Critical Reviews in Plant Sciences, 24*(4), 267–281. https://doi.org/10.1080/07352680500196017.

Maheshwari, D. K., Dubey, R. C., Aeron, A., Kumar, B., Kumar, S., Tewari, S., & Arora, N. K. (2012). Integrated approach for disease management and growth enhancement of *Sesamum indicum* L. utilising *Azotobacter chroococcum* TRA2 and chemical fertiliser. *World Journal of Microbiology and Biotechnology, 28*(10), 3015–3024. https://doi.org/10.1007/s11274-012-1112-4.

Maxton, A., Singh, P., Andy, A., Prasad, S. M., & Masih, S. A. (2018). PGPR: A boon in stress tolerance and bio control. *Research Journal of Biotechnology, 13*, 105–111.

Miransari, M., & Smith, D. L. (2014). Plant hormones and seed germination. *Environmental and Experimental Botany, 99*, 110–121. https://doi.org/10.1016/j.envexpbot.2013.11.005.

Nadeem, S. M., Naveed, M., Zahir, Z. A., & Asghar, H. N. (2013). Plant-microbe interactions for sustainable agriculture: Fundamentals and recent advances. In N. K. Arora (Ed.), *Plant microbe symbiosis: Fundamentals and advances* (pp. 51–103). India: Springer.

Naznin, H. A., Kimura, M., Miyazawa, M., & Hyakumachi, M. (2013). Analysis of volatile organic compounds emitted by plant growth promoting fungus *Phoma* sp. GS83 for growth promotion effects on tobacco. *Microbes and Environments, 28*(1), 42–49. https://doi.org/10.1264/jsme2.me12085.

Okon, Y., & Kapulnik, Y. (1986). Development and function of *Azospirillum*-inoculated roots. *Plant and Soil, 90*(1–3), 3–16. https://doi.org/10.1007/BF02277383.

Ong, K. S., Aw, Y. K., Lee, L. H., Yule, C. M., Cheow, Y. L., & Lee, S. M. (2016). *Burkholderia paludis* sp. nov., an antibiotic-siderophore producing novel *Burkholderia cepacia* complex species, isolated from Malaysian tropical peat swamp soil. *Frontiers in Microbiology, 7*, 2046. https://doi.org/10.3389/fmicb.2016.02046.

Ongena, M., & Jacques, P. (2008). *Bacillus lipopeptides*: Versatile weapons for plant disease biocontrol. *Trends in Microbiology, 16*(3), 115–125. https://doi.org/10.1016/j.tim.2007.12.009.

Pahari, A., & Mishra, B. B. (2017). Characterisation of siderophore producing rhizobacteria and its effect on growth performance of different vegetables. *International Journal of Current Microbiology and Applied Sciences, 6*(5), 1398–1405. https://doi.org/10.20546/ijcmas.2017.605.152.

Pandey, P., & Maheshwari, D. K. (2007). Two sp. microbial consortium for growth promotion of *Cajanus cajan*. *Current Science, 92*, 1137–1142.

Parmar, P., & Sindhu, S. S. (2013). Potassium solubilisation by rhizosphere bacteria: Influence of nutritional and environmental conditions. *Journal of Microbiology Research, 3*(1), 25–31.

Piao, C. G., Tang, W. H., & Chen, Y. X. (1992). Study on the biological activity of yield-increasing bacteria. *Chinese Journal of Microecology, 4*, 55–62.

Rabeendran, N., Moot, D. J., Jones, E. E., & Stewart, A. (2000). Inconsistent growth promotion of cabbage and lettuce from *Trichoderma* isolates. *New Zealand Journal of Plant Protection, 53*, 143–146.

Radjacommare, R., Kandan, A., Nandakumar, R., & Samiyappan, R. (2004). Association of the hydrolytic enzyme chitinase against *Rhizoctonia solani* in rhizobacteria treated rice plants. *Journal of Phytopathology, 152*(6), 365–370. https://doi.org/10.1111/j.1439-0434.2004.00857.x.

Rivas, R., Martens, M., de Lajudie, P., & Willems, A. (2009). Multilocus sequence analysis of the genus *Bradyrhizobium. Systematic and Applied Microbiology, 32*(2), 101–110. https://doi.org/10.1016/j.syapm.2008.12.005.

Roy Chowdhury, A., Kundu, S., & SenGupta, C. (2017). Plant growth promoting rhizobacteria (PGPR): One step ahead to sustainable agriculture. *International Journal of Innovative Science Engineering and Technology, 4*(7), 41–48.

Ryder, M. H., Yan, Z., Terrace, T. E., Rovira, A. D., & Tang, W. (1999). Uses of *Bacillus* isolated in China to suppress take all and *Rhizoctonia* root rot, and promote seedling growth of glasshouse grown wheat in Australian soils. *Soil Biology and Biochemistry, 31*, 19–29.

Sabry, S. R. S., Saleh, S. A., Batchelor, C. A., Jones, J., Jotham, J., Webster, G., ... Cocking, E. C. (1997). Endophytic establishment of *Azorhizobium caulinodans* in wheat. *Proceedings of the Royal Society of London. Series B, 264*(1380), 341–346. https://doi.org/10.1098/rspb.1997.0049.

Sacherer, P., Défago, G., & Haas, D. (1994). Extracellular protease and phospholipase C are controlled by the global regulatory gene gacA in the biocontrol strain Pseudomonas fluorescens CHA0. *FEMS Microbiology Letters, 116*(2), 155–160. https://doi.org/10.1111/j.1574-6968.1994.tb06694.x.

Schwyn, B., & Neilands, J. B. (1987). Universal chemical assay for the detection and determination of siderophores. *Analytical Biochemistry, 160*(1), 47–56. https://doi.org/10.1016/0003-2697(87)90612-9.

Shamsuddin, H. Z., TanZuan, K., Radziah, O., Halimi, M. S., Khairuddin, R. A., & Sheikh, H. (2014). Isolation and characterisation of rhizobia and plant growth-promoting rhizobacteria and their effects on growth of rice seedlings. *American Journal of Agricultural and Biological Sciences, 9*(3), 342–360. https://doi.org/10.3844/ajabssp.2014.342.360.

Shanmugaiah, V., Karmegham, N., Harikrishnan, H., Jayaprakashvel, M., & Natesan, B. (2015). Biocontrol mechanisms of siderophores against bacterial plant pathogens. Sustainable approaches to controlling plant pathogenic bacteria, Edition: First [Chapter]. In V. R. Kannan & K. K. Bastas (Eds.), *Biocontrol mechanisms of siderophores against bacterial plant pathogens* (pp. 167–186). CRC Press.

Shenker, M., Oliver, I., Helmann, M., Hadar, Y., & Chen, Y. (1992). Utilisation by tomatoes of iron mediated by a siderophore produced by *Rhizopus arrhizus. Journal of Plant Nutrition, 15*(10), 2173–2182. https://doi.org/10.1080/01904169209364466.

Shilev, S. (2013). Soil rhizobacteria regulating the uptake of nutrients and undesirable elements by plants. In N. K. Arora (Ed.), *Plant microbe symbiosis: Fundamentals and advances* (pp. 147–150). India: Springer.

Singh, S., & Kapoor, K. K. (1999). Inoculation with phosphate-solubilising microorganisms and a vesicular arbuscular mycorrhizal fungus improves dry matter yield and nutrient uptake by wheat grown in a sandy soil. *Biology and Fertility of Soils, 28*, 139–144.

Sokolova, M. G., Akimova, G. P., & Vaishlia, O. B. (2011). Effect of phytohormones synthesised by rhizosphere bacteria on plants. *Prikladnaia Biokhimiia i Mikrobiologiia, 47*, 302–307.

Spaepen, S., & Vanderleyden, J. (2011). Auxin and plant-microbe interactions. *Cold Spring Harbor Perspectives in Biology, 3*(4), a001438. https://doi.org/10.1101/cshperspect.a001438.

Strzelczyk, E., Kampert, M., & Li, C. Y. (1994). Cytokinin-like substances and ethylene production by *Azospirillum* in media with different carbon sources. *Microbiological Research, 149*(1), 55–60. https://doi.org/10.1016/S0944-5013(11)80136-9.

Sujatha, N., & Ammani, K. (2013). Siderophore production by the isolates of fluorescent pseudomonads. *International Journal of Current Research and Review, 5*, 1–7.

Sharma, S. K., Ramesh, A., & Johri, B. N. (2013). Isolation and characterisation of plant growth promoting *Bacillus amyloliquefaciens* strain Sks_bnj_1and its influence on rhizosphere soil properties and nutrition of soybean (Glycine max L. Merrill). *Journal of Virology & Microbiology, 2013*, 1–19.

Tilak, K. V. B. R., Singh, G., & Mukerji, K. G. (1999). Biocontrol—Plant growth promoting rhizobacteria: Mechanism of action. In K. G. Mukerji, B. P. Chamola, & R. K. Upadhyay (Eds.), *Biotechnological approaches in biocontrol of plant pathogens* (Vol. 10, pp. 114–115). Kluwer Academic/Plenum Publishers.

Timmusk, S., Abd El-Daim, I. A., Copolovici, L., Tanilas, T., Kännaste, A., Behers, L., ... Niinemets, Ü. (2014). Drought-tolerance of wheat improved by rhizosphere bacteria from harsh environments: Enhanced biomass production and reduced emissions of stress volatiles. *PLoS One, 9*(5), e96086. https://doi.org/10.1371/journal.pone.0096086.

Upadyay, S. K., Maurya, S. K., & Singh, D. P. (2012). Salinity tolerance in free living plant growth promoting rhizobacteria. *Indian Journal of Scientific Research, 3*, 73–78.

Vejan, P., Abdullah, R., Khadiran, T., Ismail, S., & Nasrulhaq Boyce, A. N. (2016). Role of plant growth promoting rhizobacteria in agricultural sustainability—A review. *Molecules, 21*(5), 573. https://doi.org/10.3390/molecules21050573.

Verma, V. C., Singh, S. K., & Prakash, S. (2011). Bio-control and plant growth promotion potential of siderophore producing endophytic Streptomyces from *Azadirachta indica* A. Juss. *Journal of Basic Microbiology, 51*(5), 550–556. https://doi.org/10.1002/jobm.201000155.

Vessey, J. K. (2003). Plant growth promoting rhizobacteria as biofertilisers. *Plant and Soil, 255*(2), 571–586. https://doi.org/10.1023/A:1026037216893.

Wani, S. A., Chand, S., & Ali, T. (2013). Potential use of *Azotobacter chroococcum* in crop production: An overview. *Current Agriculture Research Journal, 1*(1), 35–38. https://doi.org/10.12944/CARJ.1.1.04.

Wu, S. C., Cheung, K. C., Luo, Y. M., & Wong, M. H. (2006). Effects of inoculation of plant growth promoting rhizobacteria on metal uptake by *Brassica juncea*. *Environmental Pollution, 140*(1), 124–135. https://doi.org/10.1016/j.envpol.2005.06.023.

Yanni, Y. G., Rizk, R. Y., Corich, V., Squartini, A., Ninke, K., ... Dazzo, F. B. (1997). Natural endophytic association between *Rhizobium leguminoserum* bv. *trifolii* and rice roots and assessment of its potential to promote rice growth. *Plant and Soil, 194*(1/2), 99–114. https://doi.org/10.1023/A:1004269902246.

Yazdani, M., Bahmanyar, M. A., Pirdashti, H., & Esmaili, M. A. (2009). Effect of phosphate solubilization microorganisms (PSM) and plant growth promoting rhizobacteria (PGPR) on yield and yield components of corn (*Zea mays* L.). *International Journal of Agricultural and Biosystems Engineering, 3*, 1.

Zahran, H. H. (2001). Rhizobia from wild legumes: Diversity, taxonomy, ecology, nitrogen fixation and biotechnology. *Journal of Biotechnology, 91*(2–3), 143–153. https://doi.org/10.1016/s0168-1656(01)00342-x.

Zaidi, A., Khan, M. S., Ahemad, M., & Oves, M. (2009). Plant growth promotion by phosphate solubilising bacteria. *Acta Microbiologica et Immunologica Hungarica, 56*, 263–284.

Zhang, H., Sekiguchi, Y., Hanada, S., Hugenholtz, P., & Kim, H. (2003). *Gemmatimona aurantiaca*gen. nov, sp. nov., a Gram-negative, aerobic, polyphosphate accumulating microorganism, the firrst cultured representative of the new bacterial phylum Gemmatimonadetesphyl. *International Journal of Systematic and Evolutionary Microbiology, 53*, 1155–1163.

Zwanenburg, B., Pospíšil, T., & Ćavar Zeljković, S. (2016). Strigolactones: New plant hormones in action. *Planta, 243*(6), 1311–1326. https://doi.org/10.1007/s00425-015-2455-5.

Chapter 12
Exploring the Phytoremediation Potential of Macrophytes for Treating Sewage Effluent Through Constructed Wetland Technology (CWT) for Sustainable Agriculture

K. Suganya, Joneboina Easwar Kumar, S. Paul Sebastian, R. Poornima, Balaji Kannan, E. Parameswari, and P. Kalaiselvi

Abstract Sewage generation in India accounts for 61,754 million liters per day (MLD) in urban areas, where 22,963 MLD is treated and 38791MLD remains untreated. Due to the ever-increasing population explosion and urbanization, sewage effluent generation has been increasing. Sewage effluent is a kind of wastewater comprising of 99.9% water content along with TDS (total dissolved solids), TSS (total suspended solids), heavy metals, nitrogen, phosphorus, and also waterborne pathogens. Because of its enriched nutrient, supply can be utilized for irrigation, thereby reducing the water demands for sustainable agriculture. Though many conventional technologies are available for treating sewage, CWT (constructed wetland technology) with low maintenance and simple construction looks very promising. In this technique, macrophytes and the filtration medium play a dynamic role in eradicating the pollutants in sewage. Macrophytes utilized in CWT enhance the pollutant removal mechanism by creating oxygenated environments around the rhizosphere of plants. This chapter explains the role of macrophytes, their remediation potential, types, and mechanisms involved in constructed wetland technology to treat the sewage effluent for its effective utilization for sustainable agriculture.

K. Suganya (✉) · R. Poornima · E. Parameswari · P. Kalaiselvi
Department of Environmental Sciences, Tamil Nadu Agricultural University, Coimbatore, India

J. E. Kumar
CSIRO-NEERI, Hyderabad, India

S. P. Sebastian
Anbil Dharmalingham Agricultural College and Research Institute, Trichy, India

B. Kannan
Water Technology Centre, Tamil Nadu Agricultural University, Coimbatore, India

S. A. Bandh (ed.), *Sustainable Agriculture*,
https://doi.org/10.1007/978-3-030-83066-3_12

Keywords Constructed wetlands · Macrophytes · Effluents · Sustainable agriculture

12.1 Introduction

The population rise and industrialization are causing a significant pressure on land and water supplies by increasing wastewater that is mainly left untreated. It either infiltrates into the ground, polluting aquifers, or is disposed of into watercourses. Disposing of this wastewater is dependent on the region and availability of natural water in a particular area. Raw sewage water from cities is a combination of domestic, commercial, and industrial activities. It is rich in organic matter and essential nutrients, with the scopes of utilizing it for farming after adequate treatment.

12.2 Composition of Sewage Water

Sewage, also referred to as wastewater, contains water and solid waste suspension disposed of from households, buildings, and industries and conveyed in large pipes called sewers (Nowak & Imperowicz, 2016; Palamuleni, 2002). Depending on the source of water, storage, and treatment, the wastewater composition is very complex. While a vast proportion of these contain essential plant nutrients, they also have significant quantities of heavy metals and other types of pollutants. The sewage water generated in India holds more than 90% of water. The solid portion contains 30–40% inert materials, 40–50% organics, 5–8% miscellaneous substances, and 10–15% bio-resistant organic substances.

12.3 Characteristics of Sewage Effluent

The chemical content of wastewater differs from one location to another, depending on the factories that emit their effluents. Any municipal sewer system produces elevated levels of radioactive metallurgy as factory effluents are dumped into sewage systems. Many researchers have found differences in electrical conductivity, pH, organic carbon, suspended solids, carbonates, bicarbonates, calcium, magnesium, and other toxic elements in the wastewater from Indian cities. The pH of the sewage in various cities varies between 7.2 and 8.3, but the soil pH does not alter substantially due to its high buffering potential for these waters. The electrical conductivity of wastewater collected from different municipalities varies from 1.1 to 3.8 dS m^{-1}. Their constant usage in the agricultural fields causes increased soil salinity and ultimately restricted growth of plants. The wastewater organic C content in various towns of Bhatinda ranges between 59 and 480 mg L^{-1}. The waste streams contain large N, P, and K concentrations, with their levels ranging from 8 to 106, from 4.2 to 53 and from 19 to 2500 mg L^{-1}, respectively. In several towns, large concentrations of micro-nutrients and harmful metals are also found to be present in wastewater (Adhikari et al., 1997; Gupta et al., 1986; Gupta & Mitra, 2002; Narwal et al., 1993; Singh & Kansal, 1985). The wastewater also contains Cd, Cu, Fe, Mn, Ni, Pb, and Zn (Arora et al., 1985; Tiwana et al., 1987).

12.4 Types of Aquatic Plants

Aquatic plants are classified into three types, namely, free-floating, submerged, and emergent.

12.4.1 Free-Floating Hydrophytes

The aquatic plants with floating branches and underwater roots are known as free-floating hydrophytes. Some of these free-floating hydrophytes, e.g., duckweeds (*Lemna minor*, *Spirodela intermedia*), water lettuce (*Pistia stratotes*), water hyacinth (*Eichhornia crassipes*), and water ferns (*Salvinia minima*) are well known for their ability to remove metals from the contaminated waters. Heavy metals are actively transported in free-floating aquatic plants from the roots, where they are transferred to other areas of the plant. Passive transport occurs as the plant body comes into close contact with the polluted medium. Heavy metals collect mainly in the plant bodies upper parts during the passive transport.

12.4.2 Underwater (Submerged) Hydrophytes

Leaves are a critical component of metal absorption in submerged hydrophytes. The passive movement of the cuticles helps in heavy metal absorption. Due to the elevated inner charged mass, the passage of positive metal ions takes place. They extract heavy metals from sediments and water, e.g., American pondweed (*Potamogeton pectinatus*), coontail or hornwort (*Ceratophyllum demersum*), pondweed (*Potamogeton crispus*), and Parrot feather (*Myriophyllum spicatum*).

12.4.3 Emergent Hydrophytes

They are typically located on surfaces, with a water table of 0.5 m below the ground. Heavy metal deposition in emergent hydrophytes differs from plant to plant, as they are capable of bio-centering most metals in the field. Some of these plants often spread the metal pressure in air sections also. The examples of emergent hydrophytes include *Polygonum hydropiperoides* (smartweed), *Phragmites australis* (common reed), *Spartina alterniflora* (smooth cordgrass), *Scirpus* sp. (bulrush), and *Typha latifolia* (cattail).

12.5 Constructed Wetlands

Constructed wetlands are a planned and built-up system that uses natural components and procedures including soils, vegetation, and the allied microbial components for the treatment of wastewater. According to wetland hydrology, constructed wetlands are typically divided into two types, namely, FWS (free water surface) constructed wetlands and SSF (subsurface flow) constructed wetlands (Saeed & Sun, 2012). Based on the direction of flow, the SSF wetlands could further be classified into HF (horizontal flow) constructed wetlands and VF (vertical flow) constructed wetlands (Fig. 12.1).

Hybrid constructed wetlands, a blend of numerous wetland systems, were also introduced for the wastewater treatment. In these wetlands, the arrangement usually consists of two stages of parallel CWs connected in sequence, such as HF-VF CWs, VF-HF CWs, FWS-HF CWs, and HF-FWS CWs (Shelef et al., 2013).

12.5.1 Surface Flow Constructed Wetlands

A shallow basin soil or other media to protect plant roots, a flood management system that retains a shallow depth of water, and an elevated water level are all characteristics of a surface flow constructed wetland. They are similar to natural wetlands and can act as a wildlife habitat with aesthetic advantages and water treatment benefits. In surface flow wetlands, the near-surface layer is aerobic, while the deeper waters and substrates are usually anaerobic. These wetlands are often used to handle mine drainage, agricultural runoff, and stormwater. Additionally, they are referred to as free water surface wetlands and, if used for mine runoff, aerobic wetlands.

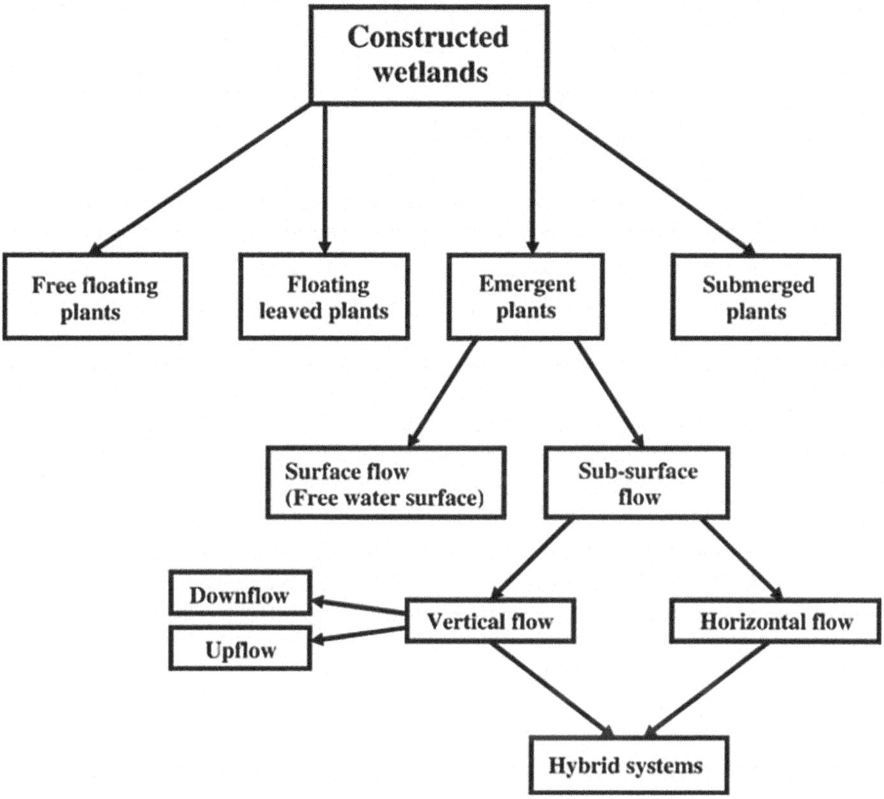

Fig. 12.1 Classification of constructed wetlands. (*Source*: Vymazal, 2007)

12.5.2 Subsurface Flow Constructed Wetlands

Subsurface flow system (SSF) wetlands also known as vegetated/reed filters submerged bed system or a root-zone system is a type of treatment system, where wastewater flows vertically or horizontally through the porous media. The common features of an SSF-CW are its bed media, vegetation type, the inlet-outlet arrangement, and the waterproof lining that prevents the groundwater contamination.

12.5.2.1 Horizontal Subsurface Flow System

Wastewater fed in the inlet of this system moves through the bed media under the bed's surface till it reaches the outlet zone in the horizontal subsurface flow system. In these types of CWs, a system of high-impact anaerobic and anoxic regions is provided where the wastewater comes into contact (Brix & Arias, 2005). The oxygen is broken into the substrate by high-impact zones surrounding the rhizomes and the roots of the wetland plants.

12.5.2.2 Vertical Subsurface Flow System

Wastewater is fed intermittently in the vertical subsurface flow systems, and it flows through the channel funnels vertically; the base has the seepage system, which gathers wastewater (Nguyen et al., 2018).

12.5.2.3 Hybrid System

Hybrid CWs treat domestic, industrial, and landfill leachate wastewater and other types of wastewater such as runoff, agricultural, and hospital wastewater. Optimum design for treating greenhouse wastewater in South Korea was reached by the pilot-scale hybrid CWs (Lee et al., 2009). Six parallel lines with different CW configurations were built in a pilot-scale hybrid CW. The highest removal efficiency was attained by the HSSF-VSSF-HSSF hybrid CW than other configurations. VSSF has greater pollutant removal efficiency than HSSF in the HSSF-VSSF-HSSF system (Fig. 12.2).

12.6 The Role of Aquatic Plants in Constructed Wetlands

In the last few decades, wastewater remediation by developed wetlands has been practiced magnificently worldwide as an adequate wastewater management alternative. The constructed wetlands (CWs) were designed to handle different wastewater types in the managed environment. In constructed wetlands, a wide variety of wastewaters such as commercial, agriculture, residential, and garbage leachates, stormwater, and industrial wastewater can be remedied. The developed wetland offers a relatively easy and affordable way to manage water pollution without damaging natural wetland supplies. Water plants are an integral component of CWs for wastewater remediation. Aquatic plants perform two critical indirect roles in CWs: (1) the leaves and stems of aquatic macrophytes increase the surface area available for microbial community attachment, and (2) aquatic plants transport gases such as oxygen to the root zone, allowing their roots to survive in anaerobic conditions. The rhizosphere supports an abundance of microbial species that are responsible for the required transformation of metallic ions, various chemicals, and nutrients. The use of aquatic macrophytes in CWs thus helps treat wastewater that has been polluted by multiple pollutants and serves as a contaminant drain. Heavy metal removal in CWs depends on nature, ionic form, season, substratum state, and plant type of metallic elements. Dense aquatic plant population in CWs has dramatically improved the performance of wastewater treatment of heavy metals.

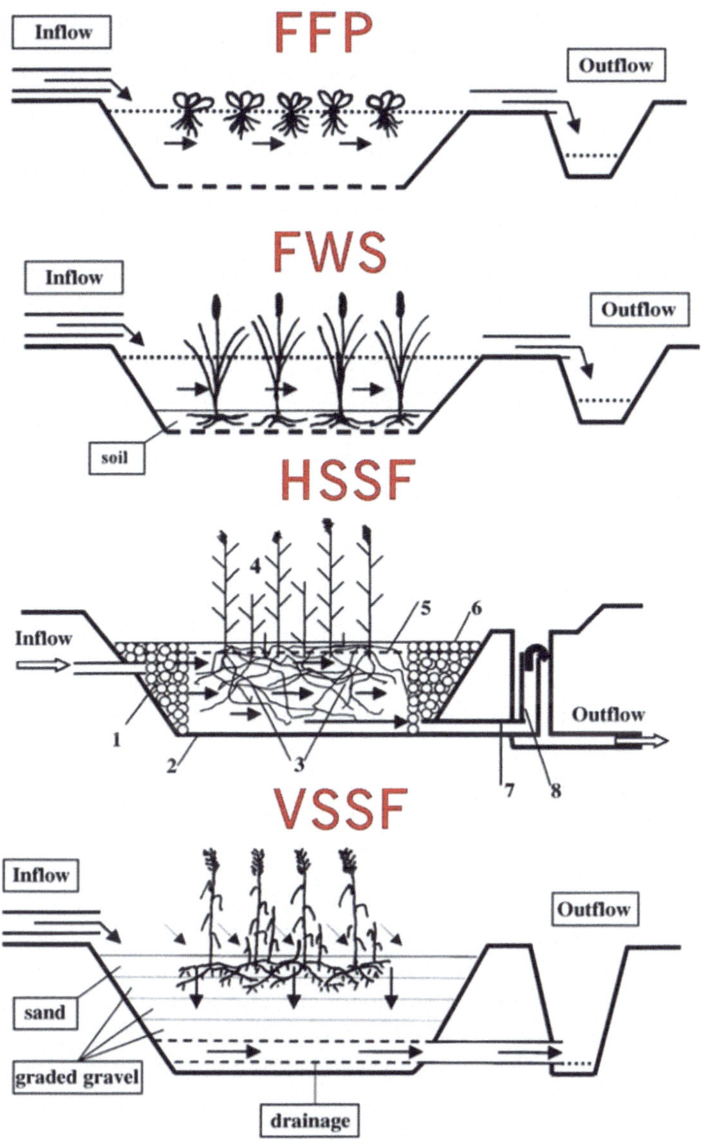

Fig. 12.2 Constructed wetlands for wastewater treatment. (*Source*: Vymazal, 2007). FFP CW with free-floating plants, FWS CW with free water surface with emergent plants, HSSF CW with horizontal subsurface flow, horizontal flow, VSSF CW with vertical subsurface flow, vertical flow

Aquatic plants contribute significantly and vigorously to the biochemistry of wetlands by the active and passive circulation of basic ingredients. In wetland aquatic macrophytes, heavy metal concentrations normally decreased in the following order: root > leaves > stems. Nonetheless, the abundance of heavy metals does not offer a sufficient measure of heavy metal absorption by aquatic plants. Heavy metal absorption in wetlands is primarily determined by the biomass of the aquatic plant (Bhattacharya et al., 2006). Recently, Rai (2019) stated that *Eichhornia crassipes* (water hyacinth) is one of the most suitable wetland plants for remediating heavy metals from wastewater. This is one of the best choices to use *Eichhornia crassipes* (water hyacinth) worldwide efficiently. Sukumaran (2013) described the potential of constructed wetland technology and the effectiveness of *E. crassipes* in remediating Pb, Cd, Ar, and Cu from industrial discharge. During a 15-day experiment, compared to *Pistia stratiotes* and *Spirodela polyrhiza*, the aquatic plant *E. crassipes* exhibited superior remediation potential for Ni, Cu, Cd, Fe, and Zn. Ladislas et al. (2015) indicated Cd, Ni, and Zn accumulation in floating hydrophytes *Juncus effusus* and *C. riparia* grown in wetlands having stormwater. The HM ratio at the roots was considerably higher than the roots. Dan et al. (2017) studied the heavy metal buildup by *Phragmites australis* and *Juncus effuses* and indicated that the quality of removal for the intended metals like Cd, Cr, Fe, Ni, Pb, and Zn was much greater for both aquatic plants in a laboratory-scale constructed wetland (Leung et al., 2017).

12.7 Rhizofiltration

Under the influence of plant roots, the rhizosphere is a region of elevated microbial activity and biomass. This rise in microbial growth and activity in the rhizosphere may explain why some xenobiotic compounds degrade at a faster rate in the rhizosphere. Rhizofiltration requires the use of the plant to remove/absorb these toxins, thus restricting their movement in underground water. Heavy metal accumulation on the surface of roots is facilitated by influences such as rhizosphere pH transfer and root exudates. Rhizofiltration systems should be capable of creating an extensive root system that accumulates high quantities of heavy metals. The rhizofiltration is ideal for both marine and terrestrial plants with long fibrous root systems. Productively, rhizofiltration is used to control and treat the factory discharge and contaminant radioactive metals in agricultural fields. Heavy metals typically taken within the soil, such as Cd, Cr, Cu, Ni, Pb, and Zn, can be sufficiently remedied through rhizofiltration.

12.8 Plant-Microbe Interactions

The rhizosphere's relationship between plants and microbial species is nuanced and has developed to support both. Plants support vast microbial communities in the rhizosphere by secreting carbohydrates and amino acids by root cells and sloughing root epidermal cells. Rhizodeposition by plants can be very extensive. Root cap cells, which shield the root from abrasion, can degrade at a rate of 10,000 cells per plant per day. Additionally, root cells secrete mucigel, a gelatinous material that acts as a lubricant, allowing roots to penetrate the soil throughout development. Along with other cell secretions, this mucigel forms root exudates. Root cap cells and exudates are critical sources of nutrients for rhizosphere microorganisms (Anderson et al., 1993). Although the alteration of the soil in the rhizosphere by plant root exudates has a significant effect on microbial populations, it is the actual composition of the plant root that provides microorganisms with a favorable surface area for colonization. For example, the grass fibrous roots have a greater surface area for colonization than taproot structures.

12.9 Root Exudates

Along with amassing biologically active molecules, plant roots actively synthesize and secrete compounds into the rhizosphere. Exudation of ions, free oxygen and water, hormones, mucilage, and a variety of primary and secondary metabolites containing carbon occurs at the root. Exudation of roots can be divided into two active cycles. The first, root excretion, involves the gradient-dependent output of unknown waste products, while the second, secretion, involves the exudation of compounds with known functions such as lubrication and protection. Root exudates are transported to the surrounding rhizosphere via the membrane of the cells. Plant products are also released from root-bound cells and root-boundary cells which, when thrive, differentiate from roots. The richness of root exudates resides primarily in low molecular compounds like carbohydrates, organic and amino acids, phenols, and other secondary metabolites. At the same time, high molecular exudates like proteins and mucilages are less diversified and comprise a more significant portion of the root exudates (Bais et al., 2006). The extent of photosynthates secreted as root exudates varies with the physiological state of the plants, nutrient availability, age of plants, and type of soil. Although not determining the root functions of any of the exudates, a variety of root exudates are essential compounds in biological processes (Walker et al., 2003).

12.9.1 Role of Root Exudates in CWs

The root exudates of CW plants and their impact on the CW's microenvironment have been a focus of many studies (Ryan et al., 2001). According to research, the rhizospheres of various wetland plants contained a diversity of microbial taxa and densities, which were related to root exudates (Brisson & Chazarenc, 2009). Organic root exudates in CWs have been shown to influence the richness and structural diversity of rhizosphere bacteria, as well as the pollutant and nutrient removal rates indirectly. This may be one of the pathways influencing the loss of nutrients from constructed wetlands (Chen et al., 2016). Many researchers have also researched microbial denitrification as a solution to root exudates.

Root exudates are key drivers of microbial activity and diversity in the rhizospheric zones, especially in the low-molecular-weight carbon (LMWC) substrates. These substrates promote nitrogen removal in the constructed wetlands. Soluble sugars and organic acids increase nitrogen removal, while amino acids decrease it (Wu et al., 2017).

12.10 The Role of Enzymes in CWs

Enzymes may function intracellularly, in the presence of or inside their originating cells or, extracellularly, in any case (Gianfreda & Rao, 2004). Enzymes are proteins that can be a valuable way to tackle most microorganism drawbacks (Gianfreda & Rao, 2004). They are the primary contributors to all biota transformations. They are the catalysts of both broad and narrow specificities; thus, they are used in mixtures with a wide variety of various compounds. They can result in comprehensive structural and toxicological transformations and even complete changes of pollutants into harmless inorganic end products. They can carry out processes that have not conceived effective chemical changes. In addition to the conventional technologies like microbial remediation, enzymes can present advantages since enzymes are more portable than microbes due to their smaller dimensions and function against a given substrate (Gianfreda & Bollag, 2002). Hydrolases, dehalogenases, transferases, and oxidation reductases are the most representative groups of enzymes in remediating toxic habitats. Many of the enzymes were examined primarily under laboratory conditions for the transformation of various xenobiotic compounds. Key groups are oxidoreductases such as mono- or dioxygenase (reductases), dehalogenase (reductases), phenoloxidases (manganese peroxidases and lignin), and cytochrome P450 monoxygenases (laccases). Proteases, amidases, and esterases can contribute to products with low or no toxicity through the breeding of esteric, amid, and peptidal associations (Coppella et al., 1990; Mulbry & Eaton, 1991; Sutherland et al., 2002).

Oxidative enzyme constitutes a vital group of enzymes (Durán & Esposito, 2000; Gianfreda et al., 2006; Rodríguez Couto & Toca Herrera, 2006; Torres et al., 2003) that play a significant role in the climate, including detoxifying contaminated surroundings. They are also engaged in the development and exchanges of humus by degradation and synthetic processes in soil between plants and soil. Additionally, they can participate in and from binding residues in reactions between humus constituents and xenobiotic molecules in water systems and humus materials in soils.

Sixteen small-scale surface flow-built wetlands measuring 2.0 m (L), 1.0 m (W), and 0.7 m (D) in length were aligned in two parallel lines and planted with four wetland plant species: *Cyperus flabelliformis*, *Hymenocallis littoralis*, *Phragmites australis*, and *Vetiveria zizanioides* were used to handle student dormitory's sewer wastewater. The sewage was routed into a septic tank and settling tank before being drained to the experimental wetlands. Urease and protease processes were found to be closely correlated with NH_4 removal in built wetlands, while phosphatase activity was found to be correlated with TP (total phosphorus) and SRP (soluble reactive phosphorus) elimination. The results suggested that urease, protease, and phosphatase may all play critical roles in the removal of NH_4 and phosphorus in built wetlands. The close association between root activity and enzyme activity showed that root activity can influence enzyme activity and that plant roots play a critical role in contaminant removal (Kong et al., 2009).

12.11 CWs for Municipal and Sewage Wastewater Treatment

Municipal wastewater is a significant source of heavy metal contamination, causing a substantial risk to the aquatic ecosystem. Cu, Zn, Pb, Ni, and Hg are possibly more harmful metals, thereby leading to phytotoxicity, bioaccumulation, and acute/chronic health implications. Aquatic plant removal of heavy metals from urban drainage, sewage water, spillage zones, and other contaminated sites has been a well-established method and experimental procedure. Aquatic plants can be used as bioaccumulators to increase the amount of heavy metals in their biomass (Bonanno et al., 2017; Bravo et al., 2017). *Typha domingensis* root and shoot tissues accumulated the maximum Cd, Fe, Ni, and Zn within the first 48 hours of the sample while planted in pots filled with urban wastewater (Mojiri, 2012). *L. gibba* has been researched as an alternate form of removal for industrial wastewater to accumulate arsenic, boron, and uranium (Pedescoll et al., 2015). Something that also demonstrated excellent in extracting heavy metals from urban wastewater was shown by *Typha latifolia* and *Phragmites australis* (Morari et al., 2015). Suganya and Sebastian (2017) reported that they treated sewage effluent using a lab-scale hybrid reed bed system using *Canna indica* and found that the system removed 68% of biological oxygen demand (BOD), 61.8% of chemical oxygen demand (COD), 71.7% of total dissolved solids (TDS), and 73.3% of total suspended solids (TSS). The root portion of the plant absorbed more heavy metals (Cr, Pb, and Ni) than the stem and leaf portion. HCWS (Hybrid constructed wetland system) planted with *Brachiaria humidicola* and *Typha angustifolia* proved to be beneficial in reducing the levels of BOD and COD of the sewage water (Suganya, 2017).

A laboratory-scale experiment was conducted at Tamil Nadu Agricultural University, Coimbatore, to evaluate the sewage treatment effluent performance of macrophytes, including Indian shot (*Canna indica*), giant reed (*Arundo donax*), arrowleaf elephant ear (*Xanthosoma sagittifolium*), cat tail (*Typha angustifolia*), and slender cyperus (*Cyperus distans*) collected from, in, and around the lakes Muthanangulam, Nagarajapuram, and Telungupalayam. Seven different retention times of 1–7 days were set to screen out the suitable macrophytes for treating the sewage effluent under CWT. Based on the experiment, it was found that pH, electrical conductivity, total dissolved solids, total suspended solids, biological oxygen demand, chemical oxygen demand, total nitrogen, and total phosphorus of the sewage effluent were expressively abridged by these plants throughout the retention time from Day 1 to Day 7. Among all the five macrophytes, three aquatic plants, namely, *Canna indica*, *Typha angustifolia*, and *Xanthosoma sagittifolium*, performed better in removing pollutants (Joneboina Easwar Kumar et al., 2019).

The three efficient aquatic plants, namely, *Canna indica*, *Xanthosoma sagittifolium*, and *Typha angustifolia*, screened out was evaluated to optimize the hydraulic loading rate (HLR) for pollutant removal efficiency by using a lab-scale model constructed wetland designed with a horizontal flow system. The experiment was conducted with three screened aquatic plants *Canna indica*, *Xanthosoma sagittifolium*, and *Typha angustifolia*, at seven different retention times of the 1st to 7th day after the beginning of experiment with two average flow of 5 ml/min and 10 ml/min. The results of the experiment state that pollutants and salt load, including heavy metals, were significantly reduced on the 7th day of retention time. *Canna indica* and *Xanthosoma sagittifolium* performed better in a model constructed wetland for treating sewage effluent with the flow rate of 5 ml/min (HLR = 0.00516 cm/day) at the retention time of the 7th day compared to the flow rate of 10 ml/min (HLR = 0.01033 cm/day). The total BOD, COD, TDS, and TSS were also removed to the tune of 44.8, 44.0, 71.3, and 94.5% in *Canna indica*; 41.3, 37.6, 62.8, and 91.0% in *Xanthosoma sagittifolium*; and 37.9, 36.7, 58.5, and 88% in *Typha angustifolia*. The plant's total growth (shoot length, root length, height, and weight) also showed a positive effect at HLR of 0.01033 cm/day with *Canna indica* and *Xanthosoma sagittifolium*. Anti-oxidant and stress enzymes, viz., peroxidase and catalase, released from these plants during the experimental period seem to be positive, thereby confirming the efficiency of aquatic plants in reducing the pollutant load. This indicates that the two plants can be used to treat the sewage effluent through CWT with high retention time and low hydraulic loading rate (HLR) for sustainable agriculture.

12.12 Conclusions

Managing and handling wastewater for its effective recycling and reusing for agriculture and other allied purposes pose a great challenge for the scientific communities. Constructed wetlands with different aquatic plants (macrophytes) seem to be the most promising techniques for recycling/treating sewage effluent that sets forth as a possible solution for wastewater treatment. Different components of the CWs, i.e., such as aquatic plants, the medium used, and microbial communities, create a suitable environment for the degradation of inorganic and organic pollutants through the secretion of enzymes and root exudates. However, the metabolism and uptake of pollutants captured by these aquatic plants need to be studied intensively. As a result, further research is needed to provide more comprehensive and compelling evidence in larger laboratory-scale, pilot-scale, or full-scale constructed wetlands.

References

Adhikari, S., Gupta, S. K., & Banerjee, S. K. (1997). Long-term effect of raw sewage application on the chemical composition of groundwater. *Journal of the Indian Society of Soil Science, 45*, 392–394.

Anderson, T. A., Guthrie, E. A., & Walton, B. T. (1993). Bioremediation in the rhizosphere. *Environmental Science and Technology, 27*(13), 2630–2636. https://doi.org/10.1021/es00049a001.

Arora, B. R., Azad, A. S., Singh, B., & Sekhon, G. S. (1985). Pollution potential of municipal wastewaters of Ludhiana, Punjab. *Indian Journal of Ecology, 12*, 1–7.

Bais, H. P., Weir, T. L., Perry, L. G., Gilroy, S., & Vivanco, J. M. (2006). The role of root exudates in rhizosphere interactions with plants and other organisms. *Annual Review of Plant Biology, 57*, 233–266. https://doi.org/10.1146/annurev.arplant.57.032905.105159.

Bhattacharya, T., Banerjee, D. K., & Gopal, B. (2006). Heavy metal uptake by *Scirpus littoralis schrad.* From fly ash dosed and metal spiked soils. *Environmental Monitoring and Assessment, 121*(1–3), 363–380. https://doi.org/10.1007/s10661-005-9133-1.

Bonanno, G., Borg, J. A., & Di Martino, V. (2017). Levels of heavy metals in wetland and marine vascular plants and their biomonitoring potential: A comparative assessment. *Science of the Total Environment, 576*, 796–806. https://doi.org/10.1016/j.scitotenv.2016.10.171.

Bravo, S., Amorós, J. A., Pérez-de-los-Reyes, C., García, F. J., Moreno, M. M., Sánchez-Ormeño, M., & Higueras, P. (2017). Influence of the soil pH in the uptake and bioaccumulation of heavy metals (Fe, Zn, Cu, Pb and Mn) and other elements (Ca, K, Al, Sr and Ba) in vine leaves, Castilla-la Mancha (Spain). *Journal of Geochemical Exploration, 174*, 79–83. https://doi.org/10.1016/j.gexplo.2015.12.012.

Brisson, J., & Chazarenc, F. (2009). Maximizing pollutant removal in constructed wetlands: Should we pay more attention to macrophyte species selection? *Science of the Total Environment, 407*(13), 3923–3930. https://doi.org/10.1016/j.scitotenv.2008.05.047.

Brix, H., & Arias, C. A. (2005). The use of vertical flow constructed wetlands for the onsite treatment of sewage wastewater: New Danish guidelines. *Ecological Engineering, 25*(5), 491–500. https://doi.org/10.1016/j.ecoleng.2005.07.009.

Chen, J., Wei, X. D., Liu, Y. S., Ying, G. G., Liu, S. S., He, L. Y., Su, H. C., Hu, L. X., Chen, F. R., & Yang, Y. Q. (2016). Removal of antibiotics and antibiotic resistance genes from domestic sewage by constructed wetlands: Optimization of wetland substrates and hydraulic loading. *Science of the Total Environment, 565*(15), 240–248.

Coppella, S. J., DelaCruz, N. D., Payne, G. F., Pogell, B. M., Speedie, M. K., Karns, J. S., … Connor, M. A. (1990). Genetic engineering approach to toxic waste management case study for organophosphate waste treatment. *Biotechnology Progress, 6*(1), 76–81. https://doi.org/10.1021/bp00001a012.

Dan, A., Oka, M., Fujii, Y., Soda, S., Ishigaki, T., & Machimura, T. (2017). Removal of heavy metals from synthetic landfill leachate in lab-scale vertical flow constructed wetlands. ke, M. *Science of the Total Environment, 584*, 742–775.

Durán, N., & Esposito, E. (2000). Potential applications of oxidative enzymes and phenoloxidase-like compounds in wastewater and soil treatment: A review. *Applied Catalysis, B: Environmental, 28*(2), 83–99. https://doi.org/10.1016/S0926-3373(00)00168-5. Enzym, B., *28*, 83–99.

Gianfreda, L., & Bollag, J.-M. (2002). Isolated enzymes for the transformation and detoxification of organic pollutants. In R. G. Burns & R. Dick (Eds.), *Enzymes in the environment: Activity, ecology and applications* (pp. 491–538). New York: Marcel Dekker.

Gianfreda, L., Iamarino, G., Scelza, R., & Rao, M. A. (2006). Oxidative catalysts for the transformation of phenolic pollutants: A brief review. *Biocatalysis and Biotransformation, 24*(3), 177–187. https://doi.org/10.1080/10242420500491938.

Gianfreda, L., & Rao, M. A. (2004). Potentials of extracellular enzymes in remediation of polluted soils: A review. *Enzyme and Microbial Technology 35*(4), 339–354.

Gupta, S. K., & Mitra, A. (2002). *Advances in land resource management for 21st. Century* (pp. 446–469). New Delhi: Soil Conservation Society of India.

Gupta, A. P., Antil, R. S., & Singh, A. (1986). CSIO. *Proceedings of the National Seminar on environmental pollution control and monitoring* (pp. 419–425). Chandigarh, India: October 22–26.

Kong, L., Wang, Y.-B., Zhao, L.-N., & Chen, Z.-H. (2009). Enzyme and root activities in surface-flow constructed wetlands. *Chemosphere, 76*(5), 601–608. https://doi.org/10.1016/j.chemosphere.2009.04.056.

Kumar, J. E., Suganya, K., Sebastian, S. P., & Kannan, T. G. (2019). Changes in physico-chemical characteristics of the sewage effluent under constructed wetland technology treatment. *The Madras Agricultural Journal, 106*(13), 58–62.

Ladislas, S., Gérente, C., Chazarenc, F., Brisson, J., & Andrès, Y. (2015). Floating treatment wetlands for heavy metal removal in highway stormwater ponds. *Ecological Engineering, 80*, 85–91. https://doi.org/10.1016/j.ecoleng.2014.09.115.

Lee, C. G., Fletcher, T. D., & Sun, G. (2009). Nitrogen removal in constructed wetland systems. *Engineering in Life Sciences, 9*(1), 11–22. https://doi.org/10.1002/elsc.200800049.

Leung, H. M., Duzgoren-Aydin, N. S., Au, C. K., Krupanidhi, S., Fung, K. Y., & Cheung, K. C. (2017). Monitoring and assessment of heavy metal contamination in a constructed wetland in Shaoguan (Guangdong Province, China): Bioaccumulation of Pb, Zn, Cu and Cd in aquatic and terrestrial components. *Environmental Science and Pollution Research International, 24*(10), 9079–9088. https://doi.org/10.1007/s11356-016-6756-4.

Mojiri, A. (2012). Phytoremediation of heavy metals from municipal wastewater by *Typha domingensis*. *African Journal of Microbiology Research, 6*, 643–647.

Morari, F., Dal Ferro, N., & Cocco, E. (2015). Municipal wastewater treatment with Phragmites australis L. and Typha latifolia L. for irrigation reuse. Boron and heavy metals. *Water, Air, and Soil Pollution, 226*(3), 56. https://doi.org/10.1007/s11270-015-2336-3.

Mulbry, W. W., & Eaton, R. W. (1991). Purification and characterization of the N-methylcarbamate hydrolase from *Pseudomonas* strain CRL-OK. *Applied and Environmental Microbiology, 57*(12), 3679–3682. https://doi.org/10.1128/aem.57.12.3679-3682.1991.

Narwal, R. P., Gupta, A. P., Singh, A., & Karwasra, S. P. S. (1993). Composition of some city waste waters and their effect on soil characteristics. *Annals of Biology, 9*, 239–245.

Nguyen, X. C., Chang, S. W., Nguyen, T. L., Ngo, H. H., Kumar, G., Banu, J. R., Vu, M. C., Le, H. S., & Nguyen, D. D. (2018). A hybrid constructed wetland for organic-material and nutrient removal from sewage: Process performance and multi-kinetic models. *Journal of Environmental Management, 222*, 378–384.

Nowak, R., & Imperowicz, A. (2016). Liquid waste from septic tanks as a source of microbiological pollution of groundwater. *Inżynieria Ekologiczna, 10*(47), 60–67. https://doi.org/10.12912/23920629/62848.

Palamuleni, L. G. (2002). Effect of sanitation facilities, domestic solid waste disposal and hygiene practices on water quality in Malawi's urban poor areas: A case study of South Lunzu Township in the city of Blantyre. *Physics and Chemistry of the Earth, Parts A/B/C, 27*(11–22), 845–850. https://doi.org/10.1016/S1474-7065(02)00079-7.

Pedescoll, A., Sidrach-Cardona, R., Hijosa-Valsero, M., & Bécares, E. (2015). Design parameters affecting metals removal in horizontal constructed wetlands for domestic wastewater treatment. *Ecological Engineering, 80*, 92–99. https://doi.org/10.1016/j.ecoleng.2014.10.035.

Rai, P. K. (2019). Heavy metals/metalloids remediation from wastewater using free floating macrophytes of a natural wetland. *Environmental Technology and Innovation, 15*. https://doi.org/10.1016/j.eti.2019.100393, PubMed: 100393.

Rodríguez Couto, S., & Toca Herrera, J. L. (2006). Industrial and biotechnological applications of laccases: A review. *Biotechnology Advances, 24*(5), 500–513. https://doi.org/10.1016/j.biotechadv.2006.04.003.

Ryan, P. R., Delhaize, E., & Jones, D. L. (2001). Function and mechanism of organic anion exudation from plant roots. *Annual Review of Plant Physiology and Plant Molecular Biology, 52*, 527–560. https://doi.org/10.1146/annurev.arplant.52.1.527.

Saeed, T., & Sun, G. (2012). A review on nitrogen and organics removal mechanisms in subsurface flow constructed wetlands: Dependency on environmental parameters, operating conditions and supporting media. *Journal of Environmental Management, 112*, 429–448. https://doi.org/10.1016/j.jenvman.2012.08.011.

Shelef, O., Gross, A., & Rachmilevitch, S. (2013). Role of plants in a constructed wetland: Current and new perspectives. *Water, 5*(2), 405–419. https://doi.org/10.3390/w5020405.

Singh, J., & Kansal, B. D. (1985). Amount of heavy metal in the waste water of different towns of Punjab and its evaluation for irrigation. *Journal of Research*. Punjab Agricultural University,, *22*, 17–24.

Suganya, K. (2017). Pollutant removal efficiency of hybrid constructed wetland system for recycling the sewage by utilizing aquatic plants. *The Madras Agricultural Journal, 104*(4–6), 121–123.

Suganya, K., & Sebastian, S. P. (2017). Phytoremediation prospective of Indian shot (*Canna indica*) in treating the sewage effluent through hybrid reed bed (HRB) technology. *International Journal of Chemical Studies, 5*(4), 102–105.

Sukumaran, D. (2013). Phytoremediation of heavy metals from industrial effluent using constructed wetland technology. *Applied Ecology and Environmental Sciences, 1*(5), 92–97. https://doi.org/10.12691/aees-1-5-4.

Sutherland, T., Russell, R., & Selleck, M. (2002). Using enzymes to clean pesticide residues. *Pesticide Outlook, 13*(4), 149–151. https://doi.org/10.1039/b206783h.

Tiwana, N. S., Panesar, R. S., & Kansal, B. D. (1987). Nanital, India. *Proceedings of the national seminar on impact of environmental protection for future development of India* (pp. 119–126).

Torres, E., Bustos-Jaimes, I., & Le Borgne, S. (2003). Potential use of oxidative enzymes for the detoxification of organic pollutants. *Applied Catalysis B: Environmental, 46*(1), 1–15. https://doi.org/10.1016/S0926-3373(03)00228-5.

Vymazal, J. (2007). Removal of nutrients in various types of constructed wetlands. *Science of the Total Environment, 380*(1–3), 48–65. https://doi.org/10.1016/j.scitotenv.2006.09.014.

Walker, T. S., Bais, H. P., Grotewold, E., & Vivanco, J. M. (2003). Root exudation and rhizosphere biology. *Plant Physiology, 132*(1), 44–51. https://doi.org/10.1104/pp.102.019661.

Wu, H., Wang, X., & He, X. (2017). Effects of selected root exudate components on nitrogen removal and development of denitrifying bacteria in constructed wetlands. *Water, 9*(6), 430. https://doi.org/10.3390/w9060430.

Chapter 13
Satellite-Based Soil Erosion Mapping

Kishor Kumar Sahu, Satyajeet Kar, and Sandeep Rout

Abstract Soil erosion has long been recognised as a significant process of soil destruction, affecting millions of hectares of land worldwide, resulting in loss of fertility and biodiversity, decreased stability of marine and terrestrial ecosystems and enhanced exposure to climate change. In semi-arid zones, the highest rate of deforestation occurred in wooded grassland, bushland and shrubland systems, while the lowest rate occurred in woodland. Satellite remote sensing technology for tracking and modelling soil erosion has exploded in popularity worldwide over the last decade. More precisely, renewed emphasis has been placed on recent advances in remote sensing technologies and the availability of these data at various resolutions, as well as on the critical need for up-to-date knowledge on soil loss levels, soil erosion monitoring and modelling, in particular, to ensure that viable agricultural fields are available to ensure food security. GIS research delivers adequate results when developing erosion surveys and risk maps using GIS data layers such as DEM, slope, aspect and land use. The Revised Universal Soil Loss Equation (RUSLE), the Water Erosion Prediction Project (WEPP) and Environmental Information Coordination are the most widely used scientific erosion prediction models that are combined with remote sensing and GIS (CORINE). Remote sensing techniques and the universal soil loss equation were established as the primary tools for mapping and tracking soil erosion in this chapter. It consists of four components: baseline sheet and rill erosion mapping, real-time rill and gully erosion monitoring, future sheet and rill erosion change forecast and long-term pattern determination.

Keywords Satellite imagery · Remote sensing · Geographic information system · Soil erosion · Erosion mapping

---

K. K. Sahu
Department of Soil Science and Agricultural Chemistry, CSK HPKV, Palampur, Himachal Pradesh, India

S. Kar
Department of Forestry, JNKVV, Jabalpur, Madhya Pradesh, India

S. Rout (✉)
Faculty of Agriculture, Sri Sri University, Cuttack, Odisha, India

S. A. Bandh (ed.), *Sustainable Agriculture*,
https://doi.org/10.1007/978-3-030-83066-3_13

13.1 Introduction

Soil is a vital natural resource because it performs critical economic, social and environmental roles. The global economy is primarily dependent on the soil as a natural resource for the supply of goods and services (Blum, 2005). However, due to the high demand for soil-generated products, commodities and services, there has been a considerable strain, especially from developing countries heavily reliant on primary sectors such as agriculture and forestry (Wessels et al., 2004).

Soil depletion by runoff is a primary ecological concern that covers 56% of the world's land. The depletion of soil is exacerbated by soil degradation triggered by human activities (Bai et al., 2008). Rill and inter-rill erosions are the recurrent forms of water erosion, including separation, transport and accumulation of soil particles into a new deposition area, deteriorating soil quality and decreasing land productivity (Fernandez et al., 2003). Soil erosion-related problems include loss of productive crop topsoil, sedimentation, infrastructure destruction and biodiversity loss contributing to global change (Morgan, 2005; Nearing et al., 2004; Onyando et al., 2005).

Although geomorphological processes can cause soil erosion, accelerated soil erosion is mainly encouraged by human activities. Accelerated soil erosion has resulted from rapid population growth, deforestation, unsuitable land production and unregulated grazing (Reusing et al., 2000; Tamene et al., 2006; Zemenu & Minale, 2014). Soil loss is often caused by an amalgamation of slope length steepness, climate change, patterns of land cover and soil's intrinsic properties, making the soil particles more vulnerable to erosion.

Owing to the lack of ability to withstand it and also to substitute the nutrients, the economic impact of soil depletion in some countries is more severe (Tamene et al., 2006). These countries have been marked by high population growth, leading to the excessive use of already harassed resources and the expansion of development on marginal and vulnerable lands. Such a mechanism exacerbates deforestation and loss of productivity, resulting in a cycle of population-poverty-land degradation.

Soil erosion is primarily influenced by topographic features, vegetation cover, soil characteristics and climatic factors. Human movements and large-scale schemes change the vegetation cover, thus affecting the rate of soil erosion. Drill and interrill erosions are primarily influenced by topographic features such as field slope, slope length and shape. The two critical climatic factors are the amount of precipitation and the intensity of precipitation, which is referred to as rainfall erosivity. Additionally, the temperature is a significant climatic aspect since it influences the vegetative materials used in mulching to manage erosion. Aggregate stability, texture, depth, organic matter and stoniness are primarily affected by soil erodibility.

Assessing the rate of soil erosion is critical for developing effective erosion prevention techniques for sustainable land and water resource management. Geographic Information System (GIS) technologies, through their advanced data storage, analysis, management and display functionality, are valuable resources in creating environmental models. Remote sensing (RS) technology, using digital image processing techniques, has provided land use/cover information. Many types of research on modelling soil erosion using RS and GIS technologies have been performed.

These capabilities of such technologies are enhanced further when paired with empirical erosion prediction models. While soil erosion models estimate soil loss and offer geographic erosion distributions, integrated erosion prediction models using RS and GIS estimate soil loss and offer spatial erosion distributions. As a result, it is critical to create accurate erosion risk maps in GIS to identify areas at high risk of erosion and implement appropriate erosion prevention techniques. Sazbo et al. (1998) successfully used RS and GIS technology to chart land loss and deterioration. Another study conducted by Bojie et al. (1995) showed that when GIS data layers such as DEM, aspect, slope and land use create erosion surveys and risk charts, GIS analysis produces adequate results.

The Revised Universal Soil Loss Equation (RUSLE), the Water Erosion Prediction Project (WEPP) and the Coordination of Information on the Environmental (CORINE) are the most frequently used empirical erosion prediction models combined with RS and GIS for mapping erosion threats. The RUSLE was established based on erosion factors like soil erodibility, topography, rainfall and vegetation cover to estimate the annual soil loss per unit area. Based on a particular erosion variable, sediment yield and erosion rates can be calculated over multiple periods in the WEPP model.

13.2 Assessing Land Degradation

1. Expert opinion: Subjective appraisal based on semi-quantitative definitions (e.g. GLASOD survey).
2. Remote sensing is the ground-based radiometry wherein satellite images and aerial photographs correlate with field measurements.
3. Field observations: This includes stratified soil sampling and analysis and long-term field studies of plants and habitats in particular locations.
4. Productivity changes: Keeping an eye on improvements in crop yields and opinions of landowners.
5. Level field criteria: Studies at the farm level are deemed necessary to ascertain the severity of deterioration and its causes and possible remedial steps.
6. Modelling: Modelling is used to estimate the danger of deterioration based on data collected from other approaches (GIS-based models), thus expanding the spectrum of applicability of observed degradation effects.
7. None of these is a singular methodology, and their synergistic applications are widespread.

13.2.1 Land Degradation Mapping and Modelling

Satellite imagery and aerial photography are highly recommended for the following purposes:

13.2.2 Assessing the spatio-temporal distribution of features associated with land degradation

13.2.3 Collecting input data for process simulation models that create maps of ground cover, plant cover and bare soil.

13.2.4 Spatio-Temporal Distribution Assessment

Surveying: To determine the land's present condition in terms of continuing erosion processes.

Identifying the spatial diversity and status of the:

- Vegetation in its natural state (structure and coverage)
- Crops used in agriculture (crop performance)
- Floor of the soil (crusting or sealing)
- Existence of soil erosion surface characteristics (rills and gullies)

Monitoring changes over time:

- Crop canopy development throughout a growing season (an indicator of erosion)
- An area's long-term growth of rill and gully formation

13.2.5 Detection and Quantification of Indicators

Numerous methods can be used to identify indicators, including:

- Field measurements
- Laboratory research
- Data gathered by remote sensing
- A combination of the above

13.2.6 Modelling Input Data

Variables that influence the process include the following:

- Interception of rainfall.
- Storage of water canopy.
- Agricultural land use changes during the growing season are deduced from airborne or satellite-borne photographs and used in process simulation models.

13.3 Soil Erosion Modelling Techniques

13.3.1 Estimation of Soil Loss

Erosion management is essential for preserving soil fertility and enhancing or sustaining the quality of water downstream. Reducing soil erosion to tolerable limits requires sufficiently designed cropping practices and soil conservation initiatives. To calculate soil loss from various land units, many methods exist, including measuring every landform and land use from drainage plots of different sizes, small unit source watersheds and sizeable mixed land use watersheds. Nevertheless, analytical and process-dependent models (equations) are used to predict soil erosion. The Universal Soil Loss Equation (USLE) is empirical. As a feature of most of the significant factors influencing sheet and rill erosion, it estimates the average annual mass of soil loss per unit area. It is considerably more challenging to evaluate soil loss than to assess runoff since several natural factors, such as soil and precipitation and human-made factors, embrace management practices. The loss of soil dramatically depends on the form of erosion.

Significant and valuable sediment yield estimates can be obtained from models for specific purposes. The best example is the Universal Soil Loss Equation (USLE) calculation of long-term average annual soil loss from a catchment.

13.3.2 Erosivity and Erodibility

Degradation of the soil is demonstrated by a drop in fertility status, a decline in the number of nutrients or physical depletion of the topsoil. The latter state is more prevalent in areas prone to soil erosion. During periods of heavy runoff, a large amount of mud, rock waste and organic matter are transferred downslope to rivers and eventually to the oceans. Soil erosion management can be accomplished by considering the susceptibility of soils and other factors. In general, the amount of erosion yield is dependent on the rain's ability to remove soil particles (rainfall erosivity) and, concurrently, on the soil's resistance to rainfall (soil erosivity). Thus, both erosivity and erodibility are essential features of soil erosion that occur when rainfall erosivity exceeds soil erodibility.

13.3.3 Erosivity of Rainfall

The word 'rainfall erosivity' refers to the soil's proclivity to be washed away from disturbed and de-vegetated regions into surface waters during storms. It is determined by the physical characteristics of precipitation, which include the size of raindrops, their propagation, their kinetic energy and their terminal velocity, among others. For a specific soil condition, the tendency of two storms to induce soil erosion may be quantitatively compared. The capacity of overland runoff flow to erode soil is determined in part by rainfall and in part by the soil's surface. The increased erosivity of the overland water flow in the presence of rain indicates a more remarkable erosive ability. Soil erosion occurs when the intensity and length of a downpour exceed the ability of the soil to absorb the rainwater. Erosion is influenced by various conditions, including the state of the soil, the slope and the amount of energy or precipitation force expected during the duration of surface disturbance.

13.4 Factors Affecting the Erosivity of Rainfall

The following variables influence the erosivity of rainstorms:

13.4.1 Intensity of Rainfall

Rainfall strength is a term that refers to the amount at which rain falls on the ground surface. It is a significant factor in the erosive aspect of rainfall. Rainfall strength is described as the force exerted by a single water droplet as it reaches the soil surface. Wischmeier and Smith (1958) suggested the following equation to equate kinetic energy to rainfall intensity:

$$\mathrm{KE} = 210.3 + 89 \log_{10} I$$

Where:

KE = Kinetic energy of the rainfall

I = Intensity of the rainfall

13.4.2 Distribution of Drop Sizes

The drop size distribution within a rainstorm has a combined effect on the rain's energy, velocity and erosivity. Increases in the median drop size result in a rise in the rainfall level. The following equation illustrates the relationship between the rainfall strength and the median drop size (D50) (Laws & Parsons, 1943):

$$D_{50} = 2.23 I^{0.182}$$

Where:

D_{50} = Median drop size in inches

I = Intensity of the rainfall (inch/h)

13.4.3 Terminal Velocity

The effect of falling raindrops' terminal velocity (a function of the drop size) is quantified in terms of their kinetic energy upon contact with the soil surface. A rainstorm with a high proportion of larger raindrops would have a higher terminal velocity and vice versa. The relationship between the kinetic energy and terminal velocity of a rainstorm is as follows:

$$E_{\mathrm{k}} = \frac{IV^2}{2}$$

Where:

E_{k} = Energy of rainfall

I = Intensity of the rainfall

V = Terminal velocity of the rainfall before impact

Ellison (1947a) developed the following empirical connection between terminal velocity, drop diameter and rainfall intensity in order to determine the volume of soil removed by rainfall:

$$E = KV^{4.33} d^{1.07} I^{0.65}$$

Where:

E = Relative amount of soil detached

K = A constant (depends upon the characteristics of the soil)

V = Velocity of the raindrops

d = Diameter of the raindrops

I = Intensity of the rainfall

13.4.4 Wind Speed

Wind speed impacts the ability of runoff to detach soil by affecting the kinetic energy of a rainstorm. As tropical areas are often subjected to windy storms, they are more potent at dislodging aggregates than predicted.

13.4.5 Slope Direction

The slope of the soil also has a significant impact on the erosivity of rainfall. Gradients in the path of the rainstorm have the effect of altering the raindrop's natural kinetic energy. It increases the raindrop's impact force as the velocity factor in the slope direction increases.

13.5 Erosivity Estimation Using Rainfall Data

The erosivity of rainfall is proportional to its kinetic energy, and the following two techniques are commonly used to determine the erosivity of rains:

1. EI_{30} Index method
2. KE > 25 Index method

13.5.1 EI_{30} Index Method

Wischmeier and Smith (1965) developed this technique because the result of the storm's kinetic energy and the maximum rainfall intensity of 30 minutes give the best estimate of soil loss. The highest average intensity encountered in any 30 minutes during the storm is determined by finding the maximum amount of rain that falls in the 30 minutes and later translating the same to intensity in mm/hour from tracking rain gauge maps. This erosivity measure is the EI_{30} index and can be measured for individual storms and weekly, monthly or annual erosivity values.

The value of the precipitation erosivity factor EI_{30} is determined as follows:

$$\mathrm{EI}_{30} = \mathrm{KE} \times I_{30}$$

Where:

KE = Kinetic energy of the rainfall

I_{30} = Maximum intensity of the rainfall for 30 minutes

Kinetic energy for the storm is computed from equation

$$KE = 210.3 + 89\log_{10} I$$

Limitation

The EI_{30} index system was developed in the United States and has been considered unsuitable for estimating erosivity in tropical and subtropical areas.

13.5.2 KE > 25 Index Method

This is a new approach proposed by Hudson for calculating the erosivity of tropical storms' rainfall. This approach is based on the premise that erosion happens only when the rainfall level reaches a specific threshold value. Studies determined that rainfall intensities less than 25 mm/h cannot result in substantial soil erosion. As a result, this approach only considers rainfall intensities more significant than 25 mm/h. That is why the process is referred to as the KE > 25 index method. It is used in the same way as the EI_{30} index and has a related measurement technique.

13.6 Procedure for Calculation

Both techniques use the same calculation practice. However, the KE > 25 approach is more beneficial since it eliminates several data points with a value less than 25 mm/h, resulting in fewer rainfall data. Both methods include data on rainfall volume and severity.

The method entails multiplying rainfall quantities by the measured kinetic energy values for each strength class. Then, all of these values are taken together to obtain the storm's overall kinetic energy. The resulting KE value is then multiplied by the actual 30-minute rainfall rate to derive the rainfall erosivity value.

13.6.1 Erodibility of the Soil

Soil erodibility is a measure of a soil's resistance to erosion depending on its physical characteristics. By and large, soils with increased penetration rates, higher organic matter levels and improved soil composition withstand erosion better. Sandy loam and soils with a loam texture are less erodible than fine sand, silt and certain clay-textured soils. A soil's erodibility can be quantitatively compared to that of other soils under a given rainfall environment. Bouyoucos (1935) proposed that soil erodibility is proportional to the mechanical composition of the soil, which includes silt, clay and sand:

$$E = \frac{\%\text{sand} + \%\text{silt}}{\%\text{clay}}$$

Where:

E = Erodibility of soil

The range of particle diameter of silt, clay and sand is

Clay = < 0.002 mm
Silt = 0.002–0.006 mm
Sand = 0.06–2.0 mm

Tillage and cropping activities that deplete soil organic matter contribute to low soil composition, soil compactness and erodibility. Compacted subsurface soil layers can have the effect of reducing penetration and increasing runoff. Additionally, forming a soil crust that appears to 'seal' the surface may result in a decrease in infiltration. A soil crust may reduce soil loss in specific locations due to sheet or rain splash erosion, but a rise in runoff water may exacerbate rill erosion problems.

There may be three different soil types with varying degrees of disturbance severity, for example:

- Low
- Moderate
- High

Stocking rates or the responses of three different soils, for example, are as follows:

- A clay
- A loam
- A sand

13.6.2 Determination of Erodibility

The term 'erodibility' refers to the soil's resistance to detachment and transport. It varies according to the aggregate stability, infiltration capability, soil texture, shear strength, infiltration performance and organic and chemical content. The soil erodibility element 'K' is used to describe the soil's erodibility. There are numerous methods for determining K, and three of the most common are discussed below.

13.6.2.1 In Situ Erosion Plots

Erosion plots allow for the determination of 'K' under field conditions. They use a normal state of bare soil with no maintenance practices and a 7° slope along the length of the plots, which is 22.13 metres. This is an expensive and time-consuming process.

13.6.2.2 Measuring K Under a Simulated Rainstorm

This technique is less time-consuming but reasonably expensive. The primary downside is that all the properties of natural rain cannot be recreated by any of the rainfall simulators designed to date. Nevertheless, in erosion research, this approach is more commonly used.

13.6.2.3 Predicting K

K can be predicted by using regression equations that describe the relationship between K and the physico-chemical properties of the soil. Wischmeier et al. (1971) developed a nomograph to express the relationship between K and soil properties. It is based on the following equation:

$$100\text{K} = 2.1 \times 10^{4} \times \left(2 - OM\right) \times m^{1.14} + 3.25 \times \left(St - 2\right) + 2.5 \times \left(Pt - 3\right)$$

Where:

OM = Organic matter content

m = Silt plus fine sand

St = Soil structure code (1 for very fine granular, 2 for fine granular, 3 for coarse granular, 4 for massive, blocky or platy)

Pt = Permeability class (1 for rapid, 2 for moderate to rapid, 3 for moderate, 4 for slow to moderate, 5 for slow, 6 for very slow).

K is predicted using the monograph devised by Wischmeier et al. (1971).

13.7 Correlation of Soil Erosion and Rainfall Energy

It is widely established that the volume of soil removed by a particular depth of rainfall is related to the pace at which it occurs. Numerous tests and various measurements of raindrop fall velocity (Ellison, 1947b) demonstrate that soil splash rate is a function of rainfall intensity:

$$S \infty V^{4.3} . D^{1.07} . I^{0.65}$$

Where:

S = Quantity of soil splashed in 30-minute duration

V = Velocity of a raindrop

D = Diameter of the raindrops

I = Intensity of rainfall

Raindrop diameters can be found in storms with different intensities within each area, resulting in regressions such as the energy of a storm being equal to the energy of each segment of rain falling at a given intensity compounded by the number of millimetres falling at this Intensity (Bisal, 1960).

The expression is given by

$$G = K.D.V^{1.4}$$

Where:

G = Weight of the soil splashed

D = Diameter of the raindrops

V = Impact velocity

K = A constant depending on the soil type

Mihara (1959) claimed that splash erosion is directly proportional to the kinetic energy of raindrops based on their mass and velocity. He established the following relationship between two distinct soil types:

For sandy soil, splash erosion ∞ K.E.0.9

For clay soil, splash erosion ∞ K.E.1.46

13.8 The Universal Soil Loss Equation (USLE)

In 1940, the United States began developing equations for estimating soil erosion. Zingg (1940) proposed that soil loss and slope length had a power-raised relationship. Later in 1947, a committee headed by Musgrave proposed a soil loss equation that bore some resemblance to the current USLE. Wischmeier and Smith (1965) developed the universal soil loss equation using data from runoff plots; the equation was later modified using more recent data from runoff plots, rainfall simulators and field observations. Controlling erosion is the most often used method for measuring soil depletion from rural watersheds. The USLE is an erosion prediction model that allows for the measurement of long-term soil erosion averages from sheet and rill erosion on a given land surface under specified conditions (Wischmeier & Smith, 1978).

It estimates the long-term average annual loss of soil from arable land segments under different cropping conditions. This estimate aims to encourage farmers and soil conservation advisors to choose combinations of land use, cropping and soil conservation practices to keep soil loss to an appropriate level. The equation (USLE) is as follows:

$$A = R \times K \times L \times S \times C \times P$$

Where:

A = Soil erosion per unit area per unit time
R = Rainfall erosivity index
K = Soil erodibility index
L = Slope length
S = Slope steepness
C = Cover management factor
P=Supporting practice factor

13.9 Parameters of Universal Soil Loss Equation

13.9.1 The Factor of Rainfall (R)

To account for the erosive force of rainfall, the volume and strength of rain over a year (erosivity index unit) are associated with the erosivity component. The word 'erosivity of runoff' refers to the ability of storms to wash the soil from disturbed and de-vegetated areas onto surface waters. Erosion is influenced by various conditions, including the state of the soil, the slope and the amount of energy or precipitation force expected during the duration of surface disturbance.

13.9.2 Factor of Soil Erodibility (K)

Soil erodibility factor is a unit of erosion index defined as the soil loss from a plot 22.1 m in length on a 9% slope under a continuous bare cultivated fallow. It varies by less than 0.1 for the least erodible soils and almost 1.0 for the most erodible soils.

13.9.3 The Factor of Topography (LS)

LS denotes the slope length-gradient factor. The topographic factor is used to calculate the slope's length and steepness. The longer the slope, the larger the amount of surface runoff; the steeper the slope, the greater the velocity of surface runoff.

13.9.4 The Factor of Crop Management (C)

C is the crop/vegetation management part and is the ratio of soil loss caused by a specific crop management strategy to the equal loss caused by continuous fallow and tilled soil. It is used to determine the relative effectiveness of soil and crop control schemes in preventing soil degradation. The C factor can be determined by choosing the crop type and tillage method.

13.9.5 The Factor of Support Practices (P)

P denotes the help practice aspect, representing the results of various activities that minimise the volume and rate of runoff, thus reducing erosion. The P factor quantifies the soil depletion caused by a support practice compared to straight row farming up and down the hill. Cross slope planting, contour forestry and strip cropping are the most often used supportive cropland activities. P should be zero in an environment with absolute support practices, suggesting no sediment loss. P should be 1.0 in an area with no support practices, indicating the highest potential sediment loss.

13.10 USLE Parameter Estimation

13.10.1 Rainfall Erosivity Factor (R)

It references the rainfall erosion index, which quantifies rainfall's tendency to erode soil particles in an exposed area. The amount of soil loss from a barren field has been determined to be directly proportional to the product of two rainfall characteristics: the storm's kinetic energy and its 30-minute maximum intensity. The outcome of these two characteristics is termed EI or EI_{30} or rainfall erosivity. It is equal to the amount of rainfall erosion index units (EI_{30}) that fell on the study site during a given time. A storm's rainfall erosion index unit (EI_{30}) is calculated as follows:

$$\mathrm{EI}_{30} = \frac{\mathrm{KE} \times I_{30}}{100}$$

Where:

KE = Kinetic energy of the rainfall
I = Intensity of the rainfall

$$\mathrm{KE} = 210.3 + 89 \log_{10} I$$

The duration of the research maybe a week, a month, a season or an entire year. Annual EI_{30} values are typically computed using data from various meteorological stations, and lines linking equivalent EI_{30} values (referred to as iso-erodent lines) are drawn for the area covered by the data stations to facilitate their use in USLE.

13.10.2 Soil Erodibility Factor (K)

The element of soil erodibility (K) in the USLE refers to the rate at which various soils erode. Due to inherent soil characteristics, some soils can erode more quickly than others under conditions of an equal slope, precipitation, vegetative cover and soil management practices. On unit runoff plots, the direct calculation of 'K' reflects the cumulative effects of all variables that substantially affect the ease with which soil is eroded or the primary slope other than 9% slope. Soil permeability, infiltration rate, soil texture, size and stability of the soil structure, organic content and soil depth are soil properties that primarily affect soil loss. These are typically calculated by unique experimental runoff plots or by using empirical erodibility equations related to factor 'K' with several soil properties. The soil erodibility factor (K) is expressed as tonnes of soil loss per hectare per unit of rainfall erosivity index, with a slope of 9% and a field length of 22 m (in some instances, 22.13 m). The soil erodibility factor (K) is calculated by taking into account, without the effect of crop cover or management, the soil loss from continuous cultivated fallow lands.

The formula used for estimating K is as follows:

$$K = \frac{AO}{S \times (\Sigma \mathrm{EI})}$$

Where:

K = Soil erodibility factor
AO = Observed soil loss
S = Slope factor
ΣEI = Total rainfall erosivity index

13.10.3 Topographic Factor (LS)

The slope length factor (L) is the ratio of soil loss under identical conditions from the field slope length under consideration to 22.13 m length plots. The size of the slope has a direct relationship with the loss of the soil, i.e. it is roughly equal to the square root of the length of the slope for soils on which the size of the slope does not affect the runoff rate (Zingg, 1940).

The gradient of the land slope factor is defined as the soil loss ratio from the field slope gradient to that from the 9% slope under otherwise identical conditions (S).

Since runoff velocity increases as field slope increases, causing more soil to be detached and carried along with the surface flow, increased slope steepness results in increased soil erosion.

Typically, the two variables L and S are merged into a single topographic component called LS. This factor is defined as the ratio of soil loss from a field with a specified steepness and slope length (i.e. 9% slope and 22.13 m length) to soil loss from the continuous fallow property. The value of LS can be determined using the formula given by Wischmeier and Smith (1962):

$$LS = \frac{\sqrt{L}}{100}\left(0.76 + 0.53S + 0.076S^2\right)$$

Where:

L = Length of field slope

S = Percent slope of the land

Wischmeier and Smith (1978) again derived the following equation for *LS* factor in MKS system, based on the observations from cropped land on slopes ranging from 3 to 18% and length from 10 to 100 m. The derived updated equation is

$$LS = \left(\frac{\lambda}{22.13}\right)^m \left[65.41\sin^2\theta + 4.56\sin\theta + 0.065\right]$$

Where:

Λ = Length of field slope

θ = Angle of slope

M = Exponent varying from 0.2 to 0.5

13.10.4 Crop Management Factor (C)

Factor C, crop management, can be described as the estimated soil loss ratio from cultivated versus fallow land. The surface form, slope and precipitation regimes are all the same. According to crops and cropping practices, soil erosion is influenced in many ways, such as the type of crop, cover quality, root growth, water use by plants, etc. The difference in rainfall distribution during the year also affects crop management, which involves the loss of soil. Given all these variables, the effectiveness of each crop and cropping practice in erosion control is assessed based on five suggested crop stages implemented by Wischmeier (1960):

- Period F (rough fallow): This period encompasses summer ploughing and seed-bed planning.
- Phase 1 (seed bed): This corresponds to the period beginning with seeding and ending 1 month later.
- Period 2 (establishment): This phase lasts between 1 and 2 months after seeding.

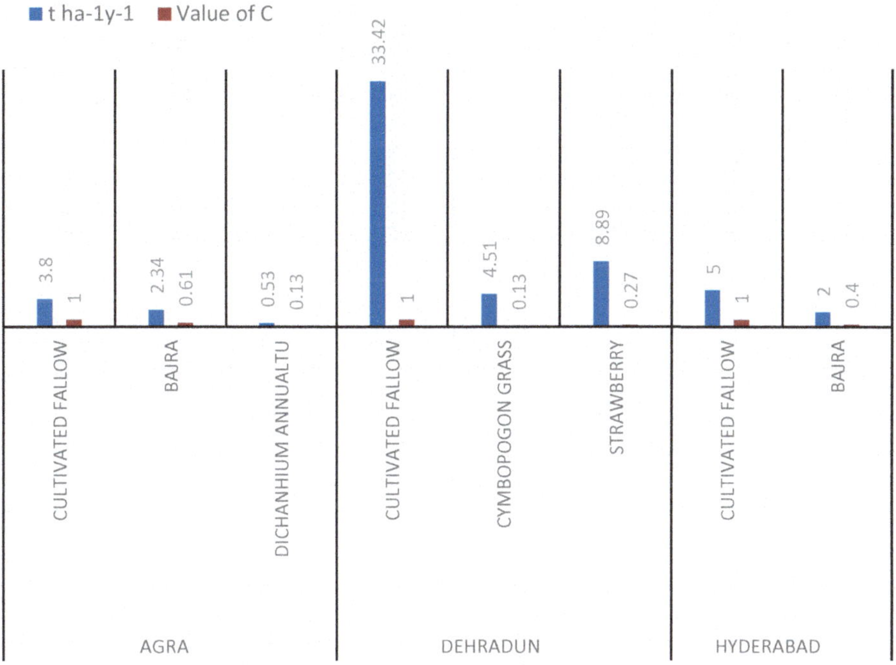

Fig. 13.1 Values of crop management factors for different stations in India. (Source: Modified from K Subramanya, 2008)

- Phase 3 (growth period): It begins with period two and ends with crop harvesting.
- Stage 4 (residue or stubble): This period encompasses everything from grain processing to summer ploughing or seedbed preparation.

The soil loss data for the above stages are collected from the runoff plot for determining the crop management factor. C is computed as the ratio of soil loss from cropped plot to the corresponding soil loss from a continuous fallow land for each of the above five crop stages separately, for a particular crop, considering various combinations of crop sequence and their productivity levels. This factor reflects the combined effect of different crop management practices. Values of factor C for some selected stations of India are shown in Fig. 13.1.

13.10.5 Support Practice Factor (P)

This element is the ratio of soil erosion caused by a support practice and straight row farming up and down the hill. Contouring, terracing and strip cropping are the primary management practices. The amount of soil lost varies according to the techniques used. The table shows the factor P for various types of support activities in different parts of India (Fig. 13.2).

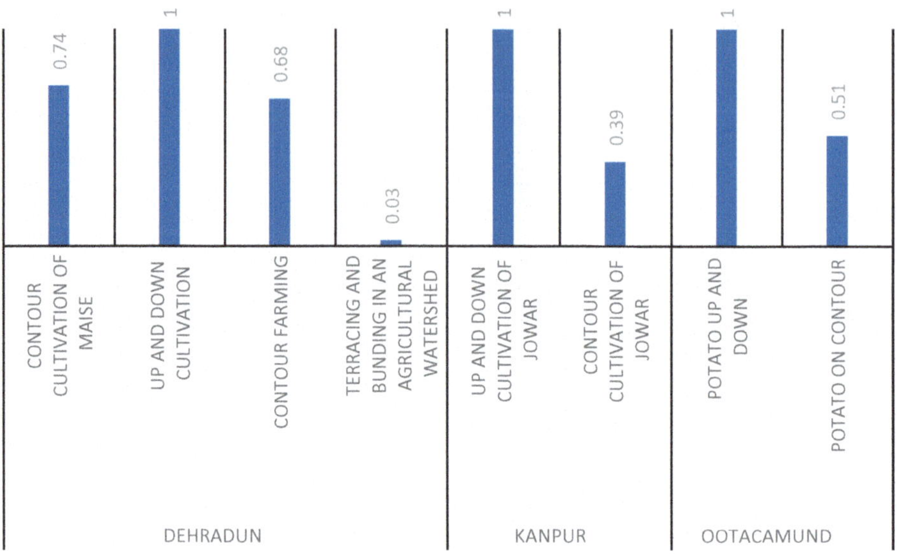

Fig. 13.2 Different values of support practice factor (P) for some Indian locations. (Source: Modified from K. Subramanya, 2008)

13.11 Applications of Universal Soil Loss Equation

USLE is an erosion prediction model, and its effectiveness is contingent upon its ability to forecast its multiple variables accurately. It is focused on a sizable experimental database relating to a variety of factors affecting USLE. The universal soil depletion equation has three critical applications:

- It forecasts land loss.
- It aids in the detection and selection of agricultural practices.
- It makes crop management recommendations.

13.12 Limitations of Universal Soil Loss Equation

The equation involves the procedure for assigning the values of different associated factors based on the practical concept. Therefore, there is a possibility to introduce some errors in selecting the appropriate values, particularly those based on the crop concept. Typically R and K factors are constant for most of the sites/regions in the catchment, whereas C and LS vary substantially with the erosion-controlled measures used.

The following are some of the limitations of the USLE:

1. *Empirical*

The USLE is an abstract equation that does not mathematically reflect the actual mechanism of soil erosion. By using observational coefficients, the probability of including predictive errors in the equation is eliminated.

2. *Prediction of Annual Soil Loss on an Average Basis*

Since this equation was constructed primarily using average annual soil loss data, its applicability is restricted to estimating the average annual soil loss for a given region. This equation produces less than the calculated value, mainly when the rainfall is intense. For each heavy flood, the storage basin whose sediment area was computed using USLE should be examined to ensure that the sedimentation amount in the storage basin remains within acceptable limits.

3. *Gully Erosion Is Not Calculated*

This equation is used to determine the extent of sheet and rill erosion but cannot forecast gully erosion. The calculation does not account for gully erosion caused by concentrated water flow, although it increases soil erosion.

4. *Non-computation of Sediment Deposition*

Only soil depletion, but not soil deposition, is calculated in the equation. Sediment accumulation at the bottom of the river is smaller than the overall loss of soil from the watershed as a whole. Nevertheless, the USLE can quantify the amount of sediment storage needed for sediment retention structures. The USLE equation can also be used as a conservative measure of potential storage needs for sediment, mainly where sediment basins usually range from 2 to 40 ha and runoff has not travelled further. The basin is intended to serve as the settlement area. Again, if the drainage is poorly managed on any site and gully erosion is in extensive form, this equation underestimates the retention structure's sediment storage requirement.

13.13 Revised Universal Soil Loss Equation (RUSLE)

Over the last few decades, a cooperative effort between scientists and users to update the USLE has resulted in the development of RUSLE. The modifications incorporated in USLE to result in the RUSLE are mentioned as under (Kenneth et al., 1991):

- Automating the equations to aid in the computations.
- A new definition for rainfall-runoff erosivity (R) in the Western United States, based on the data from over 1200 gauge locations.
- Specific revisions and additions have been made for the Eastern United States, including adjustments for regions with elevated R factors and flat slopes to account for splash erosion caused by raindrops landing on ponded water.

- The establishment of a seasonally variable definition for soil erodibility (K).
- A novel method for measuring the cover management term (C) using sub-factors for accounting for previous land usage, crop canopy, soil cover and surface roughness.
- New slope length and steepness (LS) algorithms that take into account the ratio of rill to inter-rill erosion.
- Capacity for calculating LS products for slopes with a variety of shapes.
- Rangeland restoration techniques, strip crop rotations, contour factor values and subsurface irrigation are all new conservation practices (P).

13.14 Modified Universal Soil Loss Equation (MUSLE)

Williams updated the USLE in 1975 to create the MUSLE by substituting a 'runoff factor' for the rainfall energy factor (R). The MUSLE is denoted by

$$Y = 11.8\left(Q \times q_{\mathrm{p}}\right)^{0.56} K\left(LS\right)CP$$

Where:

Y = Yield of sediment from an individual storm
Q = Volume of storm runoff
q_p *= Peak rate of runoff*
K,L,S,C,P = Different factors of universal soil loss equation

Appropriate runoff models can be used to determine Q and q_{p} values. Q is taken to reflect the detachment phase in this model, while q_{p} represents sediment transport. A sediment yield model does not require calculating the sediment delivery ratio separately, and it applies to individual storms. Additionally, it improves the precision of sediment yield estimation. From a modelling perspective, it benefits from the simulation of a watershed's constant, weekly and annual sediment yields by integrating suitable hydrological models with MUSLE.

13.15 Spatial Erosion Assessment

Three distinct methods exist for assessing the spatial extent of soil erosion:

- The first step is to determine soil erosion rates at different locations using measurement instruments or erosion plots (Hudson, 1993; Loughran, 1989). However, accurate measurements are typically expensive and time-consuming, essential equipment is scarce (Stroosnijder, 2005) and measuring results can be highly unpredictable even under comparable circumstances (Nearing et al., 1999). Field measurements are often used to determine the role of a specific erosion element, the development of models and their validity, but not for erosion spatial assessment.

- The second solution is to conduct erosion field surveys through which erosion-related characteristics such as pedestals or rills are identified (Herweg, 1996). Although quantitative data can be collected by continuously calculating the dimensions of a feature, most surveys are conducted qualitatively, with the volume of erosion classified according to the characteristics encountered. Due to management practices such as ploughing, survey timing is critical, as some features can be undetectable during the year. Surveys can map spatial erosion in small catchments of around 2 km^2 (Vigiak et al., 2005), but this becomes more complex in more expansive areas. However, systematic visual recognition of specific characteristics from aerial photos is another form of erosion survey that could be conducted for wider regions up to 50 km^2 (Bergsma, 1974).
- Integrating spatial data on erosion causes is the third and the most often used method for assessing spatial erosion. While the Universal Soil Loss Equation (Wischmeier & Smith, 1978) is frequently used, numerous other erosion models exist that allow spatial mapping of erosion (Merritt et al., 2003).

However, erosion models are designed for a specific area and size, and moving them to other scales or regions is not straightforward and may result in suboptimal or incorrect results (Brazier et al., 2000; Jetten et al., 2003; Kirkby et al., 1996; Schoorl et al., 2000). Additionally, specific erosion models include extensive data on a wide range of rainfall, soil, vegetation and slope parameters. These statistics are often unavailable or only accessible at very coarse scales in data-scarce areas such as developed countries. Qualitative data integration techniques that allow flexible selection and a combination of erosion factors can be an excellent complement to erosion models. The choice of erosion variables will be region-specific, based on the existing processes and the main parameters that account for the region's heterogeneity in these processes. Local or specialist expertise can contribute significantly to developing such qualitative approaches (De la Rosa et al., 1999; Sonneveld, 2003). The outcomes of these approaches are typically a numerical assessment of erosion risk, which is the relative likelihood of erosion occurring at a particular location compared to other sites in the mapped area.

13.16 Mapping Erosion From Space

Satellite remote sensing can provide essential input to erosion assessments at different spatial scales through various space-borne sensors currently orbiting the earth. Satellite data can aid in the rapid mapping of erosion over large areas, especially for data-poor regions. At the same time, otherwise, this could only be achieved through costly and time-consuming survey methods.

Several types of satellite images and image-derived items are available to the general public obtained from earth-observing space missions. While certain kinds of images are still costly, much data is inexpensive or free of charge, making it easier for a broader audience to use it. Therefore, satellite imagery is increasingly being used for studies of regional erosion. This can be achieved by identifying erosion characteristics and eroded areas or measuring erosion factors such as the cover or slope of vegetation.

In some instances, degraded areas, larger than 1 ha, can be distinguished from their habitat due to reduced plant cover (Pickup & Nelson, 1984), altered soil properties (Hill et al., 1995) or natural changes in the earth's surface (Lee & Liu, 2001). However, successful use of satellite remote sensing to detect degraded areas is typically limited to (semi-)arid natural and rangeland landscapes, as well as areas of extensive gully erosion (badlands). In more tropical environments, vegetation cover often obscures the visibility of the earth, and farming practices may have a direct effect on vegetation cover, soil resources and surface roughness. As a result, these variables cannot be directly linked to soil degradation in wet and agricultural fields. Along with eroded areas, satellite imaging may reveal individual erosion features such as gullies and large rills. This is partly due to distinct characteristics such as proximity to the subsoil and reduced vegetation cover but even more fundamentally due to the rills and gullies' basic spatial structure. The spatial resolution of the imagery, on the other hand, should correspond to the scale of the elements. Visual interpretation has been a widely used technique for distinguishing individual gullies from aerial photos (Martínez-Casasnovas, 2003; Nachtergaele & Poesen, 1999) and satellite imagery (Bocco & Valenzuela, 1993). Although some scholars questioned the viability of this exercise due to the spectral heterogeneity of gullies and their atmosphere, automatic gully retrieval from satellite images provides fast insight into the magnitude of gully erosion and the resulting lack of productive land for vast areas (King et al., 2005; Zinck et al., 2001).

Satellite imagery may obtain information on a range of erosion factors. Significant climate parameters for erosion studies are the volume and intensity of rainfall, measured on coarse scales, e.g. Tropical Rainfall Measuring Mission (TRMM) spaceborne data. Digital elevation models (DEMs) that can be obtained from stereo images (Toutin, 2001) or specialised techniques such as radar interferometry (Toutin & Gray, 2000) typically determine terrain attributes such as slope. Satellite data can be used to determine the spatial distribution of different soil properties, but this is mainly limited to arid or semi-arid regions due to the alarming effect of vegetation (Huete, 2004). Satellite data for the classification of land use or the extraction of continuous measurements of vegetation abundance and structure can be applied to determine vegetation cover (Hall et al., 1995). Erosion factors are not static, but over time they shift. Rainfall and vegetation are the most complex variables, while soil properties can also be altered due to, for example, tillage or crusting on short time scales. One way or another, the temporal variability of erosion variables needs to be accounted for satellite-based erosion evaluations. One alternative is to measure the variables using multi-temporal satellite imagery at various moments of the year (e.g. De Jong et al., 1999). A second choice is to decide that a mono-temporal satellite image reflects the conditions of the factor being analysed when the most significant risk of erosion is analysed. Image timing may be essential to obtain precise spatial erosion patterns, although rationales for image selection are often not established in erosion studies.

13.17 Satellites and Sensors Applied in Erosion Research

Numerous earth observation satellites orbit our earth, providing periodic images of the surface. Many of them can provide valuable information for measuring erosion, but they have been used for this purpose less often. Sensors are classified into those that measure sunlight reflections in the visible and infrared portions of the electromagnetic spectrum, thermal infrared radiance (optical systems) and those that continuously relay microwave signals and monitor the received signal (microwave systems).

In most cases, optical satellite systems have been used in erosion studies. These sensors operate in the visible and near-infrared (VNIR) range of 0.4–1.3 Am, the shortwave infrared (SWIR) range of 1.3–3.0 Am and the thermal infrared (TIR) range of 3.0–15.0 Am.

13.18 Detection of Erosion

Satellite data may be used to track erosion either directly or indirectly by the detection of erosion effects. Direct detection has been accomplished by identifying significant erosion features, the discrimination of eroded zones and the estimation of erosion rate using observational relationships. Detectable consequences include disruption caused by important erosion events and reservoir sedimentation.

13.19 Geographic Information Systems (GIS) and Simulation of Soil Erosion

Ultimately, the effectiveness of every soil erosion model is contingent upon its integration with GIS. SOMs have been implemented at the field scale as a cost-effective method for organising and managing soil protection. However, their implementation at the watershed scale has been constrained until recently by the difficulties of handling and controlling a vast amount of data and model parameters at such a spatial scale. The implementation of robust spatial hydrological tools within GIS, as well as the integration of various lumped parameter models (LPMs) and distributed parameter models (DPMs) with GIS, has allowed modellers to resolve these constraints and expand model capabilities to the watershed scale (Tim & Jolly, 1994). The ability to produce topographic parameters from digital elevation models (DEMs) enables the modelling of three-dimensional erosion in areas with complex topography (Desmet & Govers, 1995). Coupling GIS and soil erosion models has the added advantage of standardising modelling procedures in user-defined model parameters, cost and time savings associated with modelling processes and visualising modelling performance (Greene & Cruise, 1995). As a result, many existing models, such as RUSLE, WEP, EUROSEM and ANSWERS, have been successfully connected to GIS, while new models, such as LISEM and SWAT, have been built based on GIS.

13.20 Satellite Remote Sensing

Environmental factors must be observed to determine the state of the earth's wealth and monitor its dynamics. At the moment, space technology, especially satellite remote sensing, is making a significant contribution to the comprehensive and timely evaluation of large-area natural resources (Colwell, 1983). Remote sensing is primarily used to gather, store and analyse data collected by sensing systems mounted on aircraft or satellites. Currently, satellite remote sensing is a critical source of information for environmental research, including the atmosphere, seas and land surfaces. Simultaneously, military goals have fuelled its expansion. There are hundreds of artificial satellites orbiting our planet, each equipped with various sensors to capture and relay valuable data about our environment.

However, information must be gleaned from the recorded image evidence. Onboard satellites and sensors monitor electromagnetic radiation, which is sent to the ground and stored electronically. Radiation can be analysed at various wavelengths depending on the sensor's properties (e.g. visible light, infrared, thermal). The sun is the most common source of radiation. Nonetheless, in some instances, such as radar imaging, the satellite structure generates radiation by sending energy beams to the earth's surface. Thus, satellite photographs merely depict spatial differences in how electromagnetic radiation interacts with the atmosphere and the earth's crust at a given point in time. Physical models or computational methods may be used to extract information about environmental factors from the recorded imagery. In a particular study, the vector of interest dictates the image form to be used (sensor or satellite).

Along with the wavelength(s) recorded, additional sensor characteristics such as spatial and temporal resolution may be necessary. Temporal resolution refers to the frequency at which an image with the same features may be recorded and is usually inversely proportional to spatial resolution. The spatial resolution is determined by the sensor and the height of the satellite's orbit.

As a result, removing environmental variables from satellite data varies according to the image type used. Independent in situ measurements of the variable of interest are needed to create and evaluate these methods (Jensen, 2004). The variables extracted from satellite data will then be paired with additional spatial data to develop new or more accurate data (He et al., 1998; Lubczynski & Gurwin, 2005; Saha et al., 2002).

13.21 Conclusions

Soil erosion adversely affects millions of land areas, resulting in production losses, increased food insecurity, reduced ecosystem resilience and increased climate change vulnerability. In general, its spatial reach is not well known, and erosion mitigation interventions have had limited success due to the lack of adequately focused interventions, hampering progress towards preventing further property degradation. Therefore, more comprehensive and extensive work is required to evaluate the spatial variability and extent of soil erosion within given regions. Furthermore, for sustainable and efficient soil erosion control, remedial and preventive strategies are to be established, and the discrimination of soil erosion over different land management practices is needed. Although the temporal soil degradation paths and landscape innovations were examined on various aspects of soil erosion, little attention was received. An overview of the progress of remote sensing applications in mapping soil erosion over time and space is given in this chapter.

References

Bai, Z. G., Dent, D. L., Olsson, L., & Schaepman, M. E. (2008). Proxy global assessment of land degradation. *Soil Use and Management, 24*(3), 223–234. https://doi.org/10.1111/j.1475-2743.2008.00169.x.

Bergsma, E. (1974). Soil erosion sequences on aerial photographs. *ITC Journal, 3*, 342–376.

Bisal, F. (1960). The effects of raindrop size and impact velocity on sand splash. *Canadian Journal of Soil Science, 40*, 242–245. http://dx.doi.org/10.4141/cjss60-030.

Blum, W. E. H. (2005). Functions of soil for society and the environment. *Reviews in Environmental Science and Bio/Technology, 4*(3), 75–79. https://doi.org/10.1007/s11157-005-2236-x.

Bocco, G., & Valenzuela, C. R. (1993). Integrating satellite remote sensing and geographic information systems technologies in gully erosion research. *Remote Sensing Reviews, 7*(3–4), 233–240. https://doi.org/10.1080/02757259309532179.

Bojie, F., Xilin, W., & Gulinck, H. (1995). Soil erosion types in the Loess Hill and Gully area of China. *Journal of Environmental Science & Engineering, 7*, 266–272.

Bouyoucos, G. J. (1935). The clay ratio as a criterion of susceptibility of soils to erosion. *Journal of the American Society of Agronomy, 27*, 738–741.

Brazier, R. E., Beven, K. J., Freer, J. F., & Rowan, J. S. (2000). Equifinality and uncertainty in physically based soil erosion models: Application of the GLUE methodology to WEPP—the Water Erosion Prediction Project—For sites in the UK and USA. *Earth Surface Processes and Landforms, 25*(8), 825–845. https://doi.org/10.1002/1096-9837(200008)25:8<825::AID-ESP101>3.0.CO;2-3.

Colwell, R. N. (Ed.). (1983). *Manual of remote sensing* (2nd ed., p. 2240). Falls Church, VA: American Society for Photogrammetry.

De Jong, S. M., Paracchini, M. L., Bertolo, F., Folving, S., Megier, J., & De Roo, A. P. J. (1999). Regional assessment of soil erosion using the distributed model Semmed and remotely sensed data. *Catena, 37*(3–4), 291–308. https://doi.org/10.1016/S0341-8162(99)00038-7.

De la Rosa, D., Mayol, F., Moreno, J. A., Bonsón, T., & Lozano, S. (1999). An expert system/neural network model (ImpelERO) for evaluating agricultural soil erosion in Andalucia region, southern Spain. *Agriculture, Ecosystems and Environment, 73*(3), 211–226. https://doi.org/10.1016/S0167-8809(99)00050-X.

Desmet, P. J. J., & Govers, G. (1995). GIS-based simulation of erosion and deposition patterns in an agricultural landscape: A comparison of model results with soil map information. *Catena, 25*(1–4), 389–401. https://doi.org/10.1016/0341-8162(95)00019-O.

Ellison W. D. (1947a). Soil Erosion Studies-Part 5, Soil Transportation in the Splash Process. *Agricultural Engineering, 28*(8), 349–351.

Ellison W. D. (1947b). Soil Erosion Studies-Part 2, Soil Detachment Hazard by Raindrop Splash. *Agricultural Engineering, 28*(5), 197–201.

Fernandez, C., Wu, Q., McCool, D. K., & Stockle, C. O. (2003). Estimating water erosion and sediment yield with GIS, RUSLE, and SEDD. *Journal of Soil and Water Conservation. Royal Swedish Academy of Sciences, 58*(3), 128 Community Perception.

Greene, R. G., & Cruise, J. F. (1995). Urban watershed modeling using geographic information system. *Journal of Water Resources Planning and Management, 121*(4), 318–325. https://doi.org/10.1061/(ASCE)0733-9496(1995)121:4(318).

Hall, F. G., Townshend, J. R., & Engman, E. T. (1995). Status of remote-sensing algorithms for estimation of land-surface state parameters. *Remote Sensing of Environment, 51*(1), 138–156. https://doi.org/10.1016/0034-4257(94)00071-T.

He, H. S., Mladenoff, D. J., Radeloff, V. C., & Crow, T. R. (1998). Integration of GIS data and classified satellite imagery for regional forest assessment. *Ecological Applications, 8*(4), 1072–1083. https://doi.org/10.1890/1051-0761(1998)008[1072:IOGDAC]2.0.CO;2.

Herweg, K. (1996). *Field manual for assessment of current erosion damage. SCRP Ethiopia, and Centre for Development and Environment*. Berne, Switzerland: University of Berne (p. 69).

Hill, J., Mégier, J., & Mehl, W. (1995). Land degradation, soil erosion and desertification monitoring in Mediterranean ecosystems. *Remote Sensing Reviews, 12*(1–2), 107–130. https://doi.org/10.1080/02757259509532278.

Hudson, N. W. (1993). *Field measurement of soil erosion and runoff*. FAO Soils Bulletin. 68. Rome, Italy: Food and Agriculture Organization.

Huete, A. R. (2004). *Remote sensing for environmental monitoring in: Environmental monitoring and characterization*.

Jensen, J. R. (2004). *Introductory digital image processing: A remote sensing perspective* (3rd ed., p. 526). Upper Saddle River, NJ: Pearson Prentice Hall.

Jetten, V., Govers, G., & Hessel, R. (2003). Erosion models: Quality of spatial predictions. *Hydrological Processes, 17*(5), 887–900. https://doi.org/10.1002/hyp.1168.

Kenneth, R. G., Foster, G. R., Weesies, G. A., & Porter, J. P. (1991). RUSLE: Revised Universal Soil Loss Equation. *Journal of Soil and Water Conservation, 46*(1), 30–33.

King, C., Baghdadi, N., Lecomte, V., & Cerdan, O. (2005). The application of remote-sensing data to monitoring and modelling of soil erosion. *Catena, 62*(2–3), 79–93. https://doi.org/10.1016/j.catena.2005.05.007.

Kirkby, M. J., Imeson, A. C., Bergkamp, G., & Cammeraat, L. H. (1996). Scaling up processes and models from the field plot to the watershed and regional areas. *Journal of Soil and Water Conservation, 51*(5), 391–396.

Laws, J. O., & Parsons, D. A. (1943). The relation of raindrop-size to intensity. *Transactions American Geophysical Union, 24*, 452–460. https://doi.org/10.1029/TR024i002p00452.

Lee, H., & Liu, J. G. (2001). Analysis of topographic decorrelation in SAR interferometry using ratio coherence imagery. *IEEE Transactions on Geoscience and Remote Sensing, 39*(2), 223–232. https://doi.org/10.1109/36.905230.

Loughran, R. J. (1989). The measurement of soil erosion. *Progress in Physical Geography: Earth and Environment, 13*(2), 216–233. https://doi.org/10.1177/030913338901300203.

Lubczynski, M. W., & Gurwin, J. (2005). Integration of various data sources for transient groundwater modeling with spatio-temporally variable fluxes—Sardon study case, Spain. *Journal of Hydrology, 306*(1–4), 71–96. https://doi.org/10.1016/j.jhydrol.2004.08.038.

Martínez-Casasnovas, J. A. (2003). A spatial information technology approach for the mapping and quantification of gully erosion. *Catena, 50*(2–4), 293–308. https://doi.org/10.1016/S0341-8162(02)00134-0.

Merritt, W. S., Letcher, R. A., & Jakeman, A. J. (2003). A review of erosion and sediment transport models. *Environmental Modelling and Software, 18*(8–9), 761–799. https://doi.org/10.1016/S1364-8152(03)00078-1.

Mihara, H. (1959). Raindrops and soil erosion. *Bulletin of the National Institute of Agricultural, Sere A, 1*, 48–51.

Morgan, R. P. C. (2005). *Soil erosion and conservation* (3rd ed.). Malden, Oxford, Victoria, UK: Blackwell Publishing.

Nachtergaele, J., & Poesen, J. (1999). Assessment of soil losses by ephemeral gully erosion using high-altitude (stereo) aerial photographs. *Earth Surface Processes and Landforms, 24*(8), 693–706. https://doi.org/10.1002/(SICI)1096-9837(199908)24:8<693::AID-ESP992>3.0.CO;2-7.

Nearing, M. A., Govers, G., & Norton, L. D. (1999). Variability in soil erosion data from replicated plots. *Soil Science Society of America Journal, 63*(6), 1829–1835. https://doi.org/10.2136/sssaj1999.6361829x.

Nearing, M. A., Pruski, F. F., & O'Neal, M. R. (2004). Expected climate change impacts on soil erosion rates, a review. *Journal of Soil and Water Conservation, 59*(1), 43–50.

Onyando, J. O., Kisoyan, P., & Chemelil, M. C. (2005). Estimation of potential soil erosion for river perkerra catchment in Kenya. *Water Resources Management, 19*(2), 133–143. https://doi.org/10.1007/s11269-005-2706-5.

Pickup, G., & Nelson, D. J. (1984). Use of Landsat radiance parameters to distinguish soil erosion, stability, and deposition in arid Central Australia. *Remote Sensing of Environment, 16*(3), 195–209. https://doi.org/10.1016/0034-4257(84)90064-6.

Reusing, M., Schneider, T., & Ammer, U. (2000). Modeling soil loss rates in the Ethiopian highlands by integration of high-resolution MOMS-02/D2-Streaodata in a GIS. *International Journal of Remote Sensing, 21*(9), 1885–1896. https://doi.org/10.1080/014311600209797.

Saha, A. K., Gupta, R. P., & Arora, M. K. (2002). GIS-based landslide hazard zonation in the Bhagirathi (Ganga) valley, Himalayas. *International Journal of Remote Sensing, 23*(2), 357–369. https://doi.org/10.1080/01431160010014260.

Sazbo, J., Pasztor, L., Suba, Z., & Varallyay, G. (1998). Integration of remote sensing and GIS techniques in land degradation mapping. *Proceedings of the 16th international congress of soil science, Montpellier, France, August* (pp. 63–75).

Schoorl, J. M., Sonneveld, M. P. W., & Veldkamp, A. (2000). Three-dimensional landscape process modelling: The effect of DEM resolution. *Earth Surface Processes and Landforms, 25*(9), 1025–1034. https://doi.org/10.1002/1096-9837(200008)25:9<1025::AID-ESP116>3.0.CO;2-Z.

Sonneveld, B. G. J. S. (2003). Formalising expert judgments in land degradation assessment: A case study from Kenya. *Land Degradation and Development, 14*(4), 347–361. https://doi.org/10.1002/ldr.564.

Stroosnijder, L. (2005). Measurement of erosion: Is it possible? *Catena, 64*(2–3), 162–173. https://doi.org/10.1016/j.catena.2005.08.004.

Subramanya, K. (2008). *Engineering hydrology* (3rd ed., pp. 374–379). New Delhi: Tata McGraw-Hill.

Tamene, L., Park, S. J., Dikau, R., & Vlek, P. L. G. (2006). Analysis of factors determining sediment yield variability in the highlands of Northern Ethiopia. *Geomorphology, 76*(1–2), 76–91. https://doi.org/10.1016/j.geomorph.2005.10.007.

Tim, U. S., & Jolly, R. (1994). Evaluating agricultural nonpoint-source pollution using integrated geographic information-systems and hydrologic/water quality model. *Journal of Environmental Quality, 23*(1), 25–35. https://doi.org/10.2134/jeq1994.00472425002300010006x.

Toutin, T. (2001). Elevation modelling from satellite visible and infrared (VIR) data. *International Journal of Remote Sensing, 22*(6), 1097–1125. https://doi.org/10.1080/01431160117862.

Toutin, T., & Gray, L. (2000). State-of-the-art of elevation extraction from satellite SAR data. *ISPRS Journal of Photogrammetry and Remote Sensing, 55*(1), 13–33. https://doi.org/10.1016/S0924-2716(99)00039-8.

Vigiak, O., Okoba, B. O., Sterk, G., & Groenenberg, S. (2005). Modelling catchment-scale erosion patterns in the East African highlands. *Earth Surface Processes and Landforms, 30*(2), 183–196. https://doi.org/10.1002/esp.1174.

Wessels, K. J., Prince, S. D., Frost, P. E., & van Zyl, D. (2004). Assessing the effects of human-induced land degradation in the former homelands of northern South Africa with a 1km AVHRR NDVI time-series. *Remote Sensing of Environment, 91*(1), 47–67. https://doi.org/10.1016/j.rse.2004.02.005.

Wischmeier, W. H. (1960). Cropping-management factor evaluations for a universal soil-loss equation. *Soil Science Society of America Proceedings, 23*, 322–326.

Wischmeier, W. H., & Smith, D. D. (1958). Rainfall energy and its relationship to soil loss. *Transactions American Geophysical Union, 39*(2), 285–291.

Wischmeier, W. H., & Smith, D. D. (1962). Soil loss estimation as a tool in soil and water management planning. *International Association of Scientific Hydrology. Publication, 59*, 148–159.

Wischmeier, W. H., & Smith, D. D. (1965). Predicting rainfall erosion losses from crop land east of the Rocky Mountains guide for selection of practices soil and water conservation, US Department of Agriculture, Agricultural Handbook No, 282.

Wischmeier, W. H., & Smith, D. D. (1978). Predicting rainfall erosion losses—A guide to conservation planning. US Department of Agriculture Agriculture Handbook, 537.

Wischmeier, W. H., Johnson, C. B., & Crors, B. V. (1971). A soil erodibility nomograph for farmland and construction sites. *Journal of Soil and Water Conservation, 26*, 189–193.

Zemenu, D., & Minale, A. S. (2014). Adoption of soil conservation practices in North Achefer District, Northwest Ethiopia. *Chinese Journal of Population, Resources and Environment, 12*, 261–268.

Zinck, J. A., López, J., Metternicht, G. I., Shrestha, D. P., & Vázquez-Selem, L. (2001). Mapping and modeling mass movements and gullies in mountainous areas using remote sensing and GIS techniques. *International Journal of Applied Earth Observation and Geoinformation, 3*(1), 43–53. https://doi.org/10.1016/S0303-2434(01)85020-0.

Zingg, R. W. (1940). Degree and length of land slope as it effects soil loss in runoff. *Agricultural Engineering, 21*, 59–64.

Index

S. A. Bandh (ed.), *Sustainable Agriculture*,
https://doi.org/10.1007/978-3-030-83066-3

The manufacturer's authorised representative in the EU is Springer Nature Customer Service Centre GmbH, Europaplatz 3, 69115 Heidelberg, Germany. If you have any concerns regarding our products, please contact ProductSafety@springernature.com

Printed and bound by CPI Group (UK) Ltd, Croydon, CR0 4YY
30/09/2026
02225699-0006